SOIL AND WATER
CONSERVATION ENGINEERING
FOURTH EDITION

SOIL AND WATER CONSERVATION ENGINEERING
FOURTH EDITION

Glenn O. Schwab
Professor Emeritus of Agricultural Engineering
Ohio State University
Columbus, Ohio

Delmar D. Fangmeier
Professor of Agricultural and Biosystems Engineering
University of Arizona
Tucson, Arizona

William J. Elliot
Assistant Professor of Agricultural Engineering
Ohio State University
Columbus, Ohio

Richard K. Frevert
Late Professor of Agricultural and Biosystems Engineering
University of Arizona
Tucson, Arizona

WILEY

John Wiley & Sons, Inc.
New York   Chichester   Brisbane   Toronto   Singapore

Acquisitions Editor	Sally Cheney
Marketing Manager	Cathy Faduska
Copyediting Supervisor	Richard Blander
Production Manager	Linda Muriello
Senior Production Supervisor	Savoula Amanatidis
Designer	Laura Nicholls
Illustration Coordinator	Sigmund Malinowski
Manufacturing Manager	Lorraine Fumoso

Cover Photo: Courtesy of U.S. Soil Conservation Service, Ohio.

This book was set in Palatino by Digitype, Inc. and printed and
bound by Hamilton Printing Company. The cover was printed by
Phoenix Color.

Library of Congress Cataloging in Publication Data:
Soil and water conservation engineering / Glenn O.
Schwab . . . [et al.].—4th ed.
    p. cm.
    Includes index.
    ISBN 0-471-57490-2
    1. Soil conservation. 2. Water conservation. 3. Agricultural
engineering. I. Schwab, Glenn Orville, 1919– .
S623.S572 1993
631.4—dc20   92-10953
                                                       CIP

Printed in the Unitd States of America

Printed and bound by the Hamilton Printing Company.

10 9 8 7 6 5 4

Preface

In this edition the text material has been brought up to date, but it continues to emphasize engineering design of soil and water conservation practices and their impact on the environment, primarily air and water quality. Furthermore, the production of food and fiber remains an important consideration because of increasing United States and world population.

Many of the suggestions from instructors and colleagues have been included. The conversion from English to the International System of Units (SI) is nearly complete except in a few cases, such as rainfall maps, which are not available in SI units. Example problems, student problems at the end of chapters, and most illustrations are in both SI and English units. In most cases, SI and English units are given in round numbers for ease of computation, which may result in slightly different answers for the two systems of units.

As in previous editions, the purpose of this book is to provide a first-course professional text for undergraduate agricultural engineering students and for others interested in soil and water conservation. Subject matter on all the engineering phases of soil and water conservation is included, as is a limited section on hydrology. The first chapter covers general aspects with some worldwide implications; Chapters 2 through 4, hydrology; Chapters 5 through 8, erosion and its control; Chapter 9, conservation structures; Chapter 10, earth dams; Chapter 11, flood control; Chapters 12 through 15, drainage; and Chapters 16 through 21, irrigation.

We have assumed in preparing this edition that the student has taken such basic courses as calculus, surveying, mechanics, hydraulics, and soils; however, a knowledge of these subjects is not essential for understanding many portions of the text. We have attempted to emphasize the analytical approach, supplemented with sufficient field data to illustrate practical applications. Although the text material emphasizes principles rather than tables, charts, and diagrams, the book may provide practicing engineers with readily usable information as well. Many examples and student problems have been included to emphasize design principles and to facilitate an understanding of the subject matter. Computer models and software program sources have been described where applicable in the text and in Appendix I, but detailed programs have not been included because of space requirements and rapid changes and obsolescence of software programs.

February, 1992

Glenn O. Schwab
Delmar D. Fangmeier
William J. Elliot

Acknowledgments

We are deeply indebted to many individuals and organizations for the use of material. We are especially grateful to The Ferguson Foundation, Detroit, Michigan, for making the first edition possible by defraying the cost of its development. Harold E. Pinches, formerly with the Foundation, was instrumental in promoting this project. We are grateful to Massey-Ferguson Limited, Toronto, Canada, for providing funds to prepare the second edition. The following individuals have reviewed portions of the manuscript, provided information, or made valuable suggestions for this edition: C. E. Anderson, D. L. Brakensiek, A. Brate, R. S. Broughton, L. C. Brown, N. R. Fausey, W. E. Hart, C. Madramootoo, M. L. Palmer, M. A. Pelletier, C. Sandretto, P. van der Poel, A. D. Ward, L. E. Wagner, and M. Yitayew.

Special appreciation goes to the many friends and colleagues at the University of Arizona, the Ohio State University, the U.S. Soil Conservation Service, and the U.S. Department of Agriculture, who have contributed in many ways through frequent contacts. We wish also to express appreciation to our wives and families for their sympathetic understanding during the preparation of this edition.

We are dedicating this book to the memory of our colleagues, Kenneth K. Barnes, Talcott W. Edminster, and Richard K. Frevert, who contributed much to previous editions. Dr. Frevert participated in the planning and in the selection of the new authors for this edition.

G.O.S.
D.D.F.
W.J.E.

Contents

Abbreviations

Agr.	Agriculture		LR	leaching requirement
Agron.	Agronomy		mimeo.	mimeographed
ARS	Agricultural Research Service		mon.	month
ASAE	American Society of Agricultural Engineers		NOAA	National Oceanic and Atmospheric Administration
ASCE	American Society of Civil Engineers		NPSH	net positive suction head
ASTM	American Society for Testing Materials		o.d.	outside diameter
			P.C.	point of curvature
			P.I.	point of intersection
AW	available water		P.T.	point of tangency
Bull.	Bulletin		publ.	publication
cons.	conservation		PVC	polyvinyl chloride
CPT	corrugated plastic tubing		PWP	permanent wilting point
CS	crop susceptibility		RAW	readily available water
D.A.	drainage area		rep.	report
dia.	diameter		res.	research
DU	distribution uniformity		rpm	revolutions per minute
EC	electrical conductivity		RUSLE	revised universal soil loss equation
EPA	Environmental Protection Agency		SAR	sodium absorption ratio
ESSA	Environmental Science Service Administration		serv.	service
			SCS	Soil Conservation Service
ET	evapotranspiration		SDI	stress day index
EU	emission uniformity		SEW	sum excess water
expt.	experiment		SI	International System of Units
FC	field capacity			
for.	forestry		soc.	society
geophys.	geophysical		UC	uniformity coefficient
GIS	geographic information system		USBR	U.S. Bureau of Reclamation
			USDA	U.S. Department of Agriculture
GPO	Government Printing Office		USDC	U.S. Department of Commerce
HDPE	high-density polyethylene		USLE	universal soil loss equation
H.I.	horizontal interval		V.I.	vertical interval
hp	horsepower		WEPP	water erosion prediction project
i.d.	inside diameter			
L.F.	load factor			

ENGLISH UNITS

ac-ft	acre feet	h	hour
bu/ac	bushels per acre	in.	inch
cfd	cubic feet per day	ipd	inches per day
cfm	cubic feet per minute	iph	inches per hour
cfs	cubic feet per second	mi	mile
ft	foot or feet	min	minute
fpd	feet per day	mpd	miles per day
fpm	feet per minute	mph	miles per hour
fps	feet per second	ppm	parts per million
gpm	gallons per minute	t/a	tons per acre

INTERNATIONAL SYSTEM OF UNITS (SI)

cm	centimeter	m	meter
C	celsius (centigrade)	Mg	megagram (metric ton)
dS	deciSiemens	mg	milligram
g	gram	min	minute
ha	hectare	mL	milliliter
ha-m	hectare-meter	mm	millimeter
J	joule	m/s	meters per second
kg	kilogram	N	newton
km	kilometer	Pa	pascal (N/m^2)
kN	kilonewton	S	siemens
kW	kilowatt	W	watt
L	liter		

Signs and Symbols

a cross-sectional area; constant; organic matter content

A watershed area; annual soil loss; apparent specific gravity; cross-sectional area; energy

b constant; width; soil-structure code

B outside diameter; width of trench; transport coefficient

c cut; chord length; crop coefficient; profile-permeability class

C coefficient; cover-management factor; energy; constant; climatic factor; cut; coefficient of variation; coefficient of skew; correction factor; celsius

d diameter; depth; distance; dry or wet density; critical depth; equivalent depth; effective surface roughness height, ridge spacing

D diameter; depth; interrill erosion rate; runoff; degree of curvature; duration time; degree days; drainage coefficient; deflection lag factor; length of water surface exposure

DP   deep percolation

e void ratio; distance; deflection angle; vapor pressure

E efficiency; radiant energy; specific energy head; kinetic energy; degree of erosion factor; evaporation; wind erosion; elevation; modulus of elasticity

f infiltration rate; hydraulic friction; depth; fill; soil porosity

F total infiltration; fertility factor; Froude number; force or load; dimensionless force; fill; dry soil fraction; safety factor

g acceleration of gravity; gram

G sensible heat; energy

h head; wave height; height; hour; depth of channel

H total head; height; head loss; suction head; specific energy head; riser pipe length

i rainfall intensity; inflow rate; irrigation rate

I total rainfall; irrigation depth; angle of intersection; soil erodibility index; rate of application; initial rainfall extraction; impact coefficient; moment of inertia

k constant; permeability; time conversion factor; capillary conductivity; von Karman's constant; monthly evapotranspiration

K constant; hydraulic conductivity; evapotranspiration coefficient; bedding angle factor; soil erodibility; ridge roughness factor; ratio of tractive force; head loss coefficient; Scobey's coefficient of retardation; frequency factor; crop coefficient

L   dimensionless length

L length; slope length factor; slope length; liter

m meter; exponent; water content; water table height; rank order of events

M watershed area; depth of snowmelt; particle size parameter

n roughness coefficient; drainable porosity; constant; number of values; hours of sunshine; pump specific speed

N	curve number; total number of events; revolutions per minute; hours of sunshine; newton
o	outflow rate
p	wetted perimeter; percentage daytime hours of the year; atmospheric pressure; percent area shaded
P	probability; power; pressure; peak runoff rate; rainfall; conservation practice factor; water content; deep percolation
q	seepage or flow rate; sprinkler discharge rate
Q	water volume; flow rate
r	radius; rate of application, reflectance coefficient; scale ratio (prototype to model)
R	hydraulic radius; radius; rainfall and runoff erosivity index; ratio; residue cover; radiant energy; runoff
RO	surface runoff
s	slope gradient or percent; rate of water storage; distance; standard deviation; second
S	slope gradient or percent; slope steepness factor; water storage; settlement; sprinkler or drain spacing; slope erosion factor; energy; maximum difference between rainfall and runoff; seepage loss; siemens
t	time; temperature; thickness; width; Student's statistical level of significance
T	time; dimensionless time; time conversion interval; time of concentration; time of lag; recession time; time of peak; time of advance or recession; return period; concentrated surface load; tangent distance; width; tractive force; temperature; transport capacity of runoff
u	monthly evapotranspiration; volume conversion factor; wind velocity
v	velocity; rate of capillary movement; rate of soil water movement; threshold velocity
V	volume; vegetative cover factor
w	unit weight; flow conversion factor; wetted diameter
W	weight; width; watershed characteristics; water volume; furrow spacing; watt; average wind velocity; load on conduits; dry weight of soil; water depth
x	constant; variable; return period variate; distance; mean; water level; ratio
X	constant; variable; distance; horizontal coordinate
y	depth; duration variate; deviation of water depth
Y	constant; variable; minimum years of record; distance; vertical coordinate
z	sideslope ratio (horizontal to vertical); depth; height; soil roughness parameter; vertical coordinate
Z	vertical distance; infiltrated water volume or depth per unit area; depth
Δ	slope of saturated vapor pressure curve
θ	sideslope angle; slope degrees; angle of deviation; soil water content
λ	latent heat of vaporization
ρ	air density
μ	dynamic viscosity
ϕ	soil water potential (capillary)
Ψ	gravitational potential
Φ	potential
γ	psychrometric constant; water density
σ	Stefan–Boltzmann constant
τ	shear stress; hydraulic shear

Note: In the text various subscripts may be added to identify the specific definition for each symbol shown above.

CHAPTER 1

Conservation and the Environment

Soil and water conservation engineering is the application of engineering and biological principles to the solution of soil and water management problems. The conservation of natural resources implies *utilization without waste* so as to make possible a continuous high level of crop production while improving environmental quality. Engineers must develop economic systems that meet these requirements.

The engineering problems involved in soil and water conservation may be divided into the following phases: erosion control, drainage, irrigation, flood control, and water resource development and conservation. Although soil erosion takes place even under virgin conditions, the problems to be considered are caused principally by human exploitation of natural resources and the removal of the protective cover of natural vegetation. Urban–rural interface problems are even more serious because of high population density and increased runoff caused by severe changes in land use.

Drainage and irrigation involve water and its movement on the land surface or through the soil mass to provide optimum crop growth. To provide water at places and times at which it is not naturally available, surface reservoirs or other storage facilities must be developed for irrigation and domestic use. Where available, ground water supplies can be developed and maintained by recharge techniques. Flood control consists of the prevention of overflow on low land and the reduction of flow in streams during and after heavy storms. In water–short regions, soil water should be conserved by modified tillage and crop management techniques, level terracing, contouring, pitting, reservoirs, and other physical means of retaining precipitation on the land and reducing evaporative losses from the soil surface.

1.1 Engineers in Soil and Water Conservation

Sound soil and water conservation is based on the full integration of engineering, atmospheric, plant, and soil sciences. Agricultural engineers, because of their training in soils, plants, and other basic agricultural subjects, in addition to their engineering background, are well suited to integrate these sciences. To develop and execute a conservation plan, engineers must have a knowledge of the soil, including its physical and chemical characteristics, as well as a broad understanding of soil–plant–water–environment interactions. They have a unique role because their efforts are directed toward the creation of the proper environment for the optimum production of plants and animals. In addition to the agricultural application of their knowledge, they are playing an increasing role in the rural–urban sector, especially relating to air and water pollution control.

To be fully effective in applying technical training, engineers must be acquainted with the social and economic aspects that relate to soil and water conservation. They must have a full understanding of the various local, state, and federal government policies, laws, and regulations. They should also become familiar with ground and satellite mapping techniques, nationwide geographic information systems (GIS), weather records and prediction systems, soil survey reports, and other physical data. To apply the vast amount of information available, engineers should be knowledgeable and able to use computers for solving job-related problems. Some of the presently available software will be referenced later.

1.2 Conservation Ethics

The increasing world population will dictate the necessity of conserving natural resources now and in future years. Fossil fuels, soils, minerals, timber, and many other materials are being exhausted at a rapid rate. The average 70–year–old American in a lifetime will use 1 000 000 times his or her weight in water, 10 000 times in fossil fuels and construction materials, and 3000 times in metal, wood, and other manufactured products (Wolman, 1990). Recycling of paper, glass, metals, and other items is being practiced, partly because of the increasing costs of waste disposal and partly because of public support and appreciation for conservation. The decreasing population of wildlife and the disappearance of many species are evidence that much of the problem is related to air and water pollution as well as loss of habitat. In agriculture, soil erosion from farmland is not only one of the major causes of water pollution, but the loss of the land itself reduces the production of food and fiber. Government incentive programs since the 1930s have been helpful, and research efforts have developed many useful agronomic, tillage, and mechanical practices. In the short run, exploitation of the soil and other natural resources for economic benefit has been practiced since the early pioneers developed this country. Private enterprise systems encourage this concept, but it is not in the best interest of society. Natural resources should be passed on to future generations in as good or better condition than previous generations have left them. Conversion of prime farmland to urban development and other nonfarm uses without regard to future food needs continues because farmland cannot compete with other higher–value uses. Political solutions to these problems are not likely, but we can promote and teach appropriate conservation ethics.

1.3 Soil Erosion Control

The control of soil erosion by water and wind is of great importance in the maintenance of crop yields and for the mitigation of agricultural nonpoint pollution. The amount of soil eroded from fields and pastures in the United States is estimated to be the equivalent of about a depth of 1 m from 1000 farms of 200 ha each year. In addition to these losses by water, there are large losses caused by wind erosion. Not only is the soil lost in the erosion process but also a proportionally higher percentage of plant nutrients, organic matter, pesticides, and fine soil particles in the removed material is lost than in the original soil. One of the most dramatic events in the United States was the great dust storm in 1934, which swept across the country from the Great Plains to the Atlantic Coast. It blotted out the sun over a large part of the nation and sifted through the windows of New York skyscrapers (Bennett, 1939).

Erosion control is essential to maintain the crop productivity of the soil as well as to control sedimentation and pollution in streams and lakes. Federal legislation in 1978 directed all states to develop plans for controlling sediment pollution from nonpoint sources, which includes farmland. In 1985 the federal farm security act encouraged farmers to control pollution from erodible land by providing financial incentives.

Several types of erosion are shown in Fig. 1.1. The relative degree of erosion and its distribution in the United States are shown in Fig. 1.2. Although the data were obtained from a 1934 survey, erosion conditions have changed but little with time. This map indicates areas having slight, moderate, or severe erosion and does not differentiate between that caused by water and that caused by wind. Many

Fig. 1.1   Interrill, rill, and gully erosion by water together with wind erosion on cropland. (Courtesy SCS.)

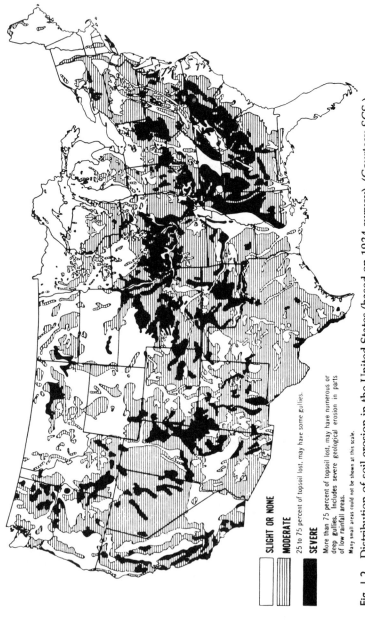

SLIGHT OR NONE

MODERATE
25 to 75 percent of topsoil lost, may have some gullies.

SEVERE
More than 75 percent of topsoil lost, may have numerous or deep gullies. Includes severe geological erosion in parts of low rainfall areas.

Many small areas could not be shown at this scale.

Fig. 1.2   Distribution of soil erosion in the United States (based on 1934 survey). (Courtesy SCS.)

small areas where severe erosion may occur locally cannot be shown on a map of this scale. Erosion caused by water from cropland in Missouri, Tennessee, and Mississippi averages in excess of 20 Mg/ha (metric tons/hectare) annually, which is more than double the national average. Wind erosion from cropland averages more than 20 Mg/ha in Texas, New Mexico, and Arizona. Conservation erosion control practices are discussed in Chapters 5 to 9.

Erosion in a serious problem worldwide. It is present in most countries where topography is rolling to steep or the climate is arid. In ancient times cities were buried in North Africa, the Middle East, and other areas by eroding soil. Some of the earliest control structures were bench terraces found in China, Japan, Italy, Peru, the Philippines, and other countries, built more than 2000 years ago (Bennett, 1939). In modern times, since the 1930s, the United States has been one of the leading countries in the development of erosion control practices. Worldwide statistics on erosion are not available.

1.4 Drainage

Agricultural drainage practices in humid areas, such as the Midwest, are essential for developing or improving much land for crop production. In arid regions under irrigation, drainage is needed to reclaim saline and sodic soils by leaching and to prevent salinity problems by maintaining a low water table. Where salinity problems are expected, land should not be developed for irrigation unless drainage can be provided. In 1985, 23 states had 96 percent of all cropland drained in the United States (Pavelis, 1987). Sixty percent of the land drained was in organized drainage districts (enterprises established under state laws). Thirty-four percent of the land drained (15 million ha) was improved by subsurface (pipe) drains. The estimated length of pipe installed is equivalent to over 2100 trips across the United States. An estimated 22 million ha could yet be drained for improved agricultural production. Of the land now drained, 75 percent will need redraining because of poor design or lack of maintenance. Drainage installations tend to fluctuate with periods of dry and wet weather and economic conditions. For example, during the years 1900 to 1910 and from 1945 to 1975 investment in drainage increased rapidly. From 1974 to 1975 pipe drainage increased 40 percent.

In the 1980s, federal land and water policy switched from financial incentives for the farmer to a policy restricting drainage of wetlands. The swampbuster provisions of the Food Security Act of 1985 specified that certain wetlands that are beneficial for wildlife and the environment should not be drained. The act is not mandatory, but noncompliance by the farmer will negate any benefits from other conservation programs.

The distribution of drained land in the United States is shown in Fig. 1.3. The leading states are Illinois, Indiana, Iowa, Ohio, Arkansas, and Louisiana. About 70 percent is cropland, 12 percent is pasture, 16 percent is woodlands, and 2 percent is other uses. A typical field needing both surface and subsurface drainage is shown in Fig. 1.4. Drainage practices are discussed in Chapters 12 to 16.

Drainage practices in the world date back thousands of years. Some notable examples of large drainage projects are the polders in Holland and the fens in England, which are low lands reclaimed from the sea. Drainage has long been recognized as essential for permanent irrigated agriculture. An estimated one third of the irrigated land of the world is affected by sodicity and salinity, which can be

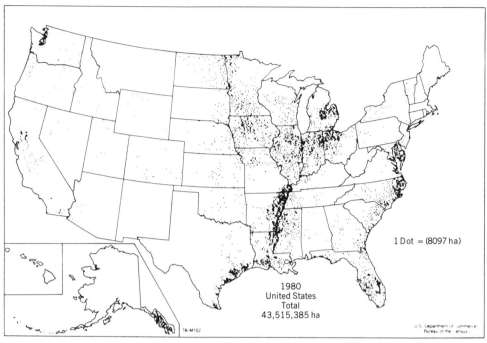

Fig. 1.3 Distribution of artificially drained agricultural land in the United States (*Source:* Pavelis, 1987.)

Fig. 1.4 Inadequate surface and subsurface drainage causes flooding following heavy rainfall. (Courtesy SCS.)

corrected by leaching with water into subsurface drains. Egypt, India, and Pakistan have much such land that needs drainage.

1.5 Irrigation

Irrigation provides one of the greatest opportunities for increasing crop production as well as improving germination, controlling air temperature, and applying chemicals with the irrigation water. In arid regions few crops can be grown without irrigation. Although only about 8 percent of the cropland in the United States is irrigated, it produces about 25 percent of the total value of farm crops. About 83 percent ("1989 Irrigation Survey," 1990) of the irrigated land is in the West, but irrigation has been increasing in the eastern states. Where the annual rainfall is less than 250 mm, irrigation is a necessity; where rainfall is from 250 to 500 mm, crop production is limited unless the land is irrigated; and where rainfall is more than 500 mm, irrigation is often required for maximum production.

Whether crop production is intensified in the humid East with modern center–pivot sprinkler system, as shown in Fig. 1.5a, or desert land is converted to lush productive land as in Fig. 1.5b, the basic needs are the same: productive soils, adequate drainage, and a reliable supply of good quality water. Relatively large quantities of water are required to satisfy the needs of the crop and to supply conveyance, evaporation, and seepage losses. The water budget shown in Fig. 1.6 illustrates where losses occur and the problems associated with the efficient use of water in an arid region.

The total irrigated area ("1989 Irrigation Survey," 1990; USDC, 1990) in the 50 states varied from 19 to 24 million ha (1987 and 1989) distributed as shown in Fig. 1.7. About 42 percent of the total is irrigated by sprinklers. Low-pressure systems and microirrigation are on the increase because of lower energy and water requirements. Irrigation practices are discussed in Chapters 18 to 21.

Worldwide expansion of irrigation after 1950 was about 3 percent per year, which paralleled the growth in population. Since 1979, the annual rate of expansion has decreased to about 1 percent per year (Postel, 1989). About two thirds of the irrigated land is in five countries as shown in Table 1.1.

Table 1.1 World and Five Countries with the Most Irrigated Land, 1986

Country	Gross Irrigated Area[a] (million ha)	Percentage of Total Irrigated[b]	Percentage of Cropland Irrigated
India	55	18	33
China	47	21	48
Soviet Union	21	7	9
United States	19	9	10
Pakistan	16	9	77
All other countries	92	36	9
World	250	100	17

[a]Equipped with irrigation facilities.
[b]Actual irrigated area.
Source: United Nations' FAO (1988) and Postel (1989).

Fig. 1.5   Irrigation has wide application from (*a*) supplementing rainfall in the East to (*b*) developing productive valleys from desert land in the West. (Courtesy SCS and Valmont Industries, Inc.)

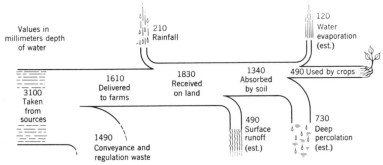

Fig. 1.6 Disposal of water diverted for irrigation where advanced irrigation practices have not yet been applied. (*Source:* USDA, 1955.)

1.6 Flood Control

Flooding in upstream watersheds (less than 2500 km² in area) is among the most significant natural phenomena in terms of causing loss of life, crops, and property as well as health hazards, water pollution, and interruption of services, such as transportation, utilities, police, fire protection, and emergency operations. Flood damage in upstream areas occurs primarily on agricultural land, whereas downstream floods cause major damage to metropolitan areas (USDA, 1989). Flood damage is estimated at $2 billion (1980 dollars) annually in rural areas (USDA, 1989). These losses will increase in future years because of continued development. Flood damage to agricultural land is shown in Fig. 1.8. Forty-one percent of the cropland prone to flooding is in the Corn Belt and Northern Plains states. The distribution of all flood-prone land in the United States and the top 10 states is

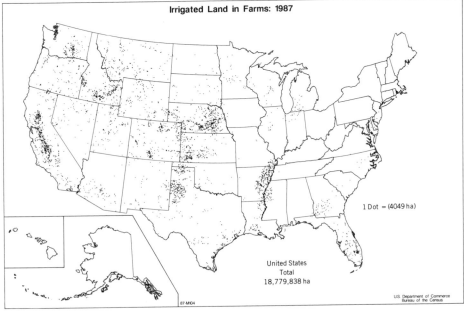

Fig. 1.7 Distribution of irrigated land in the United States. (*Source:* USDC, 1990.)

Fig. 1.8 Flood damage to agricultural land. (Courtesy SCS).

shown in Fig. 1.9. Because downstream floods on major streams are more spectacular and damage is more evident, floods in upstream areas are often neglected.

Practically all flood control projects are federally built and financed by the Department of Agriculture (upstream) and the Corps of Engineers. About 1000 projects will or have been constructed under Public Law 566 for upstream (headwater) areas (SCS, 1981). Headwater flood control measures are discussed in Chapters 10 and 11.

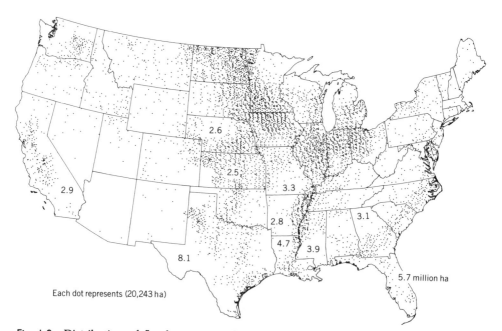

Fig. 1.9 Distribution of flood–prone rural land in the United States and the 10 highest states, with numbers in million hectares. (*Source:* USDA, 1989.)

1.7 Development, Conservation, and Quality of Water Resources

During the past 20 years about 40 percent (USDA, 1989) of the fresh water withdrawn from surface and ground water sources in the 48 contiguous states was for agricultural use, most of which was for irrigation. Agriculture consumed about 83 percent (SCS, 1981) of the total use from all sources, including precipitation. About 40 percent (Todd, 1970) of the water withdrawn for irrigation was not available to the crop. Losses include those during conveyance, seepage, water that percolates below the root zone, evaporation, and transpiration by phreatophytes. As industrial, municipal, recreational, and other nonagricultural uses of water increase their demand for a share of the limited water supplies, it is important for agriculture to improve efficiency of use and conserve its share of the nation's water supply (Fig. 1.6). Elimination of these losses is equivalent to new water supplies that could be developed. Water harvesting as shown in Fig. 1.10 is one method of increasing water supplies. In this application ground covers and sealants provide a stable and waterproof surface for collecting water for livestock and other limited uses. Recharge of ground water by water spreading and recharge wells, replenishment of irrigation with drainage water, and similar practices provide other means of water resource development.

Conservation of soil water is a critical problem in semiarid areas where irrigation is not practical and there is a recurring deficiency of soil water for crops and range production. The dry farming areas of the Great Plains, the Pacific Northwest, and many of the intermountain valleys are particularly affected. Many humid areas are also influenced by critical soil water shortages during parts of the growing season in dry years. Estimates show that two thirds of the rainfall in the Great Plains states is lost by evaporation alone. Studies have shown that if this evaporation from the soil in the 10 Great Plains states could be reduced by the equivalent of 76 mm of precipitation, these states would have an additional 37 million ha-m (1 m depth on 1 ha) of water, enough to fill Lake Mead (Hoover

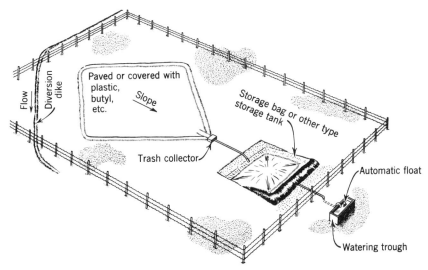

Fig. 1.10 Water harvesting as a means of making maximum use of limited precipitation for livestock water. (Courtesy Agricultural Research Service.)

Fig. 1.11   Drifting snow trapped by rows of sorghum stubble added 50 mm of soil water for the succeeding crop by holding most of the 80 mm of snow. (Courtesy Agricultural Research Service.)

dam). New practices for entrapment and storage of soil water could include more effective terrace systems, new snow trapping techniques (Fig. 1.11), tillage practices that modify the soil surface to retain more precipitation and reduce evaporation, and surface evaporation control through use of mulches and films (Fig. 1.12).

Water quality affects all humans, animals, wildlife, and crops and is becoming increasingly important as population increases and the need for food and fiber expands. Increasing use of fertilizers and pesticides, farming operations, irrigation and drainage practices, agricultural and municipal wastes, and other factors have contributed to a general decrease in water quality with time. Ground water contamination is a more serious problem than surface water in streams and lakes because of the years of delay before the problem is evident and because decades may be required to correct it. Much research will be required to detect the problem and to devise practical solutions. State and federal legislation will be required as a public cooperative effort is necessary. Much progress has been made, such as establishment of the Environmental Protection Agency and enactment of the Clean Water Act and Resources Conservation Act. Water resource development and quality are discussed in Chapters 10 and 17. The design of water conservation measures is a major thrust of this text and is discussed in many other chapters.

1.8   Impact of Conservation Practices on the Environment

Environmental quality is measured by quantitative data and/or by performance. Air, land, and water have a capacity to cleanse or regenerate themselves, biologically or mechanically (Gratto, 1971). Conservation practices can be beneficial or detrimental to the environment. For example, terracing rolling cultivated land can be beneficial by reducing erosion and sedimentation, or irrigation can be detrimental by increasing the salinity from drainage water on flow downstream. Agricultural production of food and fiber is essential, but preserving or enhancing the

Fig. 1.12 Conserving soil water with plastic film mulch which reduces the evaporation from the soil surface. (Courtesy Agricultural Research Service.)

environment is also necessary as population pressures increase. When drinking water becomes contaminated with nitrates, pesticides, sediment, and other materials, the public is and should be concerned. The reduction in wildlife and disappearance of some species on the endangered list give cause to suspect that something is seriously wrong with the environment. Changing land use will modify the quantity and quality of the habitat the land provides for wildlife. Agricultural and wildlife interests are increasingly being coordinated by local, state, and federal governments. Many conservation practices are being changed so as to improve the environment with little or no cost to agriculture. Environmental costs are paid either by taxes, when supported by the government, or by higher prices, when supported by the private sector. With the adverse economic conditions in agriculture in the 1980s, low-input, sustainable agriculture programs are being promoted. One of the many ways proposed to improve farming was to reduce fertilizers, pesticides, and other inputs, which could also enhance the environment. Others would argue that an increase in the use of fertilizers would increase production per hectare and require less land, so that erodible land could be taken out of cultivation and thus reduce erosion and pollution.

Erosion is one of the most serious problems, because it pollutes the water with sediment, nutrients, and pesticides, increasing the cost of water treatment. Considerable effort has been made to develop conservation practices since the 1930s and to encourage no tillage and similar practices to keep residues and cover on the land. Although reduced tillage practices have been effective in reducing cost of production, pesticide use is generally increased with minimum or no tillage.

In general, the impact of drainage on the environment has been beneficial. The drainage of the lakebed area in the Midwest was instigated by medical doctors, who realized that it reduced malaria long before the mosquito was the known cause. The drainage of Central Park in New York City, which was one of the earliest drainage projects, was also done for health reasons. Obviously, subsurface drainage provides a desirable environment for the roots of agricultural crops and a stable base for roads and buildings. Without drainage much land could not have been developed by humans.

During the early development of the United States, agricultural interests prevailed and the drainage of swamps and wetlands was done without question. In more recent years, drainage of these naturally wet areas was found to have some adverse effects on migrating wildlife and marine life and other environmental aspects. As much as 50 percent of the wetlands have been drained mostly in the Southeast, Midwest, and Pacific Coast states (USDA, 1989). Wetlands vary from permafrost in Alaska to everglades in Florida to the desert wetlands in Arizona. In addition to providing habitat for fish and wildlife, wetlands enhance ground water recharge, reduce flooding, trap sediment and nutrients, and provide recreational areas. Federal legislation known as the Swampbuster Act was passed to discourage wetland drainage.

Subsurface drainage water in humid areas and from irrigated lands generally contains higher concentrations of nitrate nitrogen, salt, and other chemicals than found in surface water or in the irrigation water. Wildlife interests often claim that open drainage ditches, when cleaned and straightened, reduce cover for wildlife. Because of this concern, some channel improvement projects have been redesigned to remove vegetation on only one side of the ditch and restrict cleanout and reshaping work to the opposite side. In large channels, low, small dams can be constructed to provide water pools for better fish habitat.

Irrigation projects as in the West have a major impact on the social and political structure of the entire community. Water storage and stream diversion facilities for irrigation, as well as stream flow and ground water use, influence minimum stream flow, depth of ground water, fish and wildlife population, natural vegetation and crops grown, recreation activities, roads and public utilities, and other unique and site-specific factors. One example of a serious environmental problem is the chemical pollution (selenium and boron) from irrigation drainage water, such as occurred at the Kesterson National Wildlife Refuge in California (see Chapters 14 and 17). Blockage of the drainage system was mandated by the court. The Left Bank Outfall Drain in lower Pakistan, which is a constructed channel more than 250 km in length carrying only saline drainage water to the sea, illustrates the magnitude of the pollution problem. The Salton Sea in California is an example of an evaporation disposal site for drainage water. In the construction of the Aswan dam in Egypt, many people were displaced and archaeological sites were covered with water. Similar impacts exist where large dams are built for flood control. Because of such problems as described above, environmental impact statements, which necessitate more overall planning effort, are now mandatory for most conservation projects.

1.9 Land Use and Crop Production

Since 1920 in the United States, land in crops, pasture, and forests has been diverted to nonagricultural and other uses. Land for urban development, roads, and surface water storage has increased about 56 percent from 1920 to 1982. With time, land in agriculture may change from one use to another, that is, from

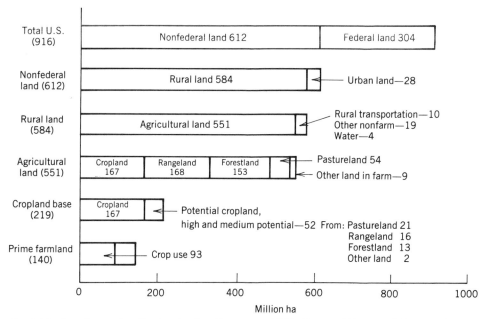

Fig. 1.13 Land–use distribution in the 50 states in 1977. (*Source:* USDA and the President's Council on Environmental Quality, 1981.)

cropland to pasture to forest and back to cropland, but urban and developed land is not likely to change back to agriculture.

The distribution of land use in the United States is shown in Fig. 1.13. Nonfederal land represents about two thirds of the total. Prime farmland, which is about 25 percent of all agricultural land, is considered our best, producing high crop yields with minimal damage or loss of soil. The loss per day of prime farmland by regions is shown in Fig. 1.14 for an 8-year period. The average prime land loss to urban development and surface water storage in the United States is

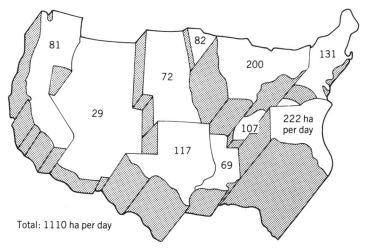

Fig. 1.14 Average loss of prime farmland in hectares per day to urban and water uses by regions for the 8-year period 1967–1975. (*Source:* Sampson, 1978.)

1110 ha per day. The Corn Belt region had 33 million ha of prime farmland or about 23 percent of the total. One of the leading causes of prime farmland loss is poorly planned urban and industrial development. Land unsuited for agriculture is often not considered for such development because of higher costs. Land for agriculture cannot compete with these more intensive uses. Many private and government agencies are concerned with agricultural land preservation, which greatly affects our long–range potential to produce food and fiber.

Crop and livestock production have been gradually increasing in the United States for more than three decades even though cropland had been nearly constant or decreasing slightly. Continued increases in yields may be threatened by soil erosion, air pollution, regulatory constraints, the increasing cost of fertilizers, water, fuel, and other resource inputs, and less productive newly cultivated land. In the future, land will have to be cultivated with greater intensity, and water-use efficiency will have to be greatly increased. Genetically improved crops and livestock will have to be developed along with more efficient use of fertilizers, pesticides, and farm equipment. Increasingly expensive and uncertain energy supplies will make it more difficult to secure higher crop yields. Greater pressure may be placed on land, water, and other inputs for production of biomass and other renewable energy resources. Proper management of soil and water resources will become increasingly important.

1.10 Population, Food and Fiber, Energy, and Pollution Problems

These problems are serious in most countries in varying degrees. Compared with developed countries (Japan, Germany, Great Britain, Australia, United States, etc.), underdeveloped countries of the world (Nigeria, India, Pakistan, Mexico, Ecuador, etc.) have about twice the population growth rate, a much lower economic growth rate per capita, and a much higher need for an increase in food and fiber production. These problems should be considered as worldwide by all countries because of humanitarian concerns, national security, long-range economic stability, energy supplies, and deterioration of the atmospheric environment. According to the United Nations' Food and Agricultural Organization (1988), the 1984 world population of 4.7 billion is expected to grow to more than 6 billion by the year 2000. Presently, no critical worldwide food shortage exists, but critical nutritional problems arise from uneven distribution of food among countries, within countries, and among families with different levels of income.

One approach to solving the energy problem is to stabilize or reverse population growth. It is seldom treated as a part of energy policy because the effective means required would offend tradition and religious beliefs. Public attitude toward population was expressed appropriately by Davis (1981). Favoring fewer people is considered inhumane, but is confused with the greater inhumanity of bringing into the world children who probably will die or suffer from starvation. Past history and social practices favored having many children. This concept is now considered inappropriate. Linked to the new attitude is the problem of energy shortage. Energy policy has been to produce or save energy for as many people as we have, not to have fewer people and give each as much energy as needed. Davis stated that reversing or stopping population growth could play a major role in solving the energy problem.

Stabilization of the population growth rate in China prior to 1990 has been quite successful, but government efforts in many other countries have been disappointing. By the year 2000 an estimated 88 percent of the world's population will be in three continents: Asia, South America, and Africa (Lamm, 1981). After government planning policies were adopted in nine developing countries, birth rates dropped from 10 to 25 percent, indicating that some progress has been made (IIED, 1987). Although the growth rate in the United States has stabilized, about one third of our population growth is due directly to immigration (on the order of one million per year).

REFERENCES

Bennett, H. H. (1939). *Soil Conservation.* McGraw-Hill, New York.

Council for Agricultural Science and Technology (CAST) (1981). *Preserving Agricultural Land: Issues and Policy Alternatives.* Report 90. Ames, IA.

Davis K. (1981). "It Is People Who Use Energy (editorial)." *Science* **211**, 441. (4481).

Gratto C. P. (1971). "Issues in Environmental Quality." *J. Soil Water Cons.* **26** (Mar.–Apr.), 44–45.

International Institute for Environment and Development and The World Resources Institute (IIED) (1987). Basic Books, New York.

"1989 Irrigation Survey" (1990). *Irrig. J.*, Jan.–Feb, 34.

Jordan W. R. (ed.) (1987). *Water and Water Policy in World Food Supplies.* Proceedings of a Conference, May 26–30, 1985. Texas A&M Univ. Press, College Station, TX.

Lamm, R. D. (1981). "Guest Editorial." *Energy Educ.* **4** (3), 1.

Pavelis, G. A. (ed.) (1987). *Farm Drainage in the United States: History, Status, and Prospects.* USDA Economic Res. Serv. Misc. Publ. 1455. GPO, Washington, DC.

Postel, S. (1989). *Water for Agriculture Facing Limits.* Paper 93. World Watch Inst., Washington, DC.

Sampson, N. (1978). *Preservation of Prime Agricultural Land, Environmental Comment.* Urban Land Inst., Washington, DC.

Todd, D. K. (1970). *The Water Encyclopedia.* Water Information Center, Port Washington, NY.

United Nations' Food and Agriculture Organization (FAO) (1988). *FAO Production Yearbook.* Statistics Series 70. Rome, Italy.

U.S. Department of Agriculture (USDA) (1955). "Water." *Yearbook of Agriculture.* GPO, Washington, DC.

———— (1989). *The Second RCA Appraisal. Soil, Water, and Related Resources on Nonfederal Land in the United States.* GPO, Washington, DC.

U.S. Department of Agriculture (USDA) and the President's Council on Environmental Quality (1981). *Final Report, National Agricultural Lands Study.* GPO, Washington, DC.

U.S. Department of Commerce (USDC) (1990). *1987 Census of Agriculture*, Vol. 2, Pt. 1: "*Subject Series*". U.S. Bureau of the Census, Washington, DC.

U.S. Soil Conservation Service (SCS) (1981). *America's Soil and Water: Condition and Trends.* Washington, DC.

Wolman, M. G. (1990). "The Impact of Man." *EOS, Trans. Am. Geophys. Union* **71**(52), 1884–1886.

CHAPTER 2

Precipitation

Precipitation, along with the atmospheric phenomena of heat, water, and air movement, is a part of the science of meteorology. This science is of particular interest to those concerned with the effective use of soil and water. The weather is often the controlling factor in problems of preventing excessive movement of soil, or retaining needed water, of increasing the intake of surface water, of adding needed water by irrigation, and of removing excess water by drainage. Water, whether too much, too little, or poorly distributed, is one of the major limitations in agricultural production.

2.1 The Hydrologic Cycle

The science of meteorology is a part of the much broader field of hydrology, which includes the study of water as it occurs in the atmosphere as well as on and below the surface of the earth. One representation of the hydrologic cycle is given in Fig. 2.1. It shows the formation of precipitation, which may occur as rain, snow, sleet, or hail. Some of this precipitation evaporates partially or completely before reaching the ground. Precipitation reaching the earth's surface may be intercepted by vegetation, it may infiltrate the surface of the ground, it may evaporate, or it may run off the surface. Evaporation may be from the surface of the ground, from free water surfaces, or from the leaves of plants through transpiration. A portion of the total rainfall moves over the earth's surface as runoff; another portion moves into the soil surface, is used by vegetation, becomes part of the deep ground water supply, or seeps slowly to streams and to the ocean.

Figure 2.1 shows also the measurements commonly made of those portions of the hydrologic cycle of special interest to agricultural engineers. These include the measurements of precipitation by rain or snow gages, the measurement of accumulated snow by snow surveys over established ranges, the measurement of runoff by gaging stream channels, and the measurement of ground water levels. These ground water levels may be measurements either of the deep water tables as indicated by the height the water rises in wells or of the shallower perched water tables of particular interest in analyzing drainage problems.

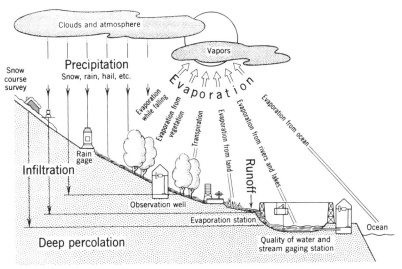

Fig. 2.1 The hydrologic cycle.

2.2 Forms of Precipitation

Precipitation may occur in any of a number of forms and may change from one form to another during its descent. The forms of precipitation consisting of falling water droplets may be classified as drizzle or rain. Drizzle consists of quite uniform precipitation with drops less than 0.5 mm in diameter. Rain consists of generally larger particles.

Precipitation may also occur as frozen water particles including snow, sleet, and hail. Snow is composed of a grouping of small ice crystals known as snowflakes. Sleet forms when raindrops are falling through air having a temperature below freezing; a hail stone is an accumulation of many thin layers of ice over a snow pellet. Of the forms of precipitation, rain and snow make the greatest contribution to our water supply.

Water at the soil surface is also made available by direct condensation and absorption from the atmosphere, commonly referred to as dew. Studies of dew formation show that 30 mm per year condenses on bare soil, but only 25 mm on a grass cover. About 15 mm is collected on corn leaves during the summer, and 33 mm is condensed on soybean leaves. Although dew is normally evaporated by noon, it is effective in reducing the rate of soil water depletion.

2.3 Characteristics of Raindrops

Since by far the largest portion of precipitation occurs as rain, and since rainfall directly affects soil erosion, the characteristics of raindrops are of interest. Raindrops include water particles as large as 7 mm in diameter. The size distribution in any one storm covers a considerable range and this size distribution varies with the rainfall intensity. Figure 2.2 gives the raindrop diameter for three of the intensities studied. Not only does the higher-intensity storm have more large-diameter raindrops, but it also has a wider range of raindrop diameters.

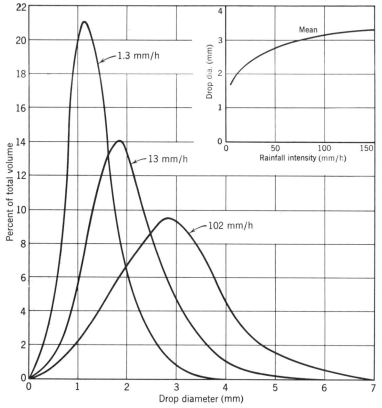

Fig. 2.2 Effect of rainfall intensity on raindrop size and its contribution to total rainfall. (Redrawn from Laws and Parsons, 1943, and Wischmeier and Smith, 1958.)

Raindrops are not necessarily spherical or even streamlined. Falling raindrops are deformed from spherical shape by unequal pressures, as a result of air resistance, developing over their surfaces. Large raindrops divide in the air, drops over 5 mm in diameter being generally unstable.

In studies of raindrops, scientists found that the velocity of fall depends on the size of the particle, and that large drops fall more rapidly. As the height of fall is increased, the velocity increases only to a height of about 11 m; the drops then approach a terminal velocity, which varies from about 5 m per second for a 1-mm drop to about 9 m per second for a 5-mm drop.

2.4 Weather Maps

The weather picture is commonly depicted by weather maps showing the position of the isobars, the ground position of the fronts, and the areas of precipitation. Such weather maps are shown in Figs. 2.3a and 2.4. Official maps show air temperature, dew point, wind direction and velocity, barometric pressure in millibars, and pressure change during the last 3 hours. The function of the weather forecaster is to prognosticate the movement of these frontal areas, their development, and the probable precipitation. As weather maps are now commonly in-

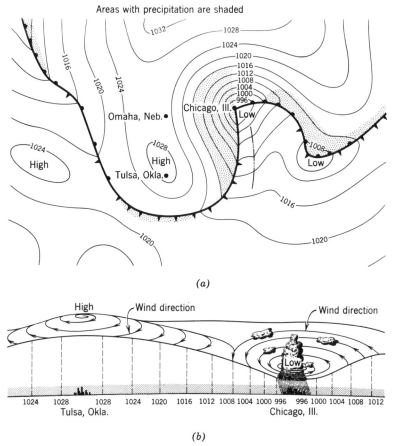

Fig. 2.3   (*a*) Portion of a weather map in April showing cloudy weather in the East, rain in Chicago, and clear skies in the Southwest. (*b*) Wind circulation around a high-pressure center at Tulsa and a low-pressure center at Chicago.

cluded in daily newspapers and on TV, the individual is provided with an opportunity to practice forecasting and to compare his or her predictions with those of the Weather Bureau. The cloud picture shown in Fig. 2.4*a* shows a typical meteorological satellite weather map.

MEASUREMENT OF PRECIPITATION

Since most estimates of runoff rates are based on precipitation data, information regarding the amounts and intensity of precipitation is of great importance.

2.5   Gaging Rainfall

The purpose of the rain gage is to measure the depth and intensity of rain falling on a flat surface. The many problems of measurements with gages include effects of topography and nearby vegetation as well as the design of the gage itself. Rain

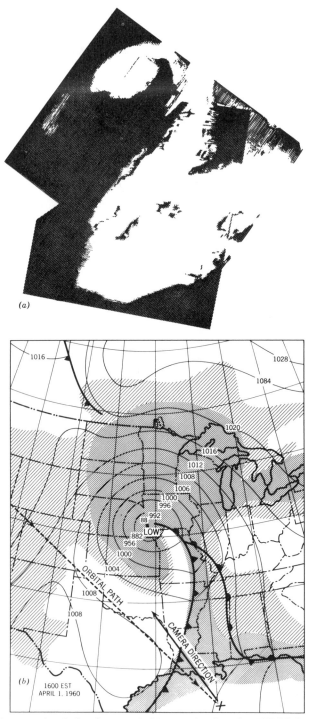

Fig. 2.4 (*a*) Photograph of cloud cover (white area) taken by TIROS I, the first photographic meteorological satellite. (*b*) Corresponding synoptic chart of the frontal system with crosshatching to indicate cloud cover. (Courtesy U.S. Weather Bureau.)

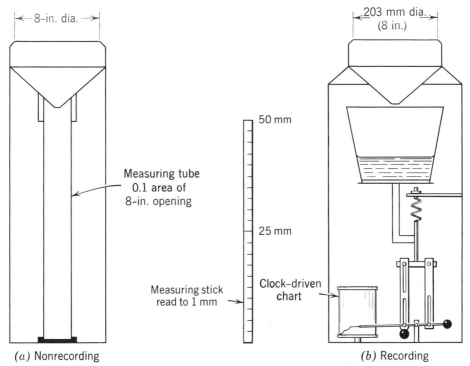

Fig. 2.5 (*a*) Weather Bureau nonrecording (standard) rain gage. (*b*) Fergusson weighing (recording) rain gage.

gages generally used in the United States are vertical, cylindrical containers with top openings 203 mm in diameter. A funnel-shaped hood is inserted to minimize evaporation losses.

Rain gages may be classified as recording or nonrecording. Nonrecording rain gages, as shown in Fig. 2.5*a*, are economical, require servicing only after rains, and are relatively free of maintenance. The gage illustrated here is the Weather Bureau type. The water is funneled into an inner cylinder one tenth of the cross section of the catch area. This provides a magnification of 10 times the depth of the water and makes it possible to measure to the nearest 0.25 mm.

Recording rain gages may be of several types. The type shown in Fig. 2.5*b* is the Fergusson weighing rain gage. Water is caught in a bucket placed above the recording mechanism. The weight of the water places a tension on the spring. The amount of displacement is recorded through an appropriate linkage on a chart placed on a clock-driven drum. The recording mechanism shown, which allows the pen to traverse the chart three times, gives a large vertical scale and makes possible a more accurate reading of the chart. Some weighing gages do not have the reversing mechanism. There are other types of recording rain gages, most of them using either the tipping bucket or the float-and-siphon principle.

For inaccessible locations, rain gages may be equipped with telemetry equipment. Radar measurement of cloud density and rainfall rate is also possible.

2.6 Measuring Snowfall

Since the water content of freshly fallen snow varies from less than 40 mm to over 400 mm of water per meter of snow, snowfall is much more difficult to measure than rainfall. Although this wide variation in density makes it hazardous to indicate the amount of snow by simple depth measurements, a water equivalent depth of 10 percent of snow depth is a commonly accepted mean. Water content of compacted snow, however, is often 30 to 50 percent of snow depth.

Snowfall measurements are often made with regular rain gages, the evaporation hood having been removed. A measured quantity of some noncorrosive, nonevaporative, antifreeze material is generally placed in the rain gage to cause the snow to melt upon entrance. Errors caused by wind are more serious in measuring snowfall than in measuring rain. Snow may also be measured by sampling the depth on a level surface with a metal sampling tube or with the top of the rain gage.

Another method of measuring snowfall is by determining the depth of snow by a snow survey. Such surveys are particularly useful in mountainous areas. These snow courses consist of ranges that are sampled at specified intervals. The sampling equipment consists of specially designed tubes that take a sample of the complete depth of the snow. The sample is then weighed and the equivalent depth of water recorded. Liquid-filled plastic snow pillows may also be placed on the soil surface before the snow season. Water equivalent is determined from a manometer gage that records the snow pressure.

By measuring these snow courses for a period of years and comparing the equivalent water depth with the observed runoff from the snow field, one can make predictions of the amount of runoff. Aerial snow surveys can be made by photographing depth gages or by picking up radio-transmitted signals from gamma-ray depth-measuring equipment. Such devices are set up on snow ranges at suitable locations. These predictions are of particular value in planning for the most effective use of the quantities of irrigation water available during the following summer, as well as in forecasting the probability of spring floods.

2.7 Errors in Measurement

Many errors in measurement result from carelessness in handling the equipment and in analyzing data. Errors characteristic of the nonrecording rain gage of the Weather Bureau type include water creeping up on the measuring stick, evaporation, leaks in the funnel or can, and denting of the cans. The volume of water displaced by the measuring stick is about 2 percent and may be taken as the correction for evaporation.

Another class of errors is due to obstructions such as trees, buildings, and uneven topography. These errors can be minimized by proper location of the rain gages. The gages are normally placed with the opening about 760 mm above the surface of the ground. They should be located so as to minimize turbulence in the wind passing across the gage. A practical rule is to have a clearance of 45 degrees from the vertical center line through the gage, but a safer rule is to be sure that the distance from the obstruction to the gage is equal to at least two times the height of the obstruction.

The wind velocity also affects the amount of water caught. A wind of 16 km/h would cause a deficit catch of about 17 percent, but at 48 km/h the deficit is increased to about 60 percent. Whenever possible, the gage should be located on level ground as the upward or downward wind movement often found on uneven topography may easily affect the amount of precipitation caught.

2.8   The Gaging Network in the United States

Precipitation records have been kept in the United States ever since it was settled; however, only since about 1890 have recording rain gages, giving the intensity of precipitation, been used. Rain gages have steadily increased in number; the gaging network in the United States now consists of about 11 000 nonrecording and 3500 recording instruments. Many of the nonrecording gages are serviced by volunteer personnel; most of the recording equipment is connected with either local, state, or federal installations. The results of these extensive gaging activities are given in the various publications of the Environmental Science Service Administration (ESSA, formerly U.S. Weather Bureau), and in reports of other federal and state agencies.

The ESSA maintains a central data processing center at Asheville, North Carolina. Precipitation as well as other climatological data are placed on computer tapes for rapid processing and evaluation.

ANALYSIS OF PRECIPITATION DATA

Rainfall data are of interest both in a specific locality and over considerable areas. Since a rain gage gives the precipitation at a given point, it is easier to make a point rainfall analysis than to study rainfall over an area.

2.9   Intensity, Duration, and Frequency of Rainfall

One of the most important rainfall characteristics is rainfall intensity, usually expressed in millimeters per hour. Very intense storms are not necessarily more frequent in areas having a high total annual rainfall. Storms of high intensity generally last for fairly short periods and cover small areas. Storms covering large areas are seldom of high intensity but may last several days. The infrequent combination of relatively high intensity and long duration gives large total amounts of rainfall. These storms do much erosion damage and may cause devastating floods. These unusually heavy storms are generally associated with warm-front precipitation. They are most apt to occur when the rate of frontal movement has decreased, when other fronts may pass by at close intervals, when stationary fronts persist in an area for a considerable period, or when tropical cyclones move into the area.

Intense rainstorms of varying duration occur from time to time over almost all portions of the United States; however, the probability of these heavy rainfalls varies with the locality. The first step in designing a water-control facility is to determine the probable recurrence of storms of different intensity and duration so that an economically sized structure can be provided. For most purposes it is not feasible to provide a structure that will withstand the greatest rainfall that has ever occurred. It is often more economical to have a periodic failure than to design for a very intense storm. Where human life is endangered, however, the design should handle runoff from storms even greater than have been recorded. For these purposes, data providing return periods of storms of various intensities and durations are essential. This return period, sometimes called recurrence interval, is defined as the period within which the depth of rainfall for a given duration will be equaled or exceeded once on the average.

A general expression for rainfall intensity is given by

$$i = \frac{KT^x}{t^n} \qquad (2.1)$$

where i = rainfall intensity,
$K, x,$ and n = constants for a given geographic location,
t = duration of storm in minutes,
T = return period in years.

Equation 2.1 has not been widely adopted because of the difficulty in evaluating the constants. Records from all stations in the United States have been statistically analyzed, but the relationship of the variables is not as simple as Eq. 2.1 would indicate.

A complete analysis of rainfall frequency data was prepared by Hershfield (1961). Maps of the United States with isohyet (lines of equal rainfall) include durations from 30 min to 24 h and return periods from 1 to 100 years. The data were obtained primarily from 200 long-record first-order stations that have recording gages. Records were processed through 1958, and the average length of record was 48 years. In addition, more than 8400 other station records were evaluated.

Records of intense precipitation have been published by the Weather Bureau since 1895. All storms with a duration of 5 min or longer were recorded if the amount of rainfall in millimeters exceeded $5 + 0.25t$, where t is the duration in minutes. Such a storm is often referred to as an excessive storm.

Rainfall frequency maps for 1- and 24-h-long storms and for return periods of 2 and 100 years are given in Fig. 2.6. By linearizing the return period and the duration, Weiss (1962) developed the following equation for the rainfall amount at a given location for partial duration series values:

$$I = 0.0256(C - A)x + 0.000256[(D - C) - (B - A)]xy$$
$$+ 0.01(B - A)y + A \qquad (2.2)$$

where I = rainfall amount in inches,
x = return period variate from Table 2.1,
y = duration variate from Table 2.2,
A = 2-year, 1-h rainfall from Fig. 2.6a,
B = 2-year, 24-h rainfall from Fig. 2.6b,
C = 100-year, 1-h rainfall from Fig. 2.6c,
D = 100-year, 24-h rainfall from Fig. 2.6d.

Equation 2.2 is also valid if rainfall is in millimeters or other units, and is useful for programming on computers to calculate the entire array of frequency–duration values. The values of x and y should be evaluated for each region of the country (see Frederick et al., 1977).

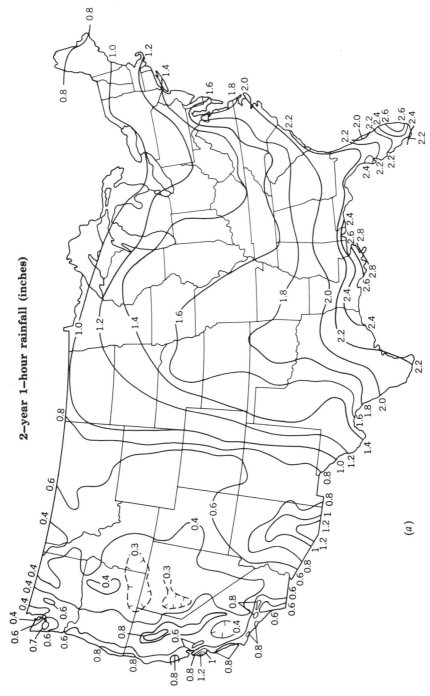

2-year 1-hour rainfall (inches)

(a)

Fig. 2.6 (a) One-hour and 24-hour rainfall in inches to be expected at return periods of 2 and 100 years. (*Source:* Hershfield, 1961.)

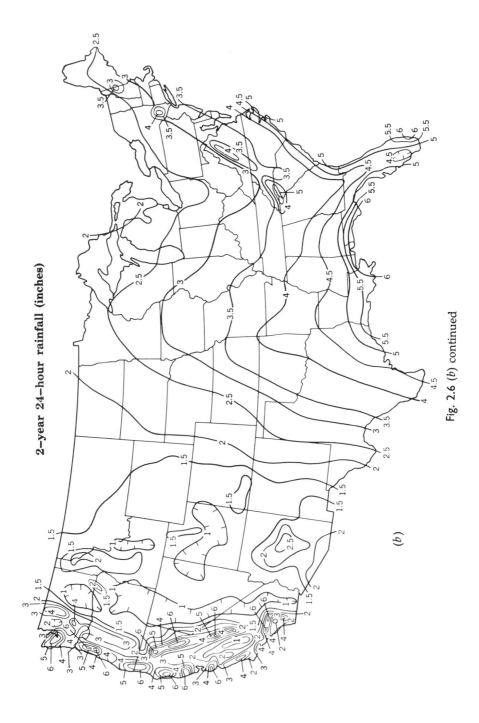

2–year 24–hour rainfall (inches)

Fig. 2.6 (*b*) continued

(*b*)

100-year 1–hour rainfall (inches)

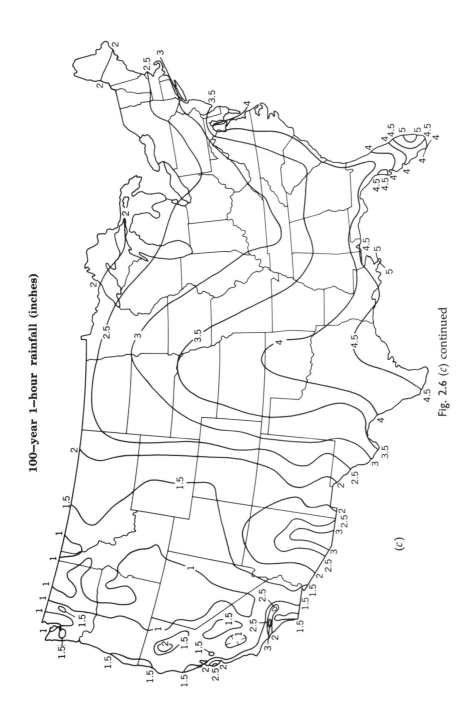

(c)

Fig. 2.6 (c) continued

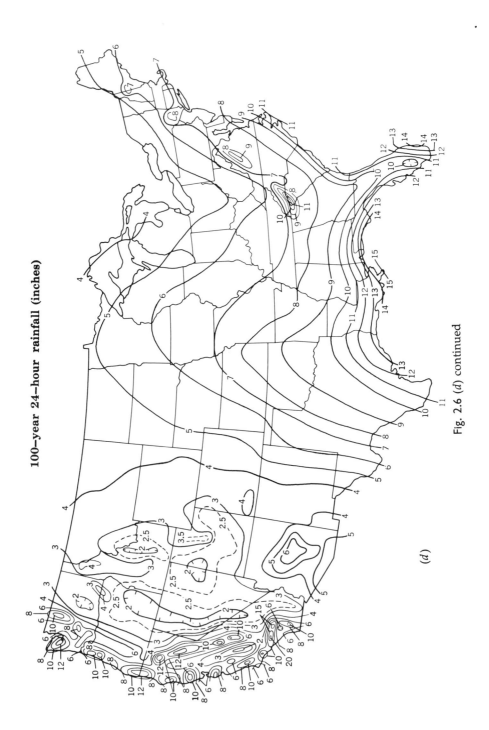

100–year 24–hour rainfall (inches)

(d)

Fig. 2.6 (d) continued

Table 2.1   Linearized Rainfall Frequency Variate for Equation 2.2

Return Period in Years	1	2	5	10	25	50	100
Linearized Variate, x	−6.93	0	9.2	16.1	25.3	32.1	39.1

Source: From Weiss (1962).

Table 2.2   Linearized Rainfall Duration Variate for Equation 2.2

Duration in Hours	0.17	0.33	0.5	0.67	1
Duration in Minutes	(10)	(20)	(30)	(40)	(60)
Linearized Variate, y	−37.0	−24.0	−15.6	−9.4	0
Duration in Hours	2	3	6	12	24
Linearized Variate, y	17.6	28.8	49.9	73.4	100.0

Source: Weiss (1962).

☐ Example 2.1

Determine the rainfall intensity for a 20-min storm and a 6-h storm that will occur once in 50 years at Chicago, Illinois.

Solution.   Read from each of the four maps in Fig. 2.6 the following values for Chicago:

Duration (h)	T (2 years)	T (100 years)
1	A = 1.43 in.	C = 2.75 in
24	B = 2.80 in.	D = 5.70 in.

Read from Table 2.1, $x = 32.1$ for $T = 50$ years, and from Table 2.2, $y = -24.0$ and 49.9, for 20-min and 6-h storm durations, respectively. Substitute the above variables in Eq. 2.2 and obtain I to compute i.

Duration (h)	T (years)	I (in.)	I (mm)	i (mm/h)
0.33 (20 min)	50	1.87	47.5	143
6	50	3.84	97.5	16.3

From Eq. 2.2 curves similar to those in Fig. 2.7 for St. Louis, Missouri, can be obtained. As will be explained in the next section, the curves in Fig. 2.7 were developed from the highest annual values for each year of record.   ☐

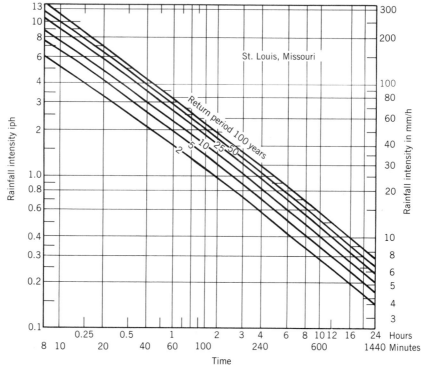

Fig. 2.7　Rainfall intensity–duration–frequency data for St. Louis, Missouri. (Adapted from Hershfield, 1961, and Weiss, 1962.)

2.10 Hydrologic Frequency Analysis

The rainfall data shown in Fig. 2.6 were developed by treating the measured values as statistical variables. Statistical methods for determining frequency distributions have been adopted for many other applications varying from the size distribution of sand grains to the distribution of flood flows. Because of this possibility for wide application, a few of the more common methods are presented.

The relationship between return period and probability of occurrence can be expressed by

$$T = 100/P \qquad (2.3)$$

where　T = return period in years,
　　　　P = probability in percent that an observed event in a given year is equal to or greater than a given event.

Selection of Data. Experience has indicated that many hydrologic events have practically no significant value in the analysis because the hydrologic design of a structure is usually governed only by a few of the extreme conditions. Therefore, the portion of the data that is of insignificant value can be excluded.

One of two methods of selecting data is adopted, either the annual series or the partial-duration series. In the annual series, only the largest single event for each year is selected for analysis. Thus, for 20 years of record, only 20 values would be analyzed. With the partial-duration series, all values above a given base are chosen regardless of the number within a given period. The partial-duration series is applicable if the second largest value (or lower) of the year would affect the design. An example is the design of drainage channels where damage may result from flooding caused largely by flows lower than the annual peak flow. The partial-duration series was selected for the analysis of rainfall given in Fig. 2.6.

The annual and partial-duration series give essentially identical results for return periods greater than 10 years. For example, if the 2-, 5-, and 10-year partial-duration series values selected from the maps in Fig. 2.6 are as given in the table below, the annual series values for corresponding return periods using conversion factors from Hershfield (1961) are as follows:

Return Period (years)	Conversion Factor	Partial Series Values (mm)	Annual Series Values (mm)
2	0.88	76	67
5	0.96	95	91
10	0.99	107	106

Regardless of the method of selecting the data, the values must satisfy two important criteria: (1) that the events be independent of a previous or subsequent event, and (2) that the data for the period of record for analysis must be representative of the long-time record. The first criterion is necessary from a statistical point of view, and the second implies that the predicted values will reflect only the pattern of occurrences from the period of record. Where the data are obtained from the highest annual values (extreme value law), the number of observations during the year should be large. Selection of a water year, such as October 1 to September 30, rather than the calendar year has proved beneficial for some types of data. Brakensiek (1959) found that starting the year on March 1 gave 15 percent better correlation than the calendar year for minimum water yields in Ohio.

Determination of Statistical Parameters.   The next step is to calculate the mean or average value and to compute the standard deviation:

$$s = \left[\frac{\Sigma X^2 - (\Sigma X)^2/n}{n-1} \right]^{1/2} \tag{2.4}$$

where X = measured value,
n = number of values.

The coefficient of variation is $C_v = s/\bar{x}$, and $s^3 = C_v^3(\bar{x})^3$. Then the unbiased estimate of

$$C_s = \frac{n}{(n-1)(n-2)} \frac{\Sigma (X^3)/n - 3(\bar{x}) \Sigma (X^2)/n + 2(\bar{x})^3}{s^3} \tag{2.5}$$

where C_s = coefficient of skew,
$\bar{x}$ = mean value $(\Sigma X/n)$.

Determination of Plotting Positions and Plotting of Data. Before proceeding, however, a brief understanding of the normal-probability curve is necessary. Figure 2.8a shows the normal curve. Deviation of the variable is plotted on the x axis and probability of occurrence on the y axis. Many phenomena in nature and some hydrologic events follow this distribution. Note that the mean occurs at zero deviation and that 34.1 percent of the values are within one standard deviation in each direction from the mean. Often the data are skewed as shown by the dashed curve in Fig. 2.8a.

Data are generally analyzed by one of two probability laws, the extreme value law or the log-probability law. The extreme value law postulates that the annual maximum values approach a definite pattern of frequency distribution when the number of observations in each year becomes large; the log-probability law states that the logarithms of the values are normally distributed.

A statistical approach for determining the plotting position as described by Chow (1951) will be followed. For an understanding of this method some hypothetical data shown in Fig. 2.8b will illustrate the method. Any point $X_c = \bar{x} + \Delta X$. The departure from the mean, ΔX, may be positive or negative and may be irregular and variable. From a statistical point of view, X_c possesses two important properties: (1) the tendency to deviate from the mean and (2) the frequency of occurrence. The first property is measured by the standard deviation defined in Eq.

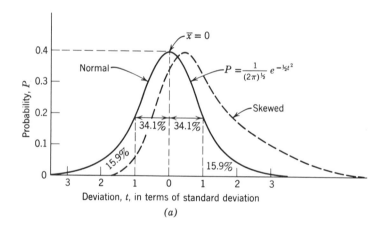

(a)

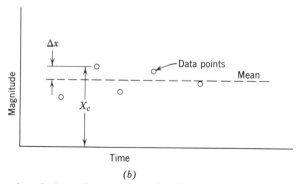

(b)

Fig. 2.8 (a) Normal and skewed probability distributions. (b) Occurrence of hydrologic events.

2.4. The second property is measured by a term called the *frequency factor, K,* which depends on the law of occurrence of a particular hydrologic event under consideration. This frequency factor is defined by the equation

$$\Delta X = sK \tag{2.6}$$

By substituting for ΔX and $s = C_v \bar{x}$,

$$X_c = \bar{x}(1 + C_v K) \tag{2.7}$$

The extreme value distribution postulates a coefficient of skewness C_s of 1.139 and $C_v = 0.363$. In this sense, it is a special case of the log-probability law, which can be applied for any coefficient of skewness. For most hydrologic data the log-probability law is suitable. The procedure for using Chow's method is illustrated in Example 2.2.

For the log-probability law the frequency factor K is a function of the return period (also P), the coefficient of variation, and the coefficient of skew. Theoretically, Chow (1954) has shown that

$$C_s = 3C_v + C_v^3 \tag{2.8}$$

These corresponding values are shown in Table 2.3. In the log-probability law the coefficient of skew is not a constant as in the extreme value law. By the selection of the appropriate scale factor for the probability graph, Chow (1954) has proposed a method whereby data with various coefficients of skew can be plotted as a straight line. When data do not fit a log-normal statistical distribution, McGuinness and Brakensiek (1964) have developed another procedure for computing a rectifying constant to obtain a straight-line plot.

Table 2.3 Theoretical Log-Probability Frequency Factors

	Return Period (years)					
	1.01	2	5	20	100	
	Probability (%) Equal to or Greater Than Given Variate					*Corresponding[a]*
C_s	99	50	20	5	1	C_v
0	−2.33	0	0.84	1.64	2.33	0
0.5	−1.98	−0.09	0.80	1.77	2.70	0.166
1.0	−1.68	−0.15	0.75	1.85	3.03	0.324
1.139[b]	−1.61	−0.16	0.73	1.86	3.11	0.363
1.4	−1.49	−0.19	0.69	1.88	3.26	0.436
1.5	−1.45	−0.20	0.68	1.89	3.31	0.462
2.0	−1.28	−0.24	0.61	1.89	3.52	0.596
3.0	−1.04	−0.28	0.51	1.85	3.78	0.818
4.0	−0.90	−0.29	0.42	1.78	3.91	1.000

[a] C_v applies for log-probability law only.
[b] For this value of C_s, the extreme value law also applies.
Source: Chow (1954).

A number of empirical equations have been developed for plotting the probability of observed events on probability paper. The following formula for the return period has been adopted by the American Society of Civil Engineers and is sometimes referred to as Gumbel's equation:

$$T = \frac{N + 1}{m}$$

(2.9)

where T = return period in years,
   N = total number of statistical events,
   m = rank of events arranged in descending order of magnitude.

In Eq. 2.9, $m = 1$ for the largest value and $m = N$ for the smallest value. This equation has a statistical basis (Gumbel, 1954) for extreme values, but its greatest merit is its simplicity. For purposes of comparison with the theoretical curve in Example 2.2, the observed data are plotted according to Eq. 2.9 in Fig. 2.9.

Adequacy of Length of Record.   The adequacy of the length of record for a given level of significance is given by Mockus (1960) as

$$Y = (4.30t \log_{10} R)^2 + 6$$

(2.10)

where Y = minimum acceptable years of record,
   t = Student's statistical value at the 90 percent level of significance with $(Y - 6)$ degrees of freedom,
   R = ratio of magnitude of the 100-year event to the 2-year event.

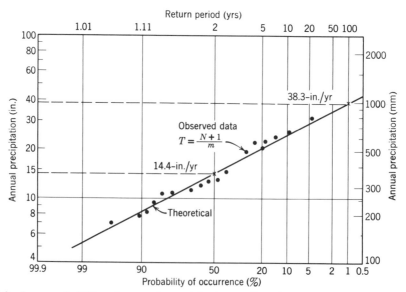

Fig. 2.9 Log–probability of annual precipitation at Los Angeles as computed in Example 2.2.

☐ *Example 2.2*

Determine the annual precipitation for return periods of 1, 2, 5, 20, and 100 years at Los Angeles by the log-probability law based on 20 years of annual rainfall shown below. For a 90 percent probability of occurrence, is the length of record adequate?

Solution.   By calculation, using Eq. 2.4,

$$s = 176 \text{ mm } (6.92 \text{ in.})$$

Year	Millimeters (in.)/year	Year	Millimeters (in.)/year
1934	371 (14.6)	1944	488 (19.2)
1935	551 (21.7)	1945	295 (11.6)
1936	307 (12.1)	1946	295 (11.6)
1937	569 (22.4)	1947	323 (12.7)
1938	594 (23.4)	1948	183 (7.2)
1939	333 (13.1)	1949	203 (8.0)
1940	488 (19.2)	1950	269 (10.6)
1941	833 (32.8)	1951	208 (8.2)
1942	284 (11.2)	1952	665 (26.2)
1943	462 (18.2)	1953	241 (9.5)

and mean annual rainfall,

$$\bar{x} = 7962/20 = 398 \text{ mm } (15.68 \text{ in.})$$

and $C_v = s/\bar{x} = 176/398 = 0.442$. From Eq. 2.7, $X_c = 398(1 + 0.442K)$.

By interpolation from Table 2.3 for $C_v = 0.442$, read values of K for each P value and compute X_c as shown below. These X_c values are plotted in Fig. 2.9 on log-probability paper.

P (%)	T (years)	K	X_c [mm (in.)]
99	1.01	−1.48	138 (5.43)
50	2	−0.19	365 (14.36)
20	5	0.69	520 (20.46)
5	20	1.88	729 (28.71)
1	100	3.27	973 (38.31)

The adequacy of length of record from Eq. 2.10 can be computed where the Student $t = 1.796$ for $(17 - 6) = 11$ degrees of freedom at the 90 percent level of significance. By trial and error, 17 for the minimum acceptable years of record was selected to agree with 16.8 as computed below. Student t values can be obtained from most statistics textbooks. From Fig. 2.9 or from X_c above,

$$R = 973/365 = 2.67$$

and substituting in Eq. 2.10,

$$Y = (4.30 \times 1.796 \times \log_{10} 2.67)^2 + 6 = 16.8 \text{ years}$$

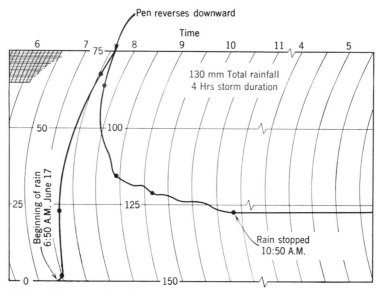

Fig. 2.10   Rain gage chart from a rain gage of the reversible, recording type.

Since the actual length of record of 20 years is greater than the minimum acceptable, the estimate of 973 mm (38.31 in.) for $T = 100$ years can be expected to be reasonably reliable. Extreme caution should be taken when the computed Y value is greater than the length of record.   □

2.11   Point Rainfall Analysis

A typical recording rain gage chart is given in Fig. 2.10. The line on the chart is a cumulative rainfall curve, the slope of the line being proportional to the intensity of the rainfall. The peak is the point of reversal of the recording gage. To analyze the chart, the time and amount of rain should be selected from representative points where the rainfall rate changes so that the data will represent the curve on the chart. These points may be tabulated as in Table 2.4, with cumulative rainfall and intensity for various periods also being recorded.

Table 2.4   Rain Gage Chart Analysis

Time (A.M.)	Time Interval (min)	Cumulative Time (min)	Rainfall during Interval[a] (mm)	Cumulative Rainfall (mm)	Rainfall Intensity for Interval (mm/h)
6.50					
7:00	10	10	1	1	6
7:10	10	20	10	11	60
7:15	5	25	11	22	132
7:35	20	45	46	68	138
7:45	10	55	19	87	114
8:25	40	95	31	118	47
9:10	45	140	6	124	8
10:50	100	240	6	130	4

[a]Corrected rainfall based on nonrecording gage depth.

To determine the highest return period for a desired duration, the time period must be selected from the most intense portion of the storm. By referring to Fig. 2.7, the appropriate return period can be determined.

☐ Example 2.3

Determine the return period for the maximum rainfall intensity occurring for any 20-min period and for the first 140 min (2.33 h) during the storm shown in Table 2.4 for St. Louis, Missouri.

Solution.   From Table 2.4 the maximum rainfall intensity for 20 min is 138 mm/h (5.43 iph). Interpolating from Fig. 2.7, read a return period of 20 years. For the 140-min storm, the average intensity is 124/2.33 = 53 mm/h, (2.1 iph), for which the return period is over 100 years.   ☐

Mass rainfall curves, required for some types of analyses, may be obtained by plotting the cumulative rainfall against time as in Fig. 2.11a. It is also often convenient to plot the rainfall intensity for increments of time as illustrated in Fig. 2.11b.

2.12   Classification of Storms

Since no two rainstorms have exactly the same time–intensity relationships, it is often convenient to group storms with regard to their characteristics. The most common characteristics used in such groupings are the intensity of the storm and the pattern of the rainfall intensity histogram.

The pattern of a storm is determined by the arrangement of the rainfall intensity histogram. Storm patterns are important because they are one of the factors determining the shape of the runoff hydrograph. Arbitrarily selected storm patterns of rainfall intensities shown in Fig. 2.12 are uniform intensity, advanced pattern, intermediate pattern, and delayed pattern. The advanced pattern of rainfall brings higher intensities when the infiltration rate is the greatest (Chapter 3), thus causing some reduction in the runoff peaks. On the other hand, the delayed pattern causes higher runoff peaks, as the high intensities occur when the infiltration is at a minimum and depression storage has been largely satisfied. In general, the cold front produces a storm of an advanced type, and the warm front a

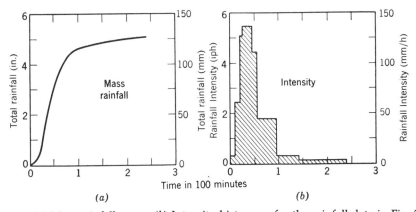

Fig. 2.11   (a) Mass rainfall curve. (b) Intensity histogram for the rainfall data in Fig. 2.10.

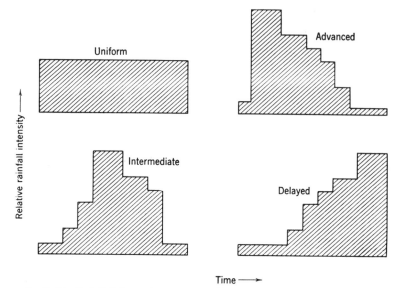

Fig. 2.12 Rainfall intensity patterns. (*Source:* Horner and Jens, 1942.)

uniform or intermediate pattern. In Ohio, in a study of 1-h storms of all intensity classes, the advanced pattern was found to be the most common.

2.13 Average Depth over Area

The rainfall depths given in Fig. 2.6 are point rainfall. Where the average depth over a watershed area must be determined, such amounts can be adjusted for different-duration storms as indicated in Fig. 2.13. The design rainfall may be

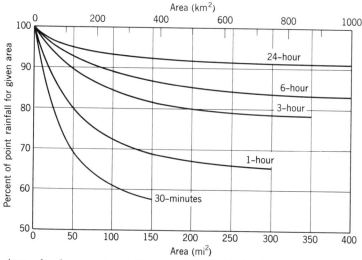

Fig. 2.13 Area–depth curves in relation to point rainfall. (Redrawn from Hershfield, 1961.)

considered as the maximum for the storm, and thus the average over a watershed will be less than the maximum, as the curves show.

Where a gaging network has been established or records are available for a given watershed, the following procedures are applicable for determining the average depth of precipitation over an area. If only one rain gage is used, the rainfall is applied over the entire area. Where several gages are available, the simplest method is to take the arithmetic mean. Since each gage may not represent equal areas, other methods often give greater accuracy.

2.14 Thiessen Method

The Thiessen method is illustrated in Fig. 2.14. The location of the rain gages is plotted on a map of the watershed. Straight lines are then drawn between the rain gages. Perpendicular bisectors are then constructed on these connecting lines in such a way that the bisectors enclose areas referred to as Thiessen polygons. All points within one polygon will be closer to its rain gage than to any of the others. The rain recorded is then considered to represent the precipitation within the appropriate polygon area.

Some difficulty may be encountered in determining which connecting lines to construct in forming the sides of the polygon. Though in general the shorter lines are used, the proper lines can best be determined by a trial-and-error procedure. Since only one set of Thiessen polygons generally needs to be drawn for a given watershed and set of rain gage locations, this procedure does not present a serious limitation. The average precipitation over a watershed can be determined by using the equation

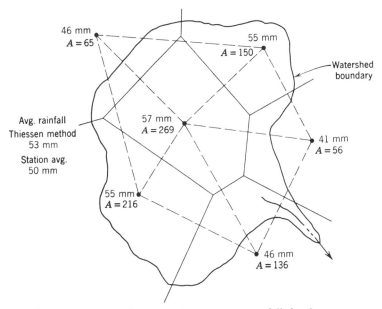

Fig. 2.14 Thiessen network for computing average rainfall depth over a watershed.

$$P = \frac{A_1 P_1 + A_2 P_2 + \cdots + A_n P_n}{A} \qquad (2.11)$$

where P represents the average depth of rainfall in a watershed of area A and P_1, $P_2, \ldots, P_n$ represent the rainfall depth in the polygon having areas A_1, $A_2, \ldots, A_n$ within the watershed.

□ *Example 2.4*
_____

A storm on the watershed illustrated in Fig. 2.14 produces rainfall at the various gage locations as indicated. Compare the average precipitation as determined by the average depth and Thiessen methods.

Solution. By the average depth method the arithmetic mean is 50 mm (1.97 in.). By the Thiessen method, the areas represented by the various rain gages are determined with a planimeter and substituted in Eq. 2.11.

$$P = \frac{(65)(46) + (150)(55) + (269)(57) + (216)(55) + (56)(41) + (136)(46)}{892}$$

$$P = 53 \text{ mm (2.08 in.)} \qquad \qquad □$$

2.15 Isohyetal Method

The isohyetal method consists of recording the depth of rainfall at the locations of the various rain gages and plotting isohyets (lines of equal rainfall) by the same methods used for locating contour lines on topographic maps. The area between isohyetals may then be measured and the average rainfall determined by the above equation.

The choice of the method of analysis will depend partly on the area of the watershed, the number of rain gages, the distribution of the rain gages, and, in some situations, the character of the rainstorm. Depth–area curves, where needed, can be constructed from isohyetal maps.

DISTRIBUTION OF PRECIPITATION IN THE UNITED STATES

2.16 Time Distribution

Diurnal. The time of day in which precipitation may be expected to occur depends on the type of precipitation. Frontal storms are not much influenced by diurnal effects. Storms of the convective type, since they are due to surface heating, are much more likely to occur in the afternoon.

Seasonal. That rainfall be distributed throughout the growing season is important. A considerable difference in the seasonal distribution of precipitation throughout the United States is shown in Fig. 2.15. Even in the areas of the West Coast, where annual precipitation is high, summertime precipitation is generally very low, making irrigation necessary. In the Middle West and South the monthly

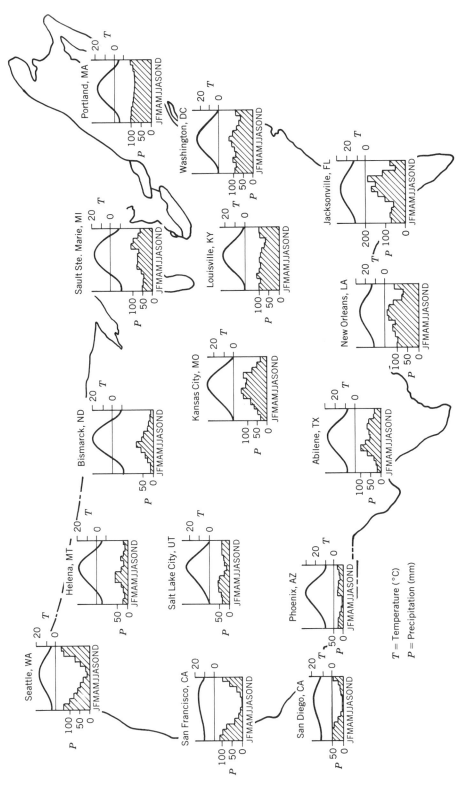

Fig. 2.15 Mean monthly precipitation (millimeters) and mean monthly temperature (degrees Centigrade) for selected locations in the United States. (Redrawn and revised from Rouse, 1950.)

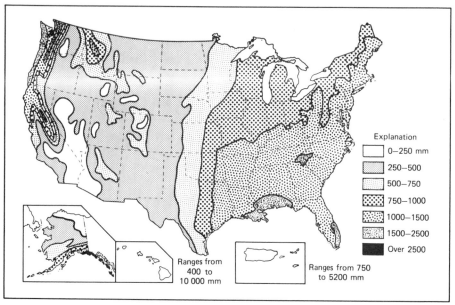

Fig. 2.16 Average annual precipitation in the United States in millimeters. (Redrawn from USDA, 1989.)

summertime precipitation is generally somewhat higher than the monthly average. In the eastern portion of the United States there is little difference between summer and winter precipitation.

Annual. The annual rainfall over the United States is shown in Fig. 2.16. Annual rainfall amounts vary from less than 100 mm to over 2500 mm in some mountainous areas. Annual precipitation is not in itself a good index of the amount of water available for plant growth because evaporation, seasonal distribution, and water-holding capacity of the soil vary with geographical location.

Cycles. That precipitation occurs in cycles has often been suggested; however, as yet there has been no statistical proof that such cycles exist or that there is any relationship between such cycles and other natural phenomena. Some evidence exists that sunspot activity is related to summer temperature and severe droughts. Thompson (1973) showed that average July–August temperatures in the Corn Belt since 1900 follow roughly about a 20-year cycle of sunspot numbers. Similar observations have been made in other countries at the same latitudes. A widely held view is that weather is a random variable.

2.17 Geographical Distribution

The geographical distribution of rainfall over the United States is largely determined by the location of large bodies of water, by the movement of the major air masses, and by changes in elevation. Figure 2.17 illustrates the effect of elevation and of moist air-mass movement on annual rainfall. It presents a section of the United States along the 40th parallel. Moving from west to east, one notes that the

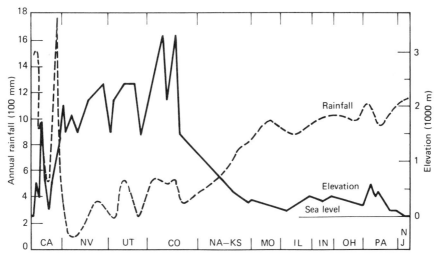

Fig. 2.17 Average annual rainfall and elevation across the United States along the 40th parallel of latitude.

highest rainfall occurs as the air is first pushed up by the mountains, with lesser rises as the drier air is pushed to higher elevations. As the air moves down the mountain slopes, lower annual rainfall is generally observed. The rainfall does not increase until the effects of the maritime tropical air moving up from the Gulf of Mexico become apparent. Then the rainfall gradually increases as the eastern boundary of the United States is approached, with the effect of the Appalachian mountains again apparent.

REFERENCES

Brakensiek, D. L. (1959). "Selecting the Water Year for Small Agricultural Water-sheds." *ASAE Trans.* **2** (1), 5–8, 10.

Chow, V. T. (1951). "General Formula for Hydrologic Frequency Analysis." *Am. Geophys. Union Trans.* **32** (April), 231–237.

——— (1954). "The Log-Probability Law and Its Engineering Applications." *ASCE* **80** (Separate No. 536).

Gumbel, E. J. (1954). *Statistical Theory of Extreme Values and Some Practical Applications.* Applied Mathematics Series 33. U.S. Bureau of Standards, Washington, DC.

Frederick, R. H., V. A. Myers, and E. P. Auciello (1977). *Five- to 60-Minute Precipitation Frequency for the Eastern and Central U.S.* NOAA Tech. Memo. HYDRO-35. Silver Spring, MD.

Haan, C. T. (1977). *Statistical Methods in Hydrology.* Iowa State University Press, Ames, IA.

Hershfield, D. N. (1961). *Rainfall Frequency Atlas of the United States.* U.S. Weather Bureau Tech. Paper 40, May. Washington, DC.

Horner, W. W., and S. W. Jens (1942). "Surface Runoff Determination from Rainfall Without Using Coefficients." *Trans. ASCE* **107**, 1039–1117.

Laws, J. O., and D. A. Parsons (1943). Hydrology Rep., *The Relation of Raindrop-Size to Intensity*. Am Geophys. Union, Pt. 2, pp. 452–460. Washington, DC.

Linsley, R. K., M. A. Kohler, and J. L. H. Paulhus (1982). *Hydrology for Engineers*. 3rd ed. McGraw-Hill, New York.

McCuen, R. H. (1989). *Hydrologic Analysis and Design*. Prentice-Hall, Englewood Cliffs, NJ.

McGuinness, J. L., and D. L. Brakensiek (1964). *Simplified Techniques for Fitting Frequency Distributions to Hydrologic Data*. ARS Agricultural Handbook 259. GPO, Washington, DC.

Mockus, V. (1960). "Selecting a Flood-Frequency Method." *ASAE Trans.* 3, 48–51, 54.

Rouse, H. (1950). *Engineering Hydraulics*. Wiley, New York.

Thompson, L. M. (1973). "Cyclical Weather Patterns in the Middle Latitudes." *J. Soil Water Cons.* **28**, 87–89.

Weiss, L. L. (1962). "A General Relation Between Frequency and Duration of Precipitation." *Mon. Weather Rev.* **90**, 87–88.

Wischmeier, W. W., and D. D. Smith (1958). "Rainfall Energy and Its Relation to Soil Loss." *Trans. Am. Geophys. Union* **39**, 285–291.

U.S. Department of Agriculture (USDA) (1989). *The Second Appraisal. Soil, Water, and Related Resources on Nonfederal Land in the United States*. Washington, DC.

PROBLEMS

2.1   Determine the total rainfall to be expected once in 5, 25, and 100 years for a 60-min storm at your present location.

2.2   Determine the maximum rainfall intensity to be expected once in 10 years for storms of durations 10, 30, 120, and 360 min, respectively, at your present location.

2.3   From rainfall data given in Table 2.4 determine the maximum rainfall intensity for any 8-, 30-, and 240-min periods. Determine the return periods for these intensities if the storm occurred in St. Louis, Missouri.

2.4   Compute the average rainfall for a given watershed by the Thiessen method from the following data. How do the weighted average and the station average compare?

Rain Gage	Area [ha (ac)]	Rainfall [mm (in.)]
A	14.0 (34.6)	58 (2.30)
B	4.5 (11.2)	41 (1.60)
C	5.3 (13.2)	51 (2.02)
D	4.9 (12.1)	43 (1.71)

2.5   During a 60-min storm the following amounts of rain fell during successive 15-min intervals: 33 mm (1.3 in.), 23 (0.9), 15 (0.6), and 5 (0.2). What is the maximum intensity for 15 min and the average intensity? If the storm had occurred in St. Louis, Missouri, how often would you expect such a 60-min storm to occur? What type of storm pattern was it? From what type of rain gage were the data obtained?

2.6 From the local rainfall data supplied by your instructor, calculate and plot the best theoretical probability curve for return periods of 2, 5, 20, and 100 years. For a 90 percent chance of occurrence, is the period of record adequate?

2.7 Plot the data in Problem 2.6 using the ASCE empirical equation for plotting positions.

2.8 If the average rainfall intensity for 50 years of record is 50 mm/h (2.0 iph) for a storm duration of 60 min and the coefficient of variation of the data is 0.324, compute the theoretical rainfall intensities for return periods of 2, 5, 20, and 100 years. Assume that the log-probability law is applicable.

2.9 Compute the mean, standard deviation, and coefficient of variation for the following maximum rainfall intensities for each year of a 5-year period: 178, 127, 126, 102, 102 mm/h (7, 5, 5, 4, 4 iph).

CHAPTER 3

Infiltration, Evaporation, and Transpiration

Three phases of the hydrologic cycle of particular interest in agriculture are infiltration, evaporation, and transpiration. Infiltration is the passage of water into the soil surface and is distinguished from percolation, which is the movement of water through the soil profile. Evaporation is the process by which water is returned to the air from a liquid to a gaseous state. Transpiration is evaporation from plants. About three fourths of the total precipitation on the land areas of the world returns directly to the atmosphere by evaporation or transpiration. Most of the balance returns to the ocean as surface or subsurface flow. Evaporation and transpiration are difficult to separate and are often considered together and called evapotranspiration.

Infiltration is of particular interest, for if water is to be conserved in the soil and made available to plants, it must first pass through the soil surface. If the infiltration rate is high, less water will pass over the soil surface and erosion will be reduced. In this way runoff quantities and peaks are lowered.

Evaporation, which may occur either from the water surface or from water on soil particles, is important for water conservation. Evapotranspiration is required for determining irrigation requirements for crops as well as water storage in ponds and reservoirs. High evapotranspiration from such crops as grass is beneficial for drainage because of the increased capacity of the soil for storing water.

INFILTRATION

The term *infiltration* refers specifically to entry of water into the soil surface. Infiltration rate has the dimensions of volume per unit of time per unit of area. These units reduce to depth per unit time. Infiltration should not be confused with hydraulic conductivity nor with soil capillary conductivity. Infiltration is the sole source of soil water to sustain the growth of vegetation and of the ground water supply of wells, springs, and streams.

The movement of water into the soil by infiltration may be limited by any restriction to the flow of water through the soil profile. Although such restriction often occurs at the soil surface, it may occur at some point in the lower ranges of the profile. The most important items influencing the rate of infiltration have to do with the physical characteristics of the soil and the cover on the soil surface, but such other factors as soil water, temperature, and rainfall intensity are also involved.

The one-dimensional flow of water through a saturated homogeneous soil can be computed by the Darcy equation

$$q = KhA/L \tag{3.1}$$

where q = the flow rate (L^3/T),
K = hydraulic conductivity of the flow medium (L/T),
h = head or potential causing flow (L),
A = cross–sectional area of flow (L^2),
L = length of the flow path (L).

The flow path may be vertically downward (as during infiltration), horizontal, or upward. The equation is valid so long as the velocity of flow and the size of soil particles are such that the Reynolds number is less than 1. Hydraulic conductivity K is a function of the effective diameter of the soil pores and the density and dynamic viscosity of the fluid.

Application of the Darcy equation is more difficult for two- and three-dimensional flow systems that have complex boundary conditions. Where infiltration is through two layers, such as the topsoil and the subsoil, the average hydraulic conductivity K can be computed from

$$K = L/(L_1/K_1 + L_2/K_2) \tag{3.2}$$

where
L = the total length of flow through all layers (L),
subscripts 1 and 2 = the soil layers.

Another L/K term should be added for each additional layer. Darcy's law is analogous to Ohm's law for electrical flow and to Fourier's law for heat flow.

Infiltration into unsaturated soil is defined by the differential equation (Klute, 1952)

$$\frac{\partial \theta}{\partial t} = \frac{\partial}{\partial z}\left(K\frac{\partial \phi}{\partial z}\right) + \frac{\partial}{\partial z}(Kg) \tag{3.3}$$

where θ = the moisture content in volume of water per unit volume of soil,
K = the unsaturated hydraulic conductivity (L/T),
ϕ = the capillary potential (L),
g = gravitational constant (L/T^2),
z = the coordinate in the vertical direction (L).

A general analytical solution to Eq. 3.3 cannot be obtained because both K and ϕ are functions of θ. Graphical solutions have been made and numerical solutions are practical with computers.

3.1 Soil Factors

Soil functions essentially as a pervious medium that provides a large number of passageways for water to move into the surface. The effectiveness of the soil as an agent for transporting water depends largely on the size and permanency of these channels. In general the size of the passageways and the infiltration into the soil are dependent on (1) the size of the particles that make up the soil, (2) the degree of aggregation between the individual particles, and (3) the arrangement of the particles and aggregates. The larger the pore size and the greater the continuity of the pores that can be maintained, the greater is the resulting infiltration rate.

The importance of maintaining permanent channels, particularly at the soil surface, is critical. The usual rapid reduction in the rate of intake of water through the surface is accompanied by the formation of a thin compact layer on the surface. This layer is a result of severe breakdown of soil structure caused in part by the beating action of raindrops and in part by an assorting action of the water flowing over the surface, fitting the fine particles around the larger ones to form a relatively impervious seal, giving the surface of the soil a slick appearance.

This surface-sealing effect can be largely eliminated when the soil surface is protected by mulch, crop residue, or by some other permeable mechanical protection. The effectiveness of such protection is illustrated in Fig. 3.1, which first shows the constant infiltration rate of soil covered by straw. After 40 min of infiltration at a constant rate, the straw was removed and the infiltration rate dropped to about one sixth of its original value. The straw had protected against the formation of the impervious surface layer, and when the straw was removed the impermeable layer developed quickly through the beating action of the raindrops. By removing the puddled surface layer of soil and protecting the newly exposed soil surface with a layer of burlap, the infiltration rate increased to a new high value. When after 40 min the burlap was again removed, the soil surface puddled, and the infiltration rate fell to a new low.

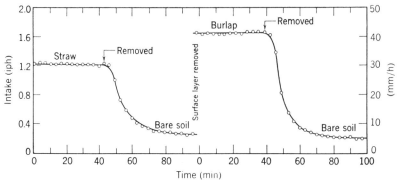

Fig. 3.1 Effect of protective cover on infiltration rates. (Redrawn from Duley, 1939.)

3.2 Vegetation

Surface sealing can be greatly reduced by vegetation. In general, vegetative cover and surface condition have more influence on infiltration rates than do the soil type and texture. The protective cover may be grasses or other close-growing vegetation as well as mulches. It has been shown that when infiltration rates are determined for soil protected by vegetation and the vegetation is removed, surface sealing occurs and infiltration drops much as illustrated in Fig. 3.1. Figure 3.2 gives a number of infiltration curves for unprotected soil and for several surface cover conditions as determined for three South Carolina soils.

3.3 Other Factors

Other factors affecting infiltration include land slope, antecedent soil moisture, and water temperature (a special case being frozen soil). The effect of slope on rate of infiltration has generally been shown to be small, and to be more important on slopes less than 2 percent than on steeper gradients. The effect of slopes steeper than 2 percent on infiltration is not significant.

Soil water generally reduces or limits the infiltration rate. The reduction is due in large part to the fact that water causes some of the colloids in the soil to swell and thereby reduces both the pore space and the rate of water movement. Consequently, in making infiltration tests in the field, it is customary to make both a dry soil run and a wet soil run, often 24 h later. Design is usually based on the minimum values obtained. In a completely saturated soil underlain with an impervious layer or layers, infiltration will be zero.

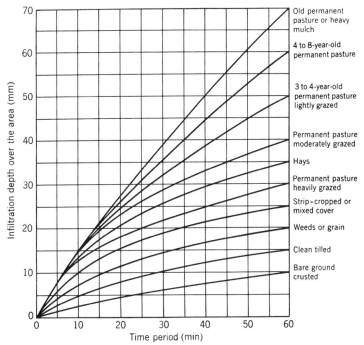

Fig. 3.2 Typical mass infiltration curves. (Redrawn from Holtan and Kirkpatrick, 1950.)

The effect of water temperature on infiltration is not significant, perhaps because the soil changes the temperature of the entering water and the size of the pore spaces may change with temperature changes. Although freezing of the soil surface greatly reduces its infiltration rate, freezing does not necessarily render the soil impervious.

3.4   Soil Additives

The physical characteristics of the soil, including the infiltration capacity, can be changed by adding chemicals. In general these additives are one of two types. The first type consists of materials that add to the permanency of the soil aggregate formations and thereby generally improve the soil structure. This improved structure causes considerable increases in both infiltration and percolation rates. The second type of additive is essentially a wetting agent that does not change the soil but instead changes the angle of contact of the soil water with the soil surface and thereby the rate at which water can move through the soil. It therefore affects water movement at depths greater than the zone of application. In general, it may be necessary to reapply these wetting agents periodically as they leach out with continued water application.

Additives are also applied that decrease infiltration rates. One group of chemical additives reduces infiltration capacity by causing soil particles to swell and to become hydrophilic. Fine clays are sometimes added to soils. These swell and seal soil pores to reduce infiltration rates. Partial or complete sealants, such as petroleum or plastic films, are applied to soil surfaces to decrease or prevent infiltration.

Decreased infiltration is desired to decrease losses of water from reservoirs or irrigation canals or to increase runoff to surface water supplies for direct use or for ground water recharge.

3.5   Predicting Infiltration

Infiltration data are commonly expressed graphically with rate as the ordinate and time as the abscissa. Figure 3.3 presents a typical infiltration curve. Here, as usual, the potential infiltration capacity initially exceeds the rate of water application;

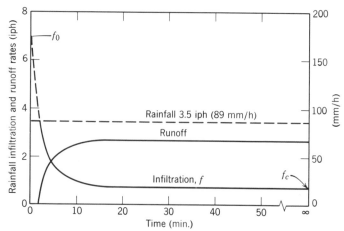

Fig. 3.3   Typical infiltration and runoff curves developed from infiltrometer data.

however, as the soil pores fill with water, and as surface sealing takes place, the rate of water intake gradually decreases. It then normally approaches a constant value which may be taken as the infiltration rate of the soil.

The infiltration curve in Fig. 3.3 can be expressed by (Horton, 1939)

$$f = f_c + (f_0 - f_c)e^{-kt} \qquad (3.4)$$

where f = infiltration capacity or the maximum rate at which soil under a given condition can take water through its surface (L/T),
f_c = the constant infiltration capacity as t approaches infinity (L/T),
f_0 = infiltration capacity at the onset of infiltration (L/T),
k = a positive constant for a given soil and initial condition,
t = time (T).

In Fig. 3.3 the infiltration capacity did not drop to equal the rainfall rate until several minutes after the onset of infiltration. During this initial period the infiltration rate was equal to the rainfall rate and less than the infiltration capacity. Up to this time the runoff rate was thus zero. As the infiltration capacity fell, below the rainfall rate, runoff occurred.

Several other methods of predicting infiltration have been developed, one of which is the Green and Ampt equation (Green and Ampt, 1911). The equation was developed by applying the Darcy equation (Eq. 3.1) to the wetted soil zone and assuming vertical flow, uniform water content, and uniform soil hydraulic conductivity (near saturation). The wetting front is considered an abrupt interface between wetted and nonwetted soil (unsaturated) and is characterized by a capillary potential. The parameters in the Green and Ampt model are effective porosity, capillary potential, and hydraulic conductivity. These can be estimated from readily measured soil properties, using equations presented by Rawls and co-workers (Rawls and Brakensiek, 1989; Rawls et al., 1982). The drainage computer model DRAINMOD incorporates the Green and Ampt infiltration equation and is an example of a practical application (see Chapter 14). This and other infiltration models have been developed, but details are beyond the scope of this text (Jury et al., 1991).

Because of the difficulty in evaluating infiltration, the SCS (1972) divided all soils into four hydrologic groups—A, B, C, and D—on the basis of infiltration rates. The procedure for applying infiltration data to obtain runoff is discussed in Chapter 4.

EVAPORATION AND TRANSPIRATION

Evaporation is the transfer of liquid water into the atmosphere. The water molecules, both in the air and in the water, are in rapid motion. Evaporation occurs when the number of moving molecules that break from the water surface and escape into the air as vapor is larger than the number that reenter the water surface from the air and become entrapped in the liquid.

Transpiration is the process through which water vapor passes into the atmosphere through the tissues of living plants. In areas of growing plants, water passes into the atmosphere by evaporation from soil surfaces and by transpiration from plants. For convenience in analyzing water transfer in this common situation, the two are combined and referred to as evapotranspiration.

The amount of water that passes through plants by the transpiration process is often a substantial portion of the total water available during the growing season. It can vary from practically nothing to as much as 635 mm in depth, depending largely on the water available, the kind of plant, the density of plant growth, the amount of sunshine, and the soil fertility and structure. Less than 1 percent is actually retained by the growing organisms. That the rate of evaporation or transpiration increases with the rise in temperature of the surface is to be expected as vapor pressure increases with increases in temperature. It has been shown that mean monthly air temperatures do not alone provide a satisfactory means of predicting mean monthly evaporation.

Wind increases the rate of evaporation, particularly as it disperses the moist layer found directly over the evaporating water surface under stagnant conditions. Because of this mixing, the characteristics of the atmosphere above the surface are of interest. As might be expected, from the decreased concentration of water molecules, evaporation increases with decreased barometric pressure. Likewise, if other conditions are unchanged, there is greater evaporation at higher elevations. Also the rate of evaporation has been found to decrease with increases in the salt content of the water.

Basic methods of predicting evaporation and evapotranspiration can be grouped into three categories:

(1) *Mass Transfer.* This approach recognizes that water moves away from evaporating and transpiring surfaces in response to the combined phenomena of turbulent mixing of the air and the vapor pressure gradient. Thornthwaite and Holzman (1942) proposed such a method. Application of methods based on mass transfer principles requires measurements of wind velocity and humidity at two or more elevations and is seldom practical.

(2) *Energy Balance.* Heat is required for evaporation of water, so if there is no change in water temperature the net radiation or heat supplied is a measure of evaporation. Energy balance methods are proving to be practical in application. The Penman (1956) equation is an example of this approach.

(3) *Empirical Methods.* Several such methods, developed from experience and field research, are based primarily on the assumption that the energy available for evaporation is proportional to the temperature. Blaney and Criddle (1950) and Thornthwaite (1948) have proposed equations of this type.

3.6 Evaporation from Water Surfaces

Many evaporation formulas for free-water surfaces are based on Dalton's law:

$$E = C(e_{\mathbf{s}} - e_{\mathbf{d}}) \tag{3.5}$$

where E = the rate of evaporation,
$\qquad C$ = a constant,

e_s = the saturated vapor pressure at the temperature of the water surface in mm Hg.

e_d = the actual vapor pressure of the air (e_s times relative humidity) in mm Hg,

Rohwer (1931) evaluated the constant in Eq. 3.5 as

$$C = (0.44 + 0.073W)(1.465 - 0.00073p) \tag{3.6}$$

where W = average wind velocity in km/h at a height of 0.15 m,
$\quad\quad p$ = atmospheric pressure in mm Hg at 0°C.

With these units E is in mm/day. To find the evaporation from reservoirs, the calculated E should be multiplied by 0.77.

Meyer (1942) evaluated the constant for pans and shallow ponds (E as mm/month),

$$C = 15 + 0.93W \tag{3.7}$$

and for small lakes and reservoirs,

$$C = 11 + 0.68W \tag{3.8}$$

where W = average wind velocity for the period in km/h at a height of 7.6 m. The vapor pressure, e_d, should be measured at 7.6 m height, and the air temperature is the average of the daily minimum and maximum.

☐ Example 3.1

Compute the evaporation for the month of June from a shallow pond if the surface water temperature is 15.6°C (60°F), the average wind speed is 4.8 km/h (3 mph), and the average temperature and relative humidity at 7.6 m (25 ft) height are 21.1°C (70°F) and 40 percent, respectively.

Solution. Substituting in Eqs. 3.5 and 3.7 (Meyer equation) and the vapor pressures from Fig. 3.4,

$$E = (15 + 0.93 \times 4.8)(13.0 - 18.5 \times 0.40)$$
$$= 109 \text{ mm/month } (4.3 \text{ in./month}) \qquad\qquad ☐$$

Evaporation measurements from free-water surfaces are commonly made using evaporation tanks or pans. The Class A pan, accepted as standard by the U.S. Weather Bureau, is 1.22 m in diameter and 254 mm deep and requires a water depth of about 190 mm. The pan is supported about 150 mm above the ground so that the air may circulate under it, and the materials and color of the pan are specified. This pan is widely used in the United States. Descriptions of other styles of pans and correction coefficients for converting evaporation data from a pan of one type to that of another are given by Meinzer (1949). These pans have higher rates of evaporation than do larger free-water surfaces, a factor of about 0.7 being recommended for converting observed evaporation rates to those for larger surface

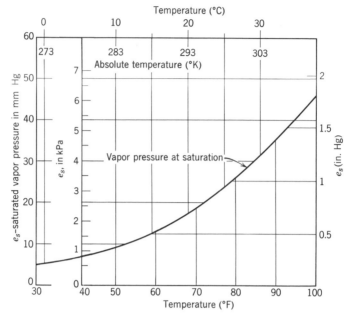

Fig. 3.4 Saturation vapor pressure as a function of temperature.

areas. Because of the "oasis" effect and likely higher water temperatures, evaporation depth is generally higher on small ponds than on large lakes. Although the vapor pressure of highly saline water is slightly less than that of pure water at the same temperature, the effect can be neglected in estimating reservoir evaporation. The geographical distribution of average annual evaporation from shallow lakes is shown in Fig. 3.5.

3.7 Evaporation from Land Surfaces

Because of differences in soil texture and in expected soil water movement, it is difficult to generalize on the amounts of evaporation from soil surfaces. For saturated soils, the evaporation may be expected to be essentially the same as from open free-water surfaces. As the water table drops, however, the evaporation rate will decrease greatly. Evaporation from the soil surface is generally unimportant at water content below field capacity, as soil water movement is very slow when the soil surface is relatively dry. Mulches are effective for several days after a rain. A mulch restricts air movement, maintains a high air vapor pressure near the soil surface, and shields the soil from solar energy, all of which reduce evaporation. Freezing of a bare soil surface causes the surface to become wet, and greatly increases the evaporation rate after thawing.

3.8 Transpiration Ratio

The effectiveness of the plant's use of water in producing dry matter is often given in terms of its transpiration ratio. This is the ratio of the weight of water transpired to the weight of dry matter in the plant. It therefore varies with the same factors as

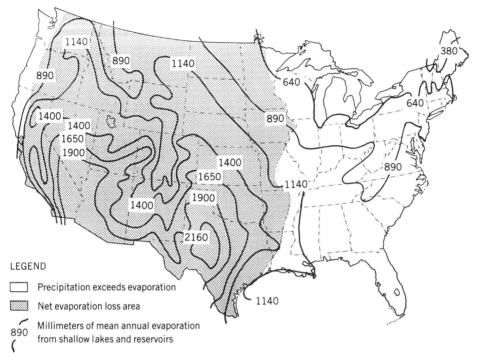

Fig. 3.5 Average annual evaporation from shallow lakes and net evaporative loss area. (Revised from USDA, 1980.)

does transpiration. Approximate transpiration ratios for several common plants are 250 for sorghum, 350 for corn, 450 for red clover, 500 for wheat, 640 for potatoes, and 900 for alfalfa. This ratio is important, especially where irrigation water is limited.

3.9 Evapotranspiration

For convenience, evaporation and transpiration are combined into evapotranspiration (ET), often referred to as consumptive use. The various methods for determining evapotranspiration include (1) tank and lysimeter experiments; (2) field experimental plots where the quantity of water applied is kept small to avoid deep percolation losses and surface runoff is measured; (3) soil water studies, a large number of samples being taken at various depths in the root zone; (4) analysis of climatological data; (5) integration methods where the water used by plants and evaporation from the water and soil surfaces are combined for the entire area involved; and (6) inflow–outflow method for large areas where yearly inflow into the area, annual precipitation, yearly outflow from the area, and the change in ground water level are evaluated.

Many practical applications can be made of evapotranspiration estimates, but the principal use is to predict soil water deficits for irrigation. Analyzing weather records and estimating evapotranspiration rates, drought frequencies, and excess water periods can show potential needs for irrigation and drainage. Similar studies to determine available tillage and harvesting days are of value in selecting opti-

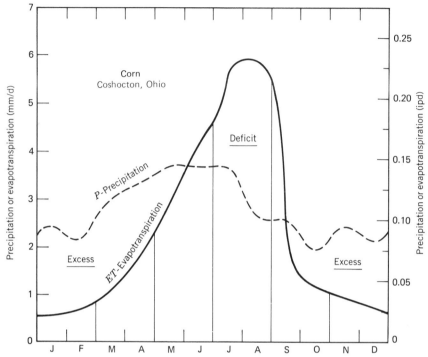

Fig. 3.6 Average precipitation and evapotranspiration from corn at Coshocton, Ohio, showing excess and deficit moisture periods.

mum size for farm machines. The average daily evapotranspiration during the year, obtained for corn from lysimeters at Coshocton, Ohio, is shown in Fig. 3.6. The greatly decreased ET at the end of the summer may delay maturation of corn and harvesting and tillage operations, should heavy rainfall occur.

EVAPOTRANSPIRATION

The basic approaches to prediction of evaporation have been developed into several usable methods for estimating evapotranspiration.

3.10 Blaney–Criddle Method

Blaney and Criddle (1950) developed an empirical method that is widely used for determining evapotranspiration from climatological and irrigation data. The procedure is to correlate existing evapotranspiration data for different crops with the monthly temperature, percent of daytime hours, and length of growing season. The correlation coefficients are then applied to determine the evapotranspiration for other areas where only climatological data are available. The monthly evapotranspiration can be computed by the formula

$$u = kp(0.46T + 8.13)$$ (3.9)

where u = monthly evapotranspiration in mm,

k = monthly evapotranspiration coefficient (function of crop type and temperature) (use local data),

T = mean monthly temperature in °C,

p = monthly percent of total daytime hours of year (monthly daytime hours × 100/total annual daytime hours).

Total evapotranspiration for the growing season or other period is the sum of the monthly totals. Mean monthly temperatures and percent of daytime hours for each month can be determined from local weather records.

☐ Example 3.2

Compute the evapotranspiration for corn at Sandusky, Ohio, latitude 41½°N, for the month of July using the Blaney–Criddle equation. Assume a crop coefficient $k = 0.6$.

Solution.   From U.S. Weather Bureau Monthly Climatological Data the long-time average monthly temperature (average of daily minimum and maximum values) is 23.9°C (75.0°F) and the average percentage of daytime hours of the year for July is 10.32. Substituting in Eq. 3.9,

$$u = 0.6 \times 10.32 \, (0.46 \times 23.9 + 8.13) = 118 \text{ mm/month (4.64 in.)} \qquad \square$$

3.11   Penman Method

Penman (1948, 1956) approached the problem of estimating the evaporation from a free-water surface by examining the energy balance at the water surface expressed by

$$R_n = E + A + S + C \qquad (3.10a)$$

where R_n = net radiant energy available at the earth's surface,

E = energy used in evaporating water,

A = energy used in heating air,

S = energy used in heating the water,

C = energy used in heating the surroundings of the water.

He reasoned that energy used in heating the water and its container could be neglected and that the evaporation of water could be predicted from the equation

$$E = R_n - A \qquad (3.10b)$$

Combination of this equation with Dalton's law (Eq. 3.5) (energy balance and mass transfer) results in an equation for evapotranspiration in which the needed data are available from meteorological records. The equation for well-watered grass or reference ET_0 (Penman, 1963) and converted to SI units (Jensen et al., 1990) is

$$\lambda ET_0 = \frac{\Delta}{\Delta + \gamma} \, (R_n - G) + \frac{\gamma}{\Delta + \gamma} \, 6.43 \, (1.0 + 0.53v_2) \, (e_s - e_d) \qquad (3.11)$$

where λET_0 = reference ET for a well-watered grass expressed as latent heat
flux density, MJ m^{-2} day $^{-1}$,

Δ = slope of the saturation vapor pressure curve in kPa/°C,

γ = psychrometric constant in kPa/°C,

R_n = net radiation in MJ m^{-2} day^{-1},

G = heat flux density to the soil in MJ m^{-2} day^{-1},

v_2 = average wind speed at a height of 2 m in m/s,

e_s = saturated vapor pressure at mean air temperature in kPa,

e_d = saturated vapor pressure at mean dew-point temperature in
kPa (also $e_s \times$ mean relative humidity).

The following equations and constants were summarized from Jensen et al. (1990).
Values for Δ can be obtained from

$$\Delta = 0.20(0.00738T + 0.8072)^7 - 0.000116 \qquad (3.12)$$

where T = mean air temperature in °C.

The psychrometric constant in kPa/°C is

$$\gamma = 0.00163\ P/\lambda \qquad (3.13)$$

$$P = 101.3 - 0.01055\ (\text{EL}) \qquad (3.14)$$

$$\lambda = 2.501 - 0.002361T \qquad (3.15)$$

where P = estimated atmospheric pressure in kPa,

EL = elevation in m,

λ = latent heat of vaporization of water in MJ/kg.

The net radiation can be calculated from

$$R_n = (1 - \alpha)\ R_s - \sigma T_a^4\ [0.34 - 0.139\ (e_d)^{0.5}]\ (0.1 + 0.9n/N) \qquad (3.16)$$

where R_s = the solar radiation received at the earth's surface in MJ m^{-2}
day^{-1},

α = the radiation reflection coefficient or albedo with values near
0.25 for green crops,

σ = Stefan–Boltzmann constant (4.903 × 10^{-9} MJ m^{-2} day^{-1} °K^{-4}),

T_a = absolute air temperature in °K (°C + 273),

n/N = ratio of actual to possible hours of sunshine.

If solar radiation is not measured, it can be obtained from

$$R_s = (0.35 + 0.61n/N)R_{so} \qquad (3.17)$$

where R_{so} is the mean solar radiation for cloudless skies in MJ m^{-2} day^{-1} from
Table 3.1.

The soil heat flux is small and often assumed as zero. The saturation vapor
pressure is calculated from

$$e_s = 3.38639[(0.00738T + 0.8072)^8 - 0.00019|1.8T + 48| + 0.001316] \qquad (3.18)$$

Table 3.1   Mean Solar Radiation for Cloudless Skies, R_{so}

Month	North Latitude				
	0	10	20	30	40
			MJ m^{-2} day^{-1}		
Jan.	28.18	25.25	21.65	17.46	12.27
Feb.	29.18	26.63	25.00	21.65	17.04
Mar.	30.02	29.43	28.18	25.96	22.90
Apr.	28.47	29.60	30.14	29.85	28.34
May	26.92	29.60	31.40	32.11	32.11
June	26.25	29.31	31.82	33.20	33.49
July	26.67	29.43	31.53	32.66	32.66
Aug.	27.76	28.76	30.14	30.44	29.18
Sept.	29.60	29.60	28.47	26.67	23.73
Oct.	29.60	28.05	25.83	22.48	18.42
Nov.	28.47	25.83	22.48	18.30	13.52
Dec.	26.80	24.41	20.50	16.04	10.76

Source: Jensen et al. (1990).

where T = the mean air temperature in °C. Equation 3.18 also can be used to determine e_d by substituting the mean dew-point temperature for T.

Since Eq. 3.11 estimates reference ET as latent heat flux density for a well-watered grass, actual ET for other crops is estimated with crop coefficients from

$$ET_c = K_c \times \lambda ET_0 / \lambda \qquad (3.19)$$

where ET_c = the estimated ET for a crop in mm/day,
  K_c = the crop coefficient for a specific crop and location.

Approximate values of crop coefficients for selected crops are given in Table 3.2. Specific values for these and other crops can be obtained from Doorenbos and Pruitt (1977), Jensen et al. (1990), Hoffman et al. (1990), or Cuenca (1989).

Table 3.2   Approximate Crop Coefficients for a Grass Reference Crop, ET_0

Date		Corn (grain) Midwest	Cotton Southwest	Potatoes Northwest	Soybeans Midwest
Apr.	1–15				
	16–30		0.1		
May	1–15	0.2	0.2	0.1	0.1
	16–31	0.4	0.3	0.4	0.3
June	1–15	0.7	0.6	0.7	0.6
	16–30	0.9	0.9	0.9	0.8
July	1–15	1.0	1.2	1.1	1.0
	16–31	1.0	1.2	1.1	1.0
Aug.	1–15	1.0	1.2	1.1	1.0
	16–31	0.9	1.2	1.0	1.0
Sept.	1–15	0.6	1.1	0.8	0.8
	16–30		0.8		0.6
Oct.	1–15		0.5		
	16–31				

☐ *Example 3.3*

Compute the estimated ET for cotton for June 16 to 30 near Bakersfield, California, 35° North Latitude, using the Penman equation. Mean maximum temperature = 36°C, mean minimum temperature = 22°C, mean dew-point temperature = 9°C, mean wind speed = 1.5 m/s, mean percentage of possible sunshine = 94 percent, elevation = 50 m, and assume G = 0.0.

Solution.

(1) Calculate the mean temperature $T = (36 + 22)/2 = 29$; then calculate Δ from Eq. 3.12.

$$\Delta = 0.2(0.00738(29) + 0.8072)^7 - 0.000116 = 0.232 \text{ kPa/°C}.$$

(2) Calculate P from Eq. 3.14.

$$P = 101.3 - 0.01055(50) = 100.8 \text{ kPa}$$

(3) Calculate the latent heat of vaporization (Eq. 3.15).

$$\lambda = 2.501 - 0.002361(29) = 2.43 \text{ MJ/kg}$$

(4) Calculate the psychrometric constant (Eq. 3.13).

$$\gamma = 0.00163(100.8)/2.43 = 0.0676 \text{ kPa/°C}$$

(5) Calculate the saturation vapor pressure at mean air temperature and at mean dew-point temperature (Eq. 3.18).

$$e_s = 3.38639[(0.00738 (29) + 0.8072)^8 - 0.00019|1.8(29) + 48| \\ + 0.0001316] = 3.947 \text{ kPa}$$

$$e_d = 3.38639[(0.00738 (9) + 0.8072)^8 - 0.00019|1.8(9) + 48| \\ + 0.0001316] = 1.112 \text{ kPa}$$

(6) Calculate the climatic constants.

$$\frac{\Delta}{\Delta + \gamma} = \frac{0.232}{0.232 + 0.0676} = 0.774$$

$$\frac{\gamma}{\Delta + \gamma} = 1.0 - \frac{\Delta}{\Delta + \gamma} = 1.0 - 0.774 = 0.226$$

(7) From Table 3.1 determine the mean cloudless solar radiation = $(33.20 + 33.49)/2 = 33.34$ MJ m^{-2} day^{-1}; then find R_s with Eq. 3.17.

$$R_s = [0.35 + 0.61 (94/100)]33.34 = 30.79 \text{ MJ m}^{-2} \text{ day}^{-1}$$

(8) Calculate net radiation (Eq. 3.16).

$$R_n = (1.0 - 0.25)30.79 - 4.903 \times 10^{-9}(29 + 273)^4[0.34 \\ - 0.139(1.112)^{0.5}](0.1 + 0.9(94/100)) = 15.63 \text{ MJ m}^{-2} \text{ day}^{-1}$$

(9)   Calculate λET_0 from Eq. 3.11.

$$\lambda ET_0 = 0.774(15.63) + 0.226(6.43)[1.0 + 0.53(1.5)]$$
$$(3.947 - 1.112) = 19.5 \text{ MJ m}^{-2} \text{ day}^{-1}$$

(10)   From Table 3.2 determine the crop coefficient $= 0.9$; then calculate the crop ET from Eq. 3.19. (*Note:* 1 kg of water $= 1$ mm $\times 1$ m^2)

$$ET_c = 0.9 \times 19.5/2.43 = 7.2 \text{ mm/day}$$

For 15 days the estimated ET for cotton is (7.2×15) 108 mm.   □

3.12   Empirical Solar Radiation Method

Jensen and Haise (1963) and Jensen (1966) presented an energy-balance approach to estimate evapotranspiration that is simpler in application than Penman's equation. Based on extensive field data, they developed a method for well-watered alfalfa reference ET_r,

$$\lambda ET_r = C_t \, (T - T_x) \, R_s \tag{3.20}$$

where $\lambda ET_r =$ alfalfa-based reference ET expressed as latent heat flux density in MJ m^{-2} day^{-1},
$\quad\quad R_s =$ solar radiation as measured or defined by Eq. 3.17 in MJ m^{-2} day^{-1},
$\quad\quad T =$ mean air temperature for the period of calculation in °C,

$$C_t = 1.0/(C_1 + 7.3C_H), \tag{3.21}$$

$$C_1 = 38 - (2 \text{ EL}/305), \text{ EL is elevation in m}, \tag{3.22}$$

$$C_H = 5.0/(e_2 - e_1) \tag{3.23}$$

$$T_x = -2.5 - 1.4 \, (e_2 - e_1) - \text{EL}/550, \tag{3.24}$$

where e_2 and e_1 are the saturation vapor pressures in kPa for the mean maximum and mean minimum temperatures, respectively, for the warmest month of the year in the area.

Crop ET_c is obtained from Eq. 3.19 using a crop coefficient for alfalfa-based reference ET and substituting ET_r for ET_0. Note that grass- and alfalfa-based crop coefficients are not identical. The coefficients vary with the stage of growth of the plant as shown in Fig. 3.7. Jensen and Haise (1963), Stegman et al. (1977), Wright (1982), and others have developed alfalfa-based coefficients for many crops and locations.

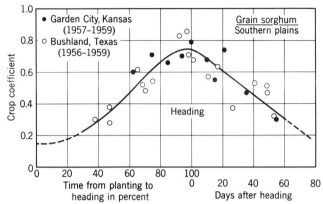

Fig. 3.7 Crop coefficients for the solar radiation method in relation to stage of growth of sorghum. (*Source:* Jensen and Haise, 1963.)

□ *Example 3.4*

Compute the estimated evapotranspiration for grain sorghum near Garden City, Kansas, for August 1 to 15 using the solar radiation method. Planting is June 1 and heading is August 1, location is 37° North Latitude, elevation is 540 m, average temperature for August 1 to 15 is 23°C, mean maximum and minimum temperatures for the warmest month are 32 and 18°C, respectively, and 80 percent of the possible sunshine is received.

Solution.

(1) Calculate C_1 from Eq. 3.22.

$$C_1 = 38 - 2(540)/305 = 34.46$$

(2) Calculate e_2 and e_1 from Eq. 3.18.

$$e_2 = 3.38639[(0.00738(32) + 0.8072)^8 - 0.00019|1.8(32) + 48| \\ + 0.001316] = 4.69$$

$$e_1 = 3.38639[(0.00738(18) + 0.8072)^8 - 0.00019|1.8(18) + 48| \\ + 0.001316] = 2.02$$

(3) Calculate C_H from Eq. 3.23.

$$C_H = 5.0/(4.69 - 2.02) = 1.873$$

(4) Calculate T_x from Eq. 3.24.

$$T_x = -2.5 - 1.4(4.69 - 2.02) - 540/550 = -7.22$$

(5) Read cloudless radiation $R_{so} = 32.66$ from Table 3.1; then determine R_s from Eq. 3.17.

$$R_s = [0.35 + 0.61(80/100)]\ 32.66 = 27.37 \text{ MJ m}^{-2} \text{ day}^{-1}$$

(6)   Calculate C_t from Eq. 3.21.

$$C_t = 1.0/(34.46 + 7.3(1.873)) = 0.0208$$

(7)   Substitute values into Eq. 3.20 to obtain λET_r.

$$\lambda ET_r = 0.0208(23 + 7.22)27.37 = 17.2 \text{ MJ m}^{-2} \text{ day}^{-1}$$

(8)   Calculate the latent heat of vaporization from Eq. 3.15.

$$\lambda = 2.501 - 0.002361(23) = 2.45 \text{ MJ/kg}$$

(9)   From Fig. 3.7 read the average crop coefficient for August 1 to 15, which is 0.7; substitute λET_r for λET_0 in Eq. 3.19 and then calculate ET_c for the period from Eq. 3.19.

$$ET_c = (0.7 \times 17.2/2.45)15 \text{ days} = 74 \text{ mm}$$

REFERENCES

Blaney, H. F., and W. D. Criddle (1950). *Determining Water Requirements in Irrigated Areas from Climatological and Irrigation Data*. (litho.) USDA SCS-TP-96. Washington, DC.

Cuenca, R. H. (1989). *Irrigation System Design: An Engineering Approach*. Prentice-Hall, Englewood Cliffs, NJ.

Doorenbos, J., and W. O. Pruitt (1977). *Guidelines for Predicting Crop Water Requirements*. FAO Irrig. and Drainage Paper No. 24, 2nd ed. FAO, Rome, Italy.

Duley, F. L. (1939). "Surface Factors Affecting the Rate of Intake of Water by Soils." *Soil Sci. Soc. Am. Proc.* **4**, 60–64.

Green, W. H., and G. A. Ampt (1911). "Studies in Soil Physics I. The Flow of Air and Water Through Soils." *J. Agr. Sci.* **4**, 1–24.

Hoffman, G. J., T. A. Howell, and K. H. Solomon (1990). *Management of Farm Irrigation Systems*. ASAE Monograph. ASAE, St. Joseph, MI.

Holtan, H. N., and M. H. Kirkpatrick, Jr. (1950). "Rainfall, Infiltration and Hydraulics of Flow in Runoff Computation." *Trans. Am. Geophys. Union* **31**, 771–779.

Horton, R. E. (1939). "Analysis of Runoff-Plot Experiments with Varying Infiltration Capacity." *Trans. Am. Geophys. Union* **20**, 693–711.

Jensen, M. E. (1966). "Empirical Methods of Estimating or Predicting Evapotranspiration Using Radiation." In *ASAE Conference Proceedings, Evapotranspiration and Its Role in Water Resources Management*, Dec., pp. 49–53, 64.

Jensen, M. E., R. D. Burman, and R. G. Allen (eds.) (1990). *Evapotranspiration and Irrigation Water Requirements*. ASCE, New York.

Jensen, M. E., and H. R. Haise (1963). "Estimating Evapotranspiration from Solar Radiation." *Proc. ASCE J. Irrig. Drainage Div.* **89** (IR4), 15–41.

Jury, W. A., W. R. Gardner, and W. H. Gardner (1991). *Soil Physics,* 5th ed. Wiley, New York.

Klute, A. (1952). "A Numerical Method for Solving the Flow Equation for Water in Unsaturated Materials." *Soil Sci.* **73**, 105–116.

Meinzer, O. E. (1949). *Hydrology: Physics of the Earth — IX.* Dover, New York.

Meyer, A. F. (1942). *Evaporation from Lakes and Reservoirs.* Minnesota Resources Commission, St. Paul.

Penman, H. L. (1948). "Natural Evapotranspiration from Open Water, Bare Soil and Grass." *Proc. R. Soc. London* **193**, 120–145.

—— (1956). "Estimating Evapotranspiration." *Trans. Am. Geophys. Union* **37**, 43–46.

—— (1963). *Vegetation and Hydrology.* Tech. Communication No. 53. Commonwealth Bureau of Soils, Harpenden, England.

Rawls, W. J., and D. L. Brakensiek (1989). "Estimation of Soil Water Retention and Hydraulic Properties." In *Unsaturated Flow in Hydrologic Modeling, Theory and Practice,* H. J. Morel-Seytoux (ed.). NATO Series C, Mathematical and Physical Science, Vol. 275, pp. 275–300. Kluwer Academic, London.

Rawls, W. J., D. L. Brakensiek, and K. E. Saxton (1982). "Estimation of Soil Water Properties." *ASAE Trans.* **25**, 1316–1320, 1328.

Rohwer, C. (1931). *Evaporation from Free Water Surfaces.* USDA Tech. Bull. 271. GPO, Washington, DC.

Stegman, E. C., et al. (1977). *Crop Curves for Water Balance Irrigation Scheduling in S.E. North Dakota.* North Dakota Agr. Expt. Sta. Res. Rep. 66. Fargo, ND.

Thornthwaite, C. W. (1948). "An Approach Toward a Rational Classification of Climate." *Geograph. Rev.* **38**, 55–94.

Thornthwaite, C. W., and B. Holzman (1942). *Measurement of Evaporation from Land and Water Surfaces.* USDA Tech. Bull. **817**. GPO, Washington, DC.

U.S. Department of Agriculture (USDA) (1980). *Part I Appraisal. Soil, Water, and Related Resources in the United States.* RCA Review Draft. Washington, DC.

U.S. Soil Conservation Service (SCS) (1972). "Hydrology." In *National Engineering Handbook,* Sect. 4. GPO, Washington, DC.

Wright, J. L. (1982). "New Evapotranspiration Crop Coefficients." *J. Irrig. Drainage Div. ASCE* **108**, 57–74.

PROBLEMS

3.1 Compute the daily evaporation from a free-water surface if the wind speed is 32 km/h (20 mph) at 0.15 m height, water and air temperature are both 29°C (80°F) and the relative humidity of air is 50 percent. Atmospheric pressure is 757 mm Hg (29.8 in. Hg). How would your computed value compare with the evaporation from a dry-soil surface? From a Class A Weather Bureau pan?

3.2 Assuming that the Horton infiltration Eq. 3.4 is valid, determine the constant infiltration rate if $f_0 = 50$ mm/h (2.0 iph), f at 10 min is 13 mm/h (0.5 iph), and $k = 12.9$. What is the infiltration rate at 20 min?

3.3 Using the Blaney–Criddle method, determine the evapotranspiration (consumptive use) for corn in May, assuming a crop coefficient k of 0.7, T of 18.3°C (65°F), and p of 10 percent.

3.4 Using Penman's equation, determine the evapotranspiration for corn during May at 40° North Latitude, assuming mean minimum, maximum, and dew-point temperatures of 17°C (63°F), 24°C (75°F), and 10°C (50°F), respectively; 60 percent of possible sunshine is received; 3.7 m/s mean wind speed; 283 m elevation; and negligible heat flux to the soil.

3.5 Using the solar radiation method, determine the evapotranspiration for grain sorghum at Bushland, Texas (37° North Latitude) for June 1 to 15. Elevation is 500 m, 90 percent of possible sunshine is received, average temperature for the period is 22°C, and mean minimum and maximum temperatures for the warmest month are 20 and 32°C, respectively. Planting date is June 1 and heading date is August 1.

3.6 Using Penman's equation, determine the evaporation from a free-water surface for the month of June of the last year and at your nearest weather station where climatological data and pan evaporation are available. How does your calculated value compare with measured pan evaporation for this month? Assume the radiation reflection coefficient for a water surface is 0.05.

CHAPTER 4

Runoff

Conservation structures and channels must be designed to handle natural flows of water from rainfall or melting snow. Runoff constitutes the hydraulic "load" that the structure or channel must withstand.

4.1 Definition

Runoff is that portion of the precipitation that makes its way toward stream channels, lakes, or oceans as surface or subsurface flow. The term *runoff* usually means surface flow. The engineer designing channels and structures to handle natural surface flows is concerned with peak rates of runoff, with runoff volumes, and with temporal distribution of runoff rates and volumes.

4.2 The Runoff Process

Before runoff can occur, precipitation must satisfy the demands of evaporation, interception, infiltration, surface storage, surface detention, and channel detention.

Interception by the vegetated canopy may be so great as to prevent a light rain from wetting the soil. Interception by dense covers of forest or shrubs commonly amounts to 25 percent of the annual precipitation. A good stand of mature corn may have a net interception storage capacity of 0.5 mm. Trees such as willows may intercept nearly 13 mm from a long, gentle storm. Interception also has a detention storage effect, delaying the progress of precipitation that reaches the soil only after running down the plant or dropping from the leaves.

Runoff will occur only when the rate of precipitation exceeds the rate at which water may infiltrate into the soil (see Chapter 3). After the infiltration rate is satisfied, water begins to fill the depressions, small and large, on the soil surface. As the depressions are filled overland flow begins. The depth of water builds up on the surface until it is sufficient to result in runoff in equilibrium with the rate of precipitation less infiltration and interception. The volume of water involved in the depth buildup is surface detention. As the flow moves into defined channels there is a similar buildup of water in channel detention. The volume of water in surface and channel detention is returned to runoff as the runoff rate subsides. The water in surface storage eventually infiltrates or evaporates.

FACTORS AFFECTING RUNOFF

The factors affecting runoff may be divided into those factors associated with the precipitation and those associated with the watershed.

4.3   Rainfall

Rainfall duration, intensity, and areal distribution (see Chapter 2) influence the rate and volume of runoff. Total runoff for a storm is clearly related to the duration for a given intensity. Infiltration will decrease with time in the initial stages of a storm. Thus, a storm of short duration may produce no runoff, whereas a storm of the same intensity but of long duration will result in runoff.

Rainfall intensity influences both the rate and the volume of runoff. An intense storm exceeds the infiltration capacity by a greater margin than does a gentle rain; thus the total volume of runoff is greater for the intense storm even though total precipitation for the two rains is the same. The intense storm actually may decrease the infiltration rate because of its destructive action on the soil structure at the surface.

Figure 4.1 shows relationships of runoff to rainfall intensity and to total rainfall per storm. The data summarize 8 years of records on bare plots at Statesville, North Carolina. In this study rainfall amounts greater than 25 mm and rainfall

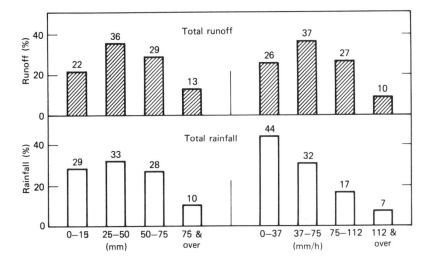

Fig. 4.1   Total storm runoff related to rainfall amounts and intensities. (Redrawn from Copley et al., 1944.)

intensities greater than 37 mm/h gave higher percentages of runoff than the corresponding percentages for the rainfall groups.

Rate and volume of runoff from a given watershed are influenced by the distribution of rainfall and of rainfall intensity over the watershed. Generally the maximum rate and volume of runoff occur when the entire watershed contributes; however, an intense storm on one portion of the watershed may result in greater runoff than a moderate storm over the entire watershed.

Runoff does not necessarily correspond to monthly precipitation, especially for the 7000-ha watershed in Fig. 4.2. Runoff from this area is high during the late winter and early spring months because of soil water that seeps below the root zone and results in interflow (ground water) into the larger streams. The runoff from the 0.8-ha watershed more nearly reflects precipitation, particularly for intensities greater than 25 mm/h.

4.4 Watershed

Watershed factors affecting runoff are size, shape, orientation, topography, geology, and surface culture. Both runoff volumes and rates increase as watershed size increases; however, both rate and volume per unit of watershed area decrease as the runoff area increases. Watershed size may determine the season at which high runoff may be expected to occur. On watersheds in the Ohio River basin 99 percent of the floods from drainage areas of 260 ha occur in May through Sep-

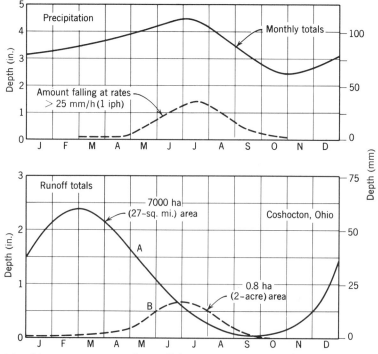

Fig. 4.2 Monthly precipitation and runoff from small and large drainage areas. (Redrawn from Harrold, 1961.)

tember; 95 percent of the floods on drainage areas of 26 million ha occur in October through April.

Long, narrow watersheds are likely to have lower runoff rates than more compact watersheds of the same size. The runoff from the former does not concentrate as quickly as it does from the compact areas, and long watersheds are less likely to be covered uniformly by intense storms. When the long axis of a watershed is parallel to the storm path, storms moving upstream cause a lower peak runoff rate than storms moving downstream. For storms moving upstream, runoff from the lower end of the watershed is diminished before the peak contribution from the headwaters arrives at the outlet; however, a storm moving downstream causes a high runoff from the lower portions coincident with high runoff arriving from the headwaters.

Topographic features, such as slope of upland areas, the degree of development and gradients of channels, and the extent and number of depressed areas affect rates and volumes of runoff. Watersheds having extensive flat areas or depressed areas without surface outlets have lower runoff than areas with steep, well-defined drainage patterns. The geologic or soil materials determine to a large degree the infiltration rate, and thus affect runoff. Vegetation and the practices incident to agriculture and forestry also influence infiltration (see Chapter 3). Vegetation retards overland flow and increases surface detention to reduce peak runoff rates. Structures such as dams, levees, bridges, and culverts all influence runoff rates.

DESIGN RUNOFF RATES

Methods of runoff estimation necessarily neglect some factors and make simplifying assumptions regarding the influence of others. Methods presented here are applicable mostly to small watersheds less than a few hundred hectares.

The capacity to be provided in a structure that must carry runoff may be termed the *design runoff rate*. Structures and channels are planned to carry runoff that occurs within a specified return period. Vegetated controls and temporary structures are usually designed for a runoff that may be expected to occur once in 10 years; expensive, permanent structures will be designed for runoffs expected only once in 50 or 100 years. Selection of the design return period, also called recurrence interval, depends on the economic balance between the cost of periodic repair or replacement of the facility and the cost of providing additional capacity to reduce the frequency of repair or replacement. In some instances the downstream damage potentially resulting from failure of the structure may dictate the choice of the design frequency.

4.5 Rational Method

The rational method of predicting a design peak runoff rate is expressed by the equation

$$q = 0.0028CiA \qquad (4.1)$$

where q = the design peak runoff rate in m³/s,
C = the runoff coefficient,
i = rainfall intensity in mm/h for the design return period and for a duration equal to the "time of concentration" of the watershed,
A = the watershed area in ha.

The time of concentration of a watershed is the time required for water to flow from the most remote (in time of flow) point of the area to the outlet once the soil has become saturated and minor depressions filled. It is assumed that, when the duration of a storm equals the time of concentration, all parts of the watershed are contributing simultaneously to the discharge at the outlet. One of the most widely accepted methods of computing the time of concentration was developed by Kirpich (1940):

$$T_c = 0.0195L^{0.77}S_g^{-0.385} \qquad (4.2)$$

where T_c = time of concentration in min (see Appendix A),
L = maximum length of flow in m,
S_g = the watershed gradient in m/m or the difference in elevation between the outlet and the most remote point divided by the length, L.

Hydrologists are not in agreement as to the best procedure for computing the time of concentration. The SCS (1972) developed a method for computing the time of concentration that considers length of the main channel, topography, vegetal cover, and infiltration rate. Horn and Schwab (1963) found that SCS's values of watershed lag time (see Eq. 4.3a) gave slightly better estimates of the actual runoff than several other methods when taken equal to the time of concentration.

For estimating time of concentration, the length of flow is divided by an estimated velocity of flow to obtain the travel time. The sum of the travel times for overland flow (sheet runoff) and for all channel flow equals the time of concentration. For such estimates the flow path is taken from the most remote point in the watershed to the outlet. In small watersheds of a few hectares, where a well-defined channel does not exist, runoff occurs mostly as overland flow.

The runoff coefficient C is defined as the ratio of the peak runoff rate to the rainfall intensity and is dimensionless. Estimates of the runoff coefficient from small single-crop watersheds at Coshocton, Ohio, showed that the primary effects were attributed to the infiltration rate, surface cover, and rainfall intensity. These estimates are presented in Table 4.1 for hydrologic soil group B. The runoff coefficient can be converted to other hydrologic soil groups by referring to Table 4.2. These soil groups are defined in Table 4.3.

The rational method assumes that the frequencies of rainfall and runoff are similar, which has been confirmed by Larson and Reich (1973). The method is a great oversimplification of a complicated process; however, the method is considered sufficiently accurate for runoff estimation in the design of relatively inexpensive structures where the consequences of failure are limited. Application of the

Table 4.1 Runoff Coefficient C for Agricultural Watersheds (Soil Group B)

Crop and Hydrologic Condition	Coefficient C for Rainfall Rates of		
	25 mm/h	*100 mm/h*	*200 mm/h*
Row crop, poor practice	0.63	0.65	0.66
Row crop, good practice	0.47	0.56	0.62
Small grain, poor practice	0.38	0.38	0.38
Small grain, good practice	0.18	0.21	0.22
Meadow, rotation, good	0.29	0.36	0.39
Pasture, permanent, good	0.02	0.17	0.23
Woodland, mature, good	0.02	0.10	0.15

Source: Horn and Schwab (1963).

rational method as presented here is normally limited to watersheds of less than 800 ha.

The rational method is developed from the assumptions that (1) rainfall occurs at uniform intensity for a duration at least equal to the time of concentration of the watershed, and (2) rainfall occurs at a uniform intensity over the entire area of the watershed. If these assumptions were fulfilled, the rainfall and runoff for the watershed would be represented graphically by Fig. 4.3a. The figure shows a rain of uniform intensity for a duration equal to the time of concentration, T_c. If a storm of duration greater than T_c occurred, the runoff rate would be less than q because the rainfall intensity would be less than i (see Chapter 2 for relationships between rainfall intensity and duration for a given return period). A rain of duration less than T_c would result in a runoff rate less than q because the entire watershed would not contribute simultaneously to the discharge at the outlet.

Table 4.2 Hydrologic Soil Group Conversion Factors

Crop and Hydrologic Condition	Factors for Converting the Runoff Coefficient C from Group B Soils to[a]		
	Group A	*Group C*	*Group D*
Row crop, poor practice	0.89	1.09	1.12
Row crop, good practice	0.86	1.09	1.14
Small grain, poor practice	0.86	1.11	1.16
Small grain, good practice	0.84	1.11	1.16
Meadow, rotation, good	0.81	1.13	1.18
Pasture, permanent, good	0.64	1.21	1.31
Woodland, mature, good	0.45	1.27	1.40

[a]Factors were computed from Table 4.3 by dividing the curve number for the desired soil group by the curve number for group B.

Table 4.3 Runoff Curve Numbers for Hydrologic Soil-Cover Complexes for Antecedent Rainfall Condition II and $I_a = 0.2S$

Land Use or Crop	Treatment or Practice	Hydrologic Condition	A	B	C	D
Fallow	Straight row	—	77	86	91	94
Row crops	Straight row	Poor	72	81	88	91
	Straight row	Good	67	78	85	89
	Contoured	Poor	70	79	84	88
	Contoured	Good	65	75	82	86
	Terraced	Poor	66	74	80	82
	Terraced	Good	62	71	78	81
Small grain	Straight row	Poor	65	76	84	88
	Straight row	Good	63	75	83	87
	Contoured	Poor	63	74	82	85
	Contoured	Good	61	73	81	84
	Terraced	Poor	61	72	79	82
	Terraced	Good	59	70	78	81
Close-seeded legumes or rotation meadow	Straight row	Poor	66	77	85	89
	Straight row	Good	58	72	81	85
	Contoured	Poor	64	75	83	85
	Contoured	Good	55	69	78	83
	Terraced	Poor	63	73	80	83
	Terraced	Good	51	67	76	80
Pasture or range		Poor	68	79	86	89
		Fair	49	69	79	84
		Good	39	61	74	80
	Contoured	Poor	47	67	81	88
	Contoured	Fair	25	59	75	83
	Contoured	Good	6	35	70	79
Meadow (permanent)		Good	30	58	71	78
Woods (farm woodlots)		Poor	45	66	77	83
		Fair	36	60	73	79
		Good	25	55	70	77
Farmsteads		—	59	74	82	86
Roads and right-of-way (hard surface)		—	74	84	90	92

Soil Group	Description	Final Infiltration Rate (mm/h)
A	*Lowest Runoff Potential.* Includes deep sands with very little silt and clay, also deep, rapidly permeable loess.	8–12
B	*Moderately Low Runoff Potential.* Mostly sandy soils less deep than A, and loess less deep or less aggregated than A, but the group as a whole has above-average infiltration after thorough wetting.	4–8
C	*Moderately High Runoff Potential.* Comprises shallow soils and soils containing considerable clay and colloids, though less than those of group D. The group has below-average infiltration after presaturation.	1–4
D	*Highest Runoff Potential.* Includes mostly clays of high swelling percent, but the group also includes some shallow soils with nearly impermeable subhorizons near the surface.	0–1

Source: SCS (1972).

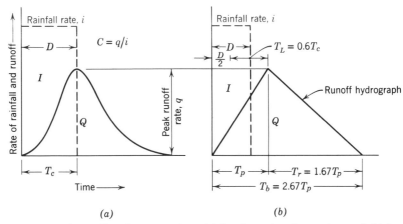

Fig. 4.3 Rainfall and runoff with assumptions (*a*) for the rational equation and (*b*) for the SCS triangular hydrograph method of runoff estimation.

The rational method is illustrated by the following problem.

☐ Example 4.1

Determine the design peak runoff rate for a 50-year return period storm from a 40-ha (100-ac) watershed near Chicago, Illinois, with the following characteristics:

(ha)	Subarea (ac)	Typography, Percent Slope	Soil Group (see Table 4.3)	Land Use, Treatment, and Hydrologic Condition
24	(60)	Flat	C	Row crop, contoured, good
16	(40)	10–30	B	Woodland, good

The maximum length of flow is 600 m (2000 ft) and the difference in elevation along this path is 3 m (10 ft).

Solution.   The watershed gradient is $(3/600)100 = 0.5$ percent. From Appendix A or Eq. 4.2, $T_c = 20$ min. From Chapter 2 for a 50-year return period near Chicago, the 20-min rainfall is 97 mm/h (3.8 iph).

The runoff coefficients C from Table 4.1 for row crop, good practice, and woodland are 0.56 and 0.10, respectively, and the factor correcting hydrologic soil group C to group B for the 24-ha subarea from Table 4.2 is 1.09.

$$C = (24/40) \times 0.56 \times 1.09 + (16/40) \times 0.10 = 0.41$$

Substituting in Eq. 4.1,

$$q = 0.0028 \times 0.41 \times 97 \times 40 = 4.45 \text{ m}^3/\text{s (157 cfs)}$$ ☐

4.6   Soil Conservation Service Method

This method described by SCS (1990) was originally developed for uniform rainfall using the assumptions for a triangular hydrograph shown in Fig. 4.3*b*. A hydrograph is a plot of the runoff rate versus time and is further described in Sections 4.14 to 4.16. The time to peak flow is

$$T_p = D/2 + T_L = D/2 + 0.6\ T_c \tag{4.3a}$$

where T_p = time to peak (T),
D = duration of excess rainfall (T),
T_L = time of lag (T),
T_c = time of concentration (T).

Time of concentration equal to $T_L/0.6$ is the longest travel time and may be obtained from the equation,

$$T_c = L^{0.8}\ [(1000/N) - 9]^{0.7}/[4407\ (S_g)^{0.5}] \tag{4.3b}$$

where T_c = time of concentration in hours,
L = longest flow length in m,
N = runoff curve number,
S_g = average watershed gradient in m/m.

The time of peak is necessary to develop a design hydrograph for routing runoff through a storage reservoir or for combining hydrographs from several subwatersheds. It is not required for peak flow estimates. The peak flow rate is calculated from the equation,

$$q = q_u AQ \tag{4.4}$$

where q = peak runoff rate (m³/s),
q_u = unit peak flow rate (m³/s per ha/mm of runoff),
A = watershed area in ha,
Q = runoff depth in mm from Eq. 4.6 (see Section 4.10).

The unit flow rate is obtained from Fig. 4.4 using the time of concentration and the ratio of initial abstraction to 24-hour rainfall. This ratio represents the fraction of rainfall that occurs before runoff begins. The initial abstraction is usually taken as 0.2S, where S is the maximum difference between rainfall and runoff calculated from Eq. 4.7. The curves in Fig. 4.4 apply only for the Type II rainfall distribution, which is applicable for the unshaded areas of the U.S. shown in Fig. 4.5. Curves for other types of rainfall are given by SCS (1990) and McCuen (1989). The SCS method is limited to rural watersheds less than 800 ha in area and average slopes greater than 0.5 percent with one main channel or two tributaries with nearly the same time of concentration. Where more than one third of the watershed has ponding or swamp conditions, peak flow rates can be adjusted (SCS, 1990 or McCuen, 1989). Application of the method is shown below.

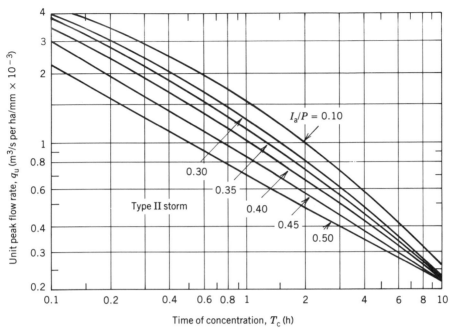

Fig. 4.4 Unit peak runoff rates for SCS Type II rainfall distribution. (Revised from SCS, 1990.)

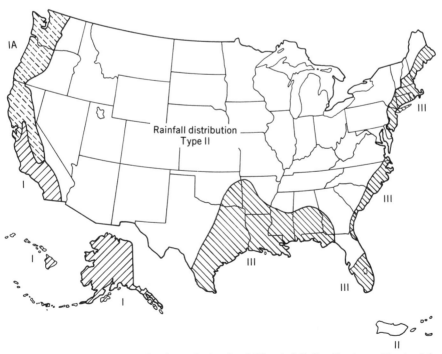

Fig. 4.5 Approximate geographic boundaries for SCS rainfall distributions. (Revised from SCS, 1990.)

□ *Example 4.2*

Determine the peak runoff rate from a 100-ha (247-ac) watershed from a 120-mm (4.7-in.), 24-hour storm that produced 10 mm (0.39 in.), depth of runoff. Assume a flow length of 460 m (1500 ft), weighted average curve number of 75, and an average watershed gradient of 0.02 m/m.

Solution.
Substitute in Eq. 4.7.

$$S = (25\ 400/75) - 254 = 85 \text{ and } 0.2\ S = 0.2 \times 85 = 17$$

From Eq. 4.3b,

$$T_c = 460^{0.8}\ [(1000/75) - 9]^{0.7}/[4407\ (0.02)^{0.5}] = 0.6 \text{ h}$$

Taking $I_a = 0.2S$,

$$I_a/P = 17/120 = 0.14$$

Read from Fig. 4.4.

$$q_u = 0.0019 \text{ m}^3/\text{s per ha/mm}$$

Substitute in Eq. 4.4.

$$q = 0.0019 \times 100 \times 10 = 1.9 \text{ m}^3/\text{s (67 cfs).}$$ □

4.7 Flood Frequency Analysis Method

One method of runoff estimation, called flood frequency analysis, depends on the existence of a number of years of record from the basin under study. These records then constitute a statistical array that defines the probable frequency of recurrence of floods of given magnitudes. Extrapolation of the frequency curves enables the hydrologist to predict flood peaks for a range of return periods. The procedure for this method is the same as that for rainfall described in Chapter 2.

4.8 Computer Prediction of Runoff

Numerous computer programs have been developed to predict storm runoff, many for special applications. The SCS has developed computer programs, which are available at all state offices (see Appendix I), that predict runoff using the procedures previously discussed. Another SCS program, TR55, was developed to be applied to small urban watersheds, but also works well for rural areas (SCS, 1986). During the late 1980s, the SCS began a major effort to develop a suite of computer software, known as the Computer Assisted Management and Planning System (CAMPS), that would not only allow runoff prediction, but would link this prediction with programs to design diversions, waterways, and terraces. The SCS is developing soil and climate databases for CAMPS that it can also link with erosion and drainage models (Rovang et al., 1990). Both the CREAMS (Knisel,

1980) and WEPP (Nearing et al., 1989) erosion models and the drainage model DRAINMOD (Skaggs, 1982) include runoff prediction within the models (see Appendix I). Many other models have been developed with runoff prediction capabilities for such applications as forests, water quality, frozen and thawing soils, wetlands, surface mines, wind erosion prediction, and plant growth applications (Goodrich and Woolhiser, 1991).

When selecting a runoff model, the user should generally select the model that best suits the purpose. Generally, more sophisticated models require larger input files, and obtaining the necessary input data can be time consuming and difficult. Models do allow the user the opportunity to compare the effects of different watershed land uses on runoff, and other hydrologic responses to runoff, and to select optimum management systems for the given situation.

4.9 Other Methods

Many other methods have been proposed for estimating flood runoff. Chow (1962) made a thorough review of 66 such methods of runoff computation.

A number of empirical formulas have been developed to describe the magnitude of extreme floods. These formulas take the form

$$q = KA^x \tag{4.5}$$

where q = the magnitude of the peak runoff (L^3/T),
  K = a coefficient dependent on various characteristics of the watershed,
  A = the watershed area (L^2),
  x = a constant for a given location.

RUNOFF VOLUME

It is often desirable to predict the total volume of runoff that may come from a watershed during a design flood. Total volume is of primary interest in the design of flood control reservoirs.

4.10 Soil Conservation Service Method

This method was developed from many years of storm flow records for agricultural watersheds in many parts of the United States. The equation, which applies to the curves in Fig. 4.6, is

$$Q = \frac{(I - 0.2S)^2}{I + 0.8S} \tag{4.6}$$

where Q = direct surface runoff depth in mm (area under hydrograph),
  I = storm rainfall in mm (see Chapter 2),
  S = maximum potential difference between rainfall and runoff in mm, starting at the time the storm begins.

On gaged watersheds I can be plotted against Q and the value of S obtained

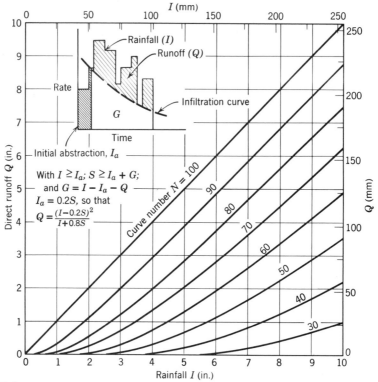

Fig. 4.6 Relationship between rainfall and runoff depth by curve numbers. (Redrawn from USDS–SCS, 1972).

directly. As pointed out in Chapter 3 and shown by Eq. 4.6, runoff decreases as S or infiltration increases. The initial abstraction, $I_a = 0.2S$, consists of interception losses, surface storage, and water that infiltrates into the soil prior to runoff.

For convenience in evaluating antecedent rainfall, soil conditions, land use, and conservation practices, USDS–SCS (1972) defines

$$S = \frac{25\ 400}{N} - 254 \tag{4.7}$$

where N is an arbitrary curve number varying from 0 to 100. Thus, if

$$N = 100, \quad \text{then } S = 0 \quad \text{and} \quad I = Q.$$

Curve numbers can be obtained from Table 4.3. These values apply to antecedent rainfall condition II, which is an average value for annual floods. Correction factors for other antecedent rainfall conditions are listed in Table 4.4.

Condition I is for low runoff potential with soil having low antecedent water content suitable for cultivation. Condition III is for wet conditions prior to the storm. As indicated in Table 4.4 no upper limit for antecedent rainfall is intended. The limits for the "dormant season" apply when the soils are not frozen and when no snow is on the ground.

Table 4.4.   Antecedent Rainfall Conditions and Curve Numbers (for $I_a = 0.2S$)

Curve Number for Condition II	Factor to Convert Curve Number for Condition II to	
	Condition I	*Condition III*
10	0.40	2.22
20	0.45	1.85
30	0.50	1.67
40	0.55	1.50
50	0.62	1.40
60	0.67	1.30
70	0.73	1.21
80	0.79	1.14
90	0.87	1.07
100	1.00	1.00

Condition	General Description	5-Day Antecedent Rainfall (mm)	
		Dormant Season	*Growing Season*
I	Optimum soil condition from about lower plastic limit to wilting point	<13	<36
II	Average value for annual floods	13–28	36–53
III	Heavy rainfall or light rainfall and low temperatures within 5 days prior to the given storm	>28	>53

Source: SCS (1972).

Since duration of a storm affects the amount of rainfall, runoff volume must be evaluated for each design application. The time of concentration is not a good criterion for the determination of storm volume since a short-duration, high-intensity storm may produce the peak flow, but not necessarily the maximum runoff volume. The SCS has established 6-h duration as the minimum storm period for flood-water-retarding structures, but this time is modified for certain conditions where a greater runoff would result.

□ Example 4.3

Determine the estimated maximum volume of runoff during the growing season for a 50-year return period that may be expected from the watershed of Example 4.1. Assume that antecedent rainfall during the last 4 of the 5 days prior to the storm was 40 mm and the critical duration of the storm is 6 h.

Solution.   From Chapter 2 the rainfall for a 6-h storm is 107 mm (4.2 in.) in Chicago. Since the percentage reduction for converting point rainfall to areal rainfall is less than 1 percent (see Chapter 2) in this case, no correction need be made. From Table 4.3 for Antecedent Rainfall Condition II, read the appropriate curve number and calculate the weighted value as follows:

Subarea, A ha (ac)	Soil Group	Land Use, Treatment, and Condition	Curve No., N	NA
24 (60)	C	Row crop, contoured, good	82	1968
16 (40)	B	Woodland, good	55	880
40 (100 ac)			Total	2848

Weighted curve number $= 2848/40 = 71.2$. Substituting in Eq. 4.7,

$$S = \frac{25\ 400}{71.2} - 254 = 103 \text{ mm}$$

$$Q = \frac{(107 - 0.2 \times 103)^2}{107 + (0.8 \times 103)} = 39 \text{ mm (1.54 in.)}$$

$$Q = \frac{39 \times 40}{1000} = 1.56 \text{ ha-m (12.6 ac-ft)}$$

□

□ *Example 4.4*

If 51 mm (2.0 in.) of rainfall occurs the day after the 50-year storm in Example 4.3, what is the expected runoff?

Solution.   From Table 4.4, Antecedent Rainfall Condition III applies as $107 + 40 = 147 > 53$. From Table 4.4, interpolate a correction factor of 1.20 for $N = 71.2$. The new curve number (71.2×1.20) is 86. Calculate from Eq. 4.7 for $N = 86$, $S = 41$ mm. Substituting in Eq. 4.6, $Q = 22$ mm (0.86 in.). Although the rainfall of 51 mm was less than half of the 107 mm in Example 4.3, the runoff was 56 percent (22/39) of the previous amount, illustrating the importance of antecedent rainfall.

□

WATER YIELD

When surface runoff is to be stored in ponds or reservoirs, the total runoff volume for a period of several months, usually the annual volume, is of more interest than the runoff from a design storm. The annual runoff is often referred to as the water yield. For gaged watersheds, frequency analysis of flow records as described in Chapter 2 usually provides the best design data. Brakensiek (1959) has shown that the selection of runoff data for a water year rather than for a calendar year can greatly improve the reliability of results. The date for beginning the water year varies with geographical location, but in general it coincides with the season of maximum runoff. For example, at Coshocton, Ohio, March 1 was the best date. Since water yield records are limited, the results from one area must often be extended to other areas of similar hydrologic conditions.

4.11   Estimation of Minimum Water Yield

Minimum water yields obtained by frequency analysis of 18 years of record at Coshocton, Ohio, are shown in Table 4.5. Since the minimum yield is the least

Table 4.5   Minimum Annual Water Yield at Coshocton, Ohio (Mixed Vegetation and for Water Year Beginning October 1)

Watershed Area (ha)	Minimum Annual Water Yield (mm) for Return Periods of			
	2 years	*10 years*	*25 years*	*50 years*
12	221	132	107	86
31	244	157	130	109
141	330	221	178	152
1041	356	234	185	160
7045	394	267	216	183

Source: Harrold (1957).

flow (dry year), it decreases with the return period, unlike rainfall and storm runoff, which increase with the return period. Minimum yield increases with watershed size because of the greater contribution of ground water flow, especially in humid areas.

☐ Example 4.5

Determine the minimum annual water yield volume from an 86-ha watershed at Coshocton, Ohio, for the driest year in 25 years.

Solution.   Read from Table 4.5, by interpolation halfway between 130 and 178 mm (86 ha is halfway between 31 and 141 ha) it is 154 mm.

$$\text{Minimum volume} = 86 \times 154 \times 10^{-3} = 13.2 \text{ ha-m } (107.4 \text{ ac-ft}) \qquad ☐$$

From water yield records in southern Iowa, Nixon and Schwab (1961) proposed a watershed rating as a means of estimating water yield from other similar watersheds. This rating included factors dependent on climate, land use, steepness of slope, soil series and type, and management practices. In this study the size of the watershed within the range of 120 to 4100 ha was of less importance than the above factors. Flow records showed that in the dry years, which are those of greatest interest, ground water contribution was small. In some geographic regions ground water flow, which is normally of long duration, may be a very important part of water yield. Springs or pipe flow can also make a major contribution.

The minimum yield for a period greater than one year is of practical interest in reservoir design. If water needs are just equal to reservoir capacity, shortage of water will likely occur if the annual water yield is not sufficient to refill the reservoir; however, if storage is greater than the need, some water can be carried over to the dry year. The water yield for southern Iowa (Fig. 4.7) illustrates this point. For a return period of 10 years the water yields for 1-, 2-, 3-, and 4-year periods were 46, 117, 198, and 318 mm, respectively. If storage is available, the annual average yields are increased from 46 to 58 (117/2), 66 (198/3), and 79 (318/4) mm per year for the 2-, 3-, and 4-year periods, respectively. For return periods greater than 10 years the average multiple-year yield becomes increasingly greater than the 1-year yield.

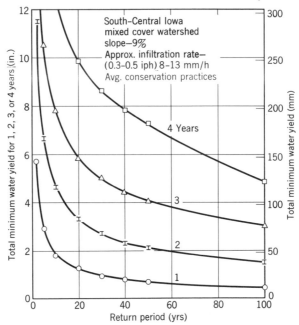

Fig. 4.7 Total minimum water yield in south-central Iowa for flow periods from 1 to 4 years. (Adapted from Nixon and Schwab, 1961.)

4.12 Other Methods

For small ponds and reservoirs the information in Fig. 10.11 in Chapter 10 will provide a conservative estimate of water yield. Although rather lengthy calculations would be required, the SCS method of estimating storm runoff given in Section 4.10 could be applied to daily rainfall. For the winter season, snowmelt may be estimated by Eq. 4.8. The annual water yield would then be the sum of the daily runoffs and snowmelt.

4.13 Snowmelt

For ungaged watersheds, snowmelt may be determined from the equation

$$M = 46KD \qquad (4.8)$$

where M = depth of snowmelt water from watershed in mm/day,
   K = a constant for watershed and climatic conditions (0.02 for low, 0.06 for average, and 0.30 for high runoff potential),
   D = degree-days for a given day, in °C.

The average temperature is the average of maximum and minimum for the day. A day with an average temperature of 15°C thus has 15 degree-days. Adjustment of temperature to the average elevation of the watershed can be made by a decrease of 0.72°C for every 100-m increase in elevation. Only temperatures above freezing are considered. Snowmelt is of most importance in the mountainous regions of the western United States.

RUNOFF HYDROGRAPHS

A hydrograph is a graphical or tabular representation of runoff rate against time. Figure 4.8 gives the hydrograph of the discharge from a 125-ha agricultural watershed. This discharge resulted from a storm of 43 mm in 30 min. For small agricultural watersheds, streamflow records from which hydrographs can be developed are generally not available. One of the characteristics of hydrographs for a given watershed is that the duration of flow is nearly constant for individual storms regardless of the peak flow. Sherman (1932) developed a hydrograph representing one inch of runoff, which is referred to as a unit hydrograph. Commons (1942) and later others developed dimensionless hydrographs.

4.14   Triangular Hydrographs

The simplest runoff hydrograph is the triangular hydrograph described in Figs. 4.3*b* and 4.9. They are applied (1) to subareas of a watershed or (2) to time increments of a rainstorm. In either case the flow rates at a given time are added from the several hydrographs, thus producing a curvilinear hydrograph. Except for the tail end, it is a good approximation of an actual hydrograph. As shown in Fig. 4.9, the time to peak and the peak flow rate are the same as for the dimensionless hydrograph. The base of the triangular hydrograph was so chosen to give the same area (volume) as the dimensionless hydrograph.

4.15   Dimensionless Hydrograph

The dimensionless hydrograph shown by the smooth curve in Fig. 4.9 has a shape that approximates the flow from an intense storm from a small watershed. It has an idealized shape and is sometimes called a synthetic hydrograph. The SCS triangular hydrograph shown by the dashed lines closely approximates the dimensionless hydrograph and greatly simplifies calculation of the variables. The dimensionless hydrograph arbitrarily has 100 units of flow for the peak and 100

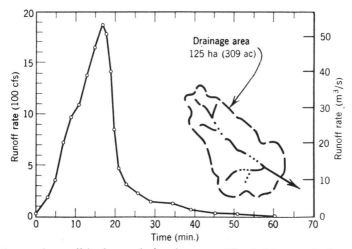

Fig. 4.8 Measured runoff hydrograph for the upper Theobold watershed, northwestern Iowa, June 1950.

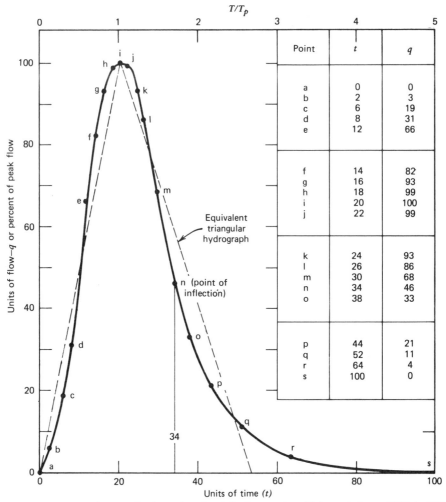

Fig. 4.9 Dimensionless and triangular flood hydrographs. (Adapted from USDA–SCS, 1972.)

units of time for the duration of flow. As described in this text, it is the hydrograph to be developed for flood routing in Chapter 11.

To develop the design hydrograph for a watershed, the peak flow and the runoff volume must be known for the desired return period storm. The design hydrograph is developed from the dimensionless hydrograph by using appropriate conversion factors. The factor u is the ratio of the total runoff volume to the area under the dimensionless hydrograph. The area under the hydrograph is 2620 square units. Thus, each square unit under the basic hydrograph has a value of

$$u = Q/2620 \qquad (4.9)$$

for the design storm having a total runoff volume Q. The factor w is the ratio of peak runoff for the design storm to the peak flow of 100 on the dimensionless hydrograph. Each unit of flow on the dimensionless hydrograph has a value of

$$w = q/100 \qquad (4.10)$$

in the hydrograph of the design storm. The factor k is the value that each unit of time on the dimensionless hydrograph represents in the design hydrograph. On the design hydrograph 1/100 of the peak flow times 1/100 of the duration of runoff must equal 1/2620 of the flood volume, just as it does on the dimensionless hydrograph. Since w is equal to 1/100 of the design peak flow, k must be equal to 1/100 of the design duration, and u is 1/2620 of the design flood volume. Therefore,

$$wk = u$$

and

$$k = u/w \qquad (4.11)$$

When runoff rate is measured in cubic meters per second, runoff volume is measured in hectare-meters, and time is measured in minutes.

$$k = \frac{u(\text{ha-m}) \times 10\ 000\ (\text{m}^2/\text{ha})}{w(\text{m}^3/\text{s}) \times 60\ (\text{s/min})} = 167\ \frac{u}{w}$$

The coordinates of the design hydrograph are obtained by multiplying the ordinates and abscissas of the dimensionless hydrograph by w and k, respectively.

☐ Example 4.6

Develop a runoff hydrograph for a design return period of 50 years for the watershed of Examples 4.1 and 4.3.

Solution.   The peak runoff is 4.45 m³/s and the flood volume is 1.56 ha-m. From Eqs. 4.9, 4.10, and 4.11,

$$u = 1.56/2620 = 0.0006 \text{ ha-m/unit}$$
$$w = 4.45/100 = 0.045 \text{ m}^3/\text{s per unit}$$
$$k = 167 \times 0.0006/0.045 = 2.25 \text{ min/unit}$$

Ordinates and abscissas of the design hydrograph are obtained by multiplying the values of q and t from Fig. 4.9 by w and k, respectively. The calculated coordinates are as follows:

Point	kt	wq	Point	kt	wq
a	0	0	j	49.5	4.46
b	4.5	0.14	k	54.0	4.19
c	13.5	0.86	l	58.5	3.87
d	18.0	1.40	m	67.5	3.06
e	27.0	2.97	n	76.5	2.07
f	31.5	3.69	o	85.5	1.49
g	36.0	4.19	p	99.0	0.95
h	40.5	4.46	q	117.0	0.50
i	45.0	4.50	r	144.0	0.18
Point i is at T_p			s	225.0	0

☐

4.16 Other Hydrographs

A large number of synthetic hydrographs have been developed, but none of these has received general acceptance. Because of limited flow records, these hydrographs as well as peak flow estimates leave much to be desired.

Several theoretical hydrographs have been proposed based on different statistical frequency distributions. Dodge (1959) developed a unit hydrograph from the Poisson probability function, Gray (1973) employed a two-parameter gamma distribution, and Reich (1962) investigated a three-parameter Pearson type III function. Linsley et al. (1982) described several methods of developing unit hydrographs and a hydrograph for overland flow.

REFERENCES

Brakensiek, D. L. (1959). "Selecting the Water Year for Small Agricultural Watersheds." *ASAE Trans.* **2,** 5–8, 10.

Chow, V. T. (1962). *Hydrologic Determination of Waterway Areas for the Design of Drainage Structures in Small Drainage Basins.* Ill. Eng. Exp. Sta. Bull. 462, March.

Commons, G. G. (1942). "Flood Hydrographs." *Civil Eng.* **12,** 571–572.

Copley, T. L., et al. (1944). *Effects of Land Use and Season on Runoff and Soil Loss.* N.C. Agr. Exp. Sta. Bull. 347.

Dodge, J. C. I. (1959). "A General Theory of the Unit Hydrograph." *J. Geophys. Res.* **64,** 241–256.

Goodrich, D. C., and D. A. Woolhiser (1991). "Catchment Hydrology." *Rev. Geophys. Suppl.,* April.

Gray, D. M. (ed.) (1973). *Handbook on the Principles of Hydrology.* 2nd print. Water Information Center, Port Washington, NY.

Harrold, L. L. (1957). "Minimum Water Yield from Small Agricultural Watersheds." *Am. Geophys. Union Trans.* **38,** 201–208.

―――― (1961). "Hydrologic Relationships on Watersheds in Ohio." *Soil Cons.* **26,** 208–210.

Horn, D. L., and G. O. Schwab (1963). "Evaluation of Rational Runoff Coefficients for Small Agricultural Watersheds." *ASAE Trans.* **6**(3), 195–198, 201.

Kirpich, P. Z. (1940). "Time of Concentration of Small Agricultural Watersheds." *Civil Eng.* **10,** 362.

Knisel, W. G. (ed.) (1980). *CREAMS: A Field-Scale Model for Chemicals, Runoff, and Erosion from Agricultural Management Systems.* Cons. Res. Rep. No. 26. USDA–Science and Education Administration, Washington, DC.

Larson, C. L., and B. M. Reich (1973). "Relationship of Observed Rainfall and Runoff Recurrence Intervals." In *Flood and Droughts,* E. F. Schulz et al. (eds.). *Proceedings of 2nd International Symposium in Hydrology, Sept. 1972.* Water Research Publ., Ft. Collins, CO.

Linsley, R. K., M. A. Kohler, and J. L. H. Paulhus (1982). *Hydrology for Engineers,* 3rd ed. McGraw–Hill, New York.

McCuen, R. H. (1989). *Hydrologic Analysis and Design.* Prentice-Hall, Englewood Cliffs, NJ.

Nearing, M. A., G. R. Foster, L. J. Lane, and S. C. Finkner (1989). "A Process-Based Soil Erosion Model for USDA-Water Erosion Prediction Project Technology." *ASAE Trans* **32,** 1587–1593.

Nixon, P. R., and G. O. Schwab (1961). "Water Yield Prediction in Southern Iowa Based on Watershed Characteristics." *Iowa State J. Sci.* **35**, 331–342.

Potter, W. D. (1961). *Peak Rates of Runoff from Small Watersheds.* U.S. Bureau Public Roads, Hydraulic Design Series No. 2, April. GPO, Washington, DC.

Reich, B. M. (1962). *Design Hydrographs for Very Small Watersheds from Rainfall.* Civil Engineering Section, July. Colorado State University, Ft. Collins, CO.

Rovang, R. M., L. P. Herndon, T. E. Radermacher, and K. E. Harward (1990). *Experience Gained in Development of SCS Prototype Engineering Software.* Paper No. 90-2620. ASAE, St. Joseph, MI.

Sherman, L. K. (1932). "Stream Flow from Rainfall by Unit Graph Method." *Eng. News-Record* **108**, 501–505.

Skaggs, R. W. (1982). "Field Evaluation of a Water Management Simulation Model." *ASAE Trans.*, **25**, 666–674.

U.S. Soil Conservation Service (SCS) (1972). "Hydrology." In *National Engineering Handbook*, Sect. 4. GPO, Washington, DC.

—— (1986). *Urban Hydrology for Small Watersheds.* Tech. Release 55. National Technical Information Service, Springfield, VA.

—— (1990). *Engineering Field Manual*, Chap. 2. (litho.) Washington, DC.

PROBLEMS

4.1 Calculate the 10-year return period peak runoff for an 80-ha (200-ac) watershed having a runoff coefficient of 0.4. The maximum length of flow of water is 610 m (2000 ft) and the fall along this path is 6.0 m (20 ft). Watershed is at your present location.

4.2 Determine the peak rate of runoff from an 8-ha (20-ac) watershed during a 30-min storm that gave the following amounts of rain during successive 5-min periods: 3, 8, 15, 15, 8, and 3 mm (0.1, 0.3, 0.6, 0.6, 0.3, and 0.1 in.). The time of concentration for the watershed is 10 min. The runoff coefficients for the 2 ha (5 ac) are 0.2 and 0.6 for the remaining 6.0 ha (15 ac). What is the approximate return period for this runoff if the watershed is at your present location?

4.3 By the rational method, determine the design peak runoff for a 50-year return period. The watershed consists of 50 ha (120 ac), one third in rotation meadow and the remainder in row crops on the contour. The hydrologic soil group is B, and the time of concentration is 30 min. The watershed is at your present location.

4.4 If the watershed in Problem 4.3 is in southern Alabama, what would be the peak runoff rate? If the return period was decreased to 2 years, what would be the rate of runoff?

4.5 Determine the depth (water) of snowmelt in one day for average conditions if a snow field is at an average elevation of 1830 m (6000 ft). The nearest weather station at an elevation of 300 m (1000 ft) had a minimum temperature of 12°C and a maximum of 16°C.

4.6 Assuming that 15 degree-days (°C) were accumulated in 4 days, what depth of snow would be melted under average conditions if the snow had a water equivalent of 10 percent?

4.7   Determine the runoff volume in millimeters (in.) of depth from a 50-year return period storm using local data at your present location assuming Antecedent Rainfall Condition III, Hydrologic Soil-Cover Complex Curve Number 70, and duration of the critical storm 3 h.

4.8   Determine the flood volume in ha-m (ac-ft) for a 50-year return period storm for the dormant season at your present location. The 360-ha (900-ac) watershed is in row crops, terraced on Hydrologic Soil Group C in good condition for one third of the area, and the remaining two thirds is in woodland on Hydrologic Soil Group D in poor condition. Assume 25 mm (1 in.) of rainfall occurred 3 days prior to the design storm and the critical storm duration is 24 h.

4.9   Determine the minimum annual water yield from a 40-ha (100-ac) watershed near Coshocton, Ohio, which will occur once in 25 years. What would be the equivalent surface area of a reservoir to store this quantity of water, if the average depth was 2.4 m (8 ft)?

4.10   From Fig. 10.11 determine the minimum size drainage area for a farm pond at your present location if the storage capacity is 1.2 ha-m (10 ac-ft) and if seepage and evaporation are neglected. What is the equivalent water yield in depth over the drainage area?

4.11   Assuming the dimensionless hydrograph is applicable, determine the duration of flow if the peak runoff is 8.5 m³/s (300 cfs) and the volume of flow is 100 mm (4 in.) for a 50-year storm from an 85-ha (210-ac) watershed. What are the coordinates for point n on the design hydrograph?

4.12   Determine q, T_p, and T_b for a triangular unit hydrograph for 25 mm (1 in.) of runoff for a 40-ha (100-ac) watershed at your present location if the storm duration is 2 h and T_c is 2 h (see Fig. 4.3b).

4.13   By frequency analysis methods discussed in Chapter 2 determine the estimated maximum annual discharge for return periods of 2 and 100 years from the following 18 years of maximum annual floods (1939–1956) from a gaged watershed; 53, 91, 109, 0.8, 0.8, 0.3, 20, 36, 0.5, 0.3, 56, 38, 0.3, 0.8, 0.3, 5, 0.3, and 3 mm (2.1, 3.6, 4.3, 0.03, 0.03, 0.01, 0.8, 1.4, 0.02, 0.01, 2.2, 1.5, 0.01, 0.03, 0.01, 0.2, 0.01, and 0.1 in.). Is the length of record adequate for these estimates to be reliable 90 percent of the time?

CHAPTER 5

Water Erosion and Control Practices

Erosion is one of the most important agricultural problems in the world. It is a primary source of sediment that pollutes streams and fills reservoirs. Some estimates in the 1970s were as high as 4 billion Mg annually in the United States. This amount represents about a 30 percent increase over that estimated in the 1930s even though government subsidies, educational programs, and new practices have been instigated.

Since the early 1970s, greater emphasis has been given to erosion as a contributor to nonpoint pollution. *Nonpoint* refers to erosion from the land surface rather than from channels and gullies. Eroded sediment can carry nutrients, particularly phosphates, to waterways, and contribute to eutrophication of lakes and streams. Adsorbed pesticides are also carried with eroded sediments, adversely affecting surface water quality.

The two major types of erosion are geological erosion and erosion from human or animal activities. Geological erosion includes soil-forming as well as soil-eroding processes that maintain the soil in a favorable balance, suitable for the growth of most plants. Human- or animal-induced erosion includes a breakdown of soil aggregates and accelerated removal of organic and mineral particles resulting from tillage and removal of natural vegetation. Geological erosion has contributed to the formation of our soils and their distribution on the surface of the earth. This long-time eroding process caused most of our present topographic features, such as canyons, stream channels, and valleys.

Water erosion is the detachment and transport of soil from the land by water, including runoff from melted snow and ice. Types of water erosion include interrill (raindrop and sheet), rill, gully, and stream channel erosion. Water erosion is accelerated by farming, forestry, and construction activities.

5.1   Factors Affecting Erosion by Water

The major variables affecting soil erosion are climate, soil, vegetation, and topography. Of these, vegetation and, to some extent, soil and topography may be controlled. The climatic factors are beyond the power of humans to control.

Climate. Climatic factors affecting erosion are precipitation, temperature, wind, humidity, and solar radiation. Temperature and wind are most evident through their effects on evaporation and transpiration; however, wind also changes raindrop velocities and the angle of impact. Humidity and solar radiation are somewhat less directly involved in that they are associated with temperature and rate of soil water depletion.

The relationship between precipitation characteristics and runoff and soil loss is complex. In a study of 19 independent variables measuring rainfall characteristics, the most important single measure of the erosion-producing factor of a rainstorm was the product, rainfall energy times maximum 30-min intensity (Wischmeier and Smith, 1958). Studies on individual erosion processes have found that interrill erosion varies with the rainfall intensity to a power varying from 1.56 to 2.09 (Watson and Laflen, 1986), with the power of 2 generally being accepted (Lane et al., 1987). Rill erosion is a function of runoff rate, which depends both on rainfall intensity and on soil infiltration rates.

Soil. Physical properties of soil affect the infiltration capacity and the extent to which particles can be detached and transported. The corresponding soil characteristics that describe the ease with which soil particles may be eroded are soil detachability and soil transportability. In general, soil detachability increases as the size of the soil particles or aggregates increase, and soil transportability increases with a decrease in the particle or aggregate size. That is, clay particles are more difficult to detach than sand, but clay is more easily transported. The properties that influence erosion include soil structure, texture, organic matter, water content, clay mineralogy, and density or compactness, as well as chemical and biological characteristics of the soil. As yet, no single soil characteristic or index has been identified as a satisfactory means of predicting erodibility.

Vegetation. The major effects of vegetation in reducing erosion are (1) interception of rainfall by absorbing the energy of the raindrops and thus reducing surface sealing and runoff, (2) retardation of erosion by decreased surface velocity, (3) physical restraint of soil movement, (4) improvement of aggregation and porosity of the soil by roots and plant residue, (5) increased biological activity in the soil, and (6) transpiration, which decreases soil water, resulting in increased storage capacity and less runoff. These vegetative influences vary with the season, crop, degree of maturity of the vegetation, soil, and climate, as well as with the kind of vegetative material, namely, roots, plant tops, and plant residues.

Residues from vegetation protect the surface from raindrop impact and improve the soil structure. Residue and tillage management practices can have a dramatic effect on soil erosion.

Topography. Topographic features that influence erosion are degree of slope, shape and length of slope, and size and shape of the watershed. On steep slopes, runoff water is more erosive, and can more easily transport detached sediment downslope. On longer slopes, an increased accumulation of overland flow tends to increase rill erosion. Concave slopes, with lower slopes at the foot of the hill, are less erosive than convex slopes.

5.2 Raindrop Erosion

Raindrop erosion is soil detachment and transport resulting from the impact of water drops directly on soil particles or on thin water surfaces. Although the impact of raindrops on shallow streams may not splash soil, it does increase turbulence, providing a greater sediment-carrying capacity.

Tremendous quantities of soil are splashed into the air, most particles more than once. The amount of soil splashed into the air as indicated by the splash losses from small elevated pans was found to be 50 to 90 times greater than the runoff losses. On bare soil, it is estimated that as much as 200 Mg/ha is splashed into the air by heavy rains. The relationship among erosion, rainfall momentum, and energy is determined by raindrop mass, size distribution, shape, velocity, and direction. The relationship between rainfall intensity and energy has been found to be (Foster et al., 1981)

$$E = 0.119 + 0.0873 \log_{10}i \tag{5.1}$$

where E = kinetic energy in MJ/ha-mm,
i = intensity of rainfall in mm/h.

The effect of a single drop as it strikes is shown in the high-speed photograph in Fig. 5.1a; the successive steps that take place in drop-crater formation are illustrated in Fig. 5.1b. Splashed particles may move more than 0.6 m in height and more than 1.5 m laterally on level surfaces.

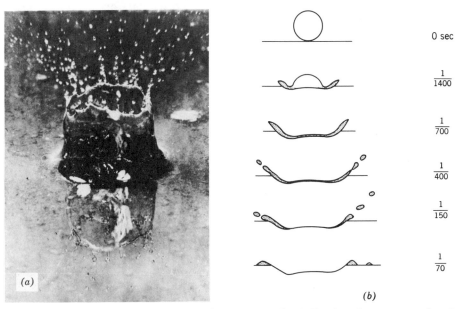

Fig. 5.1 Raindrop characteristics when striking moist soil. (a) The drop bursts upward and outward. (Courtesy Agricultural Research Service.) (b) Steps in drop-crater formation. (After Mihara, 1952).

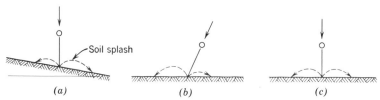

Fig. 5.2 Differential soil movement caused by raindrop splash. (*a*) Sloping land. (*b*) Inclined rainfall. (*c*) Vertical rainfall. (*Source:* Kohnke and Bertrand, 1959.)

Factors affecting the direction and distance of soil splash are slope, wind, surface condition, and impediments to splash such as vegetative cover and mulches. On sloping land, the splash moves farther downhill than uphill, not only because the soil particles travel farther, but also because the angle of impact causes the splash reaction to be in a downhill direction (Fig. 5.2). Components of wind velocity up or down the slope have an important effect on soil movement by splash. Surface roughness and impediments to splash tend to counteract the effects of slope and winds. Contour furrows and ridges break up the slope and cause more of the soil to be splashed uphill. If raindrops fall on crop residue or growing plants, the energy is absorbed and thus soil splash is reduced. Raindrop impact on bare soil not only causes splash but also decreases aggregation and causes deterioration of soil structure.

5.3 Sheet Erosion

The idealized concept of sheet erosion was the uniform removal of soil in thin layers from sloping land, resulting from sheet or overland flow. Fundamental studies of the mechanism of erosion, in which both time-lapse and high-speed photographic techniques have been used, indicate that this idealized form of erosion rarely occurs. Minute rilling takes place almost simultaneously with the first detachment and movement of soil particles. The constant meander and change of position of these microscopic rills obscure their presence from normal observation, hence establishing the false concept of sheet erosion.

The beating action of raindrops combined with surface flow causes initial microscopic rilling. Raindrops detach the soil particles, and the detached sediment can reduce the infiltration rate by sealing the soil pores. The eroding and transporting power of sheet flow is a function of the rainfall intensity, infiltration rate, and field slope, for a given size, shape, and density of soil particle or aggregate.

5.4 Interrill Erosion

Splash and sheet erosion are sometimes combined and called interrill erosion. Research has shown interrill erosion to be a function of soil properties, rainfall intensity, and slope. The relationship among these parameters is generally expressed as (Watson and Laflen, 1986)

$$D_i = K_i i^2 S_f \tag{5.2}$$

where D_i = interrill erosion rate in kg/m²-s,
  K_i = interrill erodibility of soil in kg-s/m⁴,

i = rainfall intensity in m/s,

S_f = slope factor = $1.05 - 0.85 \exp(-4 \sin \theta)$ (Liebenow et al., 1990), (5.3)

and where θ = slope in degrees.

5.5 Rill Erosion

Rill erosion is the detachment and transport of soil by a concentrated flow of water. Rills are small enough to be removed by normal tillage operations. Rill erosion is the predominant form of erosion under most conditions. It is most serious where intense storms occur on soils with high-runoff-producing characteristics and highly erodible topsoil.

Rill erosion is a function of the hydraulic shear τ of the water flowing in the rill, and two soil properties, the rill erodibility K_r and the critical shear τ_c, the shear below which soil detachment is negligible (Lane et al., 1987). Detachment rate D_r is the erosion rate occurring beneath the submerged area of the rill. The relationship among these variables is

$$D_r = K_r \, (\tau - \tau_c) \left(1 - \frac{Q_s}{T_c}\right)$$ (5.4)

where D_r = rill detachment rate in kg/m²-s,
$\quad K_r$ = rill erodibility resulting from shear in s/m,
$\quad \tau_c$ = critical shear below which no erosion occurs in Pa,
$\quad Q_s$ = rate of sediment flow in the rill in kg/m-s,
$\quad T_c$ = sediment transport capacity of rill in kg/m-s,
$\quad \tau$ = hydraulic shear of flowing water in Pa = $\rho \, g \, r \, s$, (5.5)

and where ρ = density of water in kg/m³,
$\quad g$ = acceleration resulting from gravity in m/s²,
$\quad r$ = hydraulic radius of rill in m,
$\quad s$ = hydraulic gradient of rill flow.

The interrill and rill erosion processes are used in several process-based erosion prediction computer models, including CREAMS (Knisel, 1980) and WEPP (Lane and Nearing, 1989) (see Appendix I).

5.6 Gully Erosion

Gully erosion produces channels larger than rills. These channels carry water during and immediately after rains, and, as distinguished from rills, gullies cannot be obliterated by tillage. The amount of sediment from gully erosion is usually less than from upland areas, but the nuisance from having fields divided by large gullies has been the greater problem. In tropical areas, gully growth following deforestation and cultivation has led to severe problems from soil loss, and damage to buildings, roads, and airports (Aneke, 1985).

The rate of gully erosion depends primarily on the runoff-producing characteristics of the watershed; the drainage area; soil characteristics; the alignment, size, and shape of the gully; and the slope in the channel (Bradford et al., 1973). A gully develops by processes that may take place either simultaneously or during different periods of its growth. These processes are (1) waterfall erosion or headcutting at the gully head, (2) erosion caused by water flowing through the gully or by raindrop splash on exposed gully sides, (3) alternate freezing and thawing of the exposed soil banks, and (4) slides or mass movement of soil into the gully.

Evaluation and prediction of gully development are difficult because the factors are not well defined and field records of gullying are inadequate. From aerial photographs and field topographic surveys, Beer and Johnson (1963) developed a prediction equation for the deep loess region in western Iowa based on watershed runoff characteristics and soil properties. Gully formation was found to depend on soil shear strength, infiltration, and depth of water table by Bradford et al. (1973). In many cases, an impeding layer resulted in saturated soil conditions at the floor of the gully. The saturated soils tended to be weak, leading to undercutting and side sloughing. Runoff from subsequent storms would then remove loose soil from the gully floor.

5.7 Stream Channel Erosion

Stream channel erosion consists of soil removal from stream banks or soil movements in the channel. Stream channel erosion and gully erosion are distinguished primarily in that stream channel erosion applies to the lower end of headwater tributaries and to streams that have nearly continuous flow and relatively flat gradients, whereas gully erosion generally occurs in intermittent streams near the upper ends of headwater tributaries.

Stream banks erode either by runoff flowing over the side of the stream bank or by scouring and undercutting below the water surface. Stream bank erosion, less serious than scour erosion, is often increased by the removal of vegetation, overgrazing, tilling too near the banks, or straightening of the channel. Scour erosion is influenced by velocity and direction of flow, depth and width of the channel, and soil texture. Poor alignment and the presence of obstructions such as sandbars increase meandering, the major cause of erosion along the bank.

5.8 Sediment Transport

Numerous methods for predicting sediment transport capacity of channels have been developed, based on channel hydraulic shear (Eq. 5.5), flow rate, velocity, and sediment properties. The reader should consult specific references on sediment transport, such as Vanoni (1975), for further information on these complex processes. Transport capacity of individual rills has been estimated by the relationship (Foster and Meyer, 1972)

$$T_c = B\tau^{1.5}$$ (5.6)

where T_c = transport capacity per unit width in kg/m-s,
B = transport coefficient based on soil and water properties,
τ = hydraulic shear of rill channel in Pa.

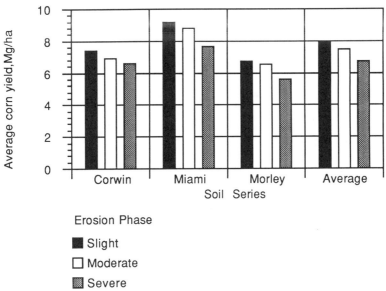

Fig. 5.3 Effect of erosion phase and soil series on crop yield. (Based on data by Schertz et al., 1989.)

SOIL LOSSES

The importance of soil losses is indicated by the effect of erosion phase on crop yield, as shown in Fig. 5.3. The researchers reported that much of the reduced yield observed on eroded soils was due to a decrease in the amount of water available to the plant on eroded soils (Schertz et al., 1989). On some soils, these crop yield decreases can be largely overcome by higher fertilization levels. On other soils, particularly more shallow soils on sloping terrain, erosion may completely destroy productivity if appropriate conservation practices are not initiated (USDA, 1989).

From watershed studies in Ohio, the annual and monthly soil losses for corn are shown in Fig. 5.4 for areas approximately 0.6 ha in size on slopes from 6 to 15 percent. The soil losses from watersheds with recommended conservation practices are compared with the soil losses from the corresponding check watersheds representing typical Ohio practices. In Fig. 5.4, the losses from the check watershed are more than double those from the corresponding conservation watershed. With a few exceptions, erosion from the conservation watersheds is considerably less than that from the check areas.

Soil losses, or relative erosion rates for different management systems, are estimated to assist farmers and government agencies in evaluating existing farming systems or in planning to decrease soil losses. In the period 1945 to 1965, a method of estimating losses based on statistical analyses of field plot data from small plots located in many states was developed, which resulted in the Universal Soil Loss Equation (USLE). A revised version of the USLE (RUSLE) has been developed for computer applications, allowing more detailed consideration of farming practices and topography for erosion prediction (Renard et al., 1991).

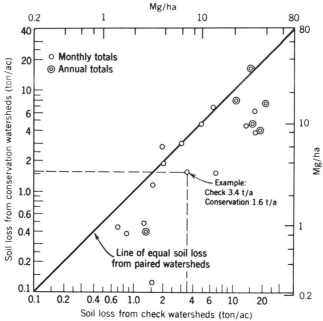

Fig. 5.4 Annual and monthly soil loss from small paired watersheds (corn years only). (Redrawn from Harrold, 1949.)

Since the mid-1960s, scientists have been developing process-based erosion computer programs that estimate soil loss by considering the processes of infiltration, runoff, detachment, transport, and deposition of sediment. Numerous research programs have been developed, and programs are being improved for field use. The process-based Water Erosion Prediction Project (WEPP) model is replacing the USLE in SCS and other U.S. government agencies (Foster, 1988).

5.9 Predicting Upland Erosion

The USLE continues to be a widely accepted method of estimating sediment loss despite its simplification of the many variables involved. It is useful for determining the adequacy of conservation measures in farm planning, and for predicting nonpoint sediment losses in pollution control programs. The average annual soil loss, as determined by Wischmeier and Smith (1978), can be estimated from the equation

$$A = RKLSCP \qquad (5.7)$$

where A = average annual soil loss in Mg/ha,
R = rainfall and runoff erosivity index for geographic location (Fig. 5.5),
K = soil erodibility factor (Fig. 5.6, Eq. 5.8),
L = slope length factor (Eq. 5.9a),
S = slope steepness factor (Eq. 5.10),
C = cover management factor (Table 5.2),
P = conservation practice factor (Eq. 5.11).

In RUSLE, K varies to account for seasonal variation in soil erodibility.

For a given storm, the rainfall and runoff erosivity index (EI) is the product of the kinetic energy of the storm (Eq. 5.1) and the maximum 30-min intensity for that storm. The EI values for all the storms occurring in a given year for a location are added to give an annual erosivity index. The average annual rainfall and runoff erosivity index R is shown in Fig. 5.5 for the continental United States. Estimates for R have been made for many other locations in the world (Hudson, 1985). Estimates of R for single storms and annual events with 2- to 20-year return periods have also been formulated, and the values for 3 sites are given in Table 5.1, with 180 other sites given in Wischmeier and Smith (1978).

The soil erodibility factor K for a series of benchmark soils was obtained by direct soil loss measurements from fallow plots located in many U.S. states. A nomograph to estimate K for a given soil on which it is not known is presented in Fig. 5.6, or can be calculated from the regression equation

$$K = 2.8 \times 10^{-7}M^{1.14}(12 - a) + 4.3 \times 10^{-3}(b - 2) + 3.3 \times 10^{-3}(c - 3) \quad (5.8)$$

where M = particle size parameter (% silt + % very fine sand) $\times$ (100 − % clay),
  a = percent organic matter,
  b = soil structure code (very fine granular, 1; fine granular, 2; medium or coarse granular, 3; blocky, platy, or massive, 4),
  c = profile permeability class (rapid, 1; moderate to rapid, 2; moderate, 3; slow to moderate, 4; slow, 5; very slow, 6).

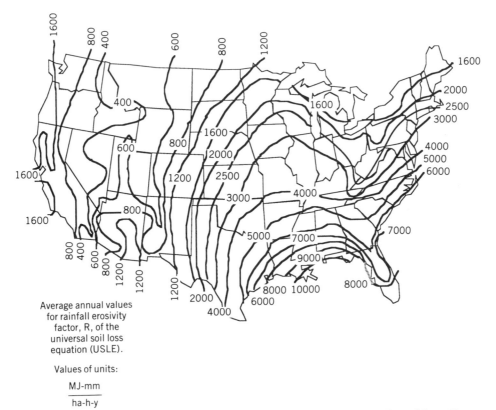

Average annual values for rainfall erosivity factor, R, of the universal soil loss equation (USLE).

Values of units:

$$\frac{\text{MJ-mm}}{\text{ha-h-y}}$$

Fig. 5.5 Rainfall and runoff erosivity index R by geographic location. (Adapted from Foster et al., 1981.)

Table 5.1   Annual and Single-Storm Erosivity Index R

Location	Return Period in Years			
	2	5	10	20
Annual R		MJ-mm/ha-h-year		
Little Rock, AR	5242	7182		9684
Indianapolis, IN	2825	3830		5140
Devils Lake, ND	953	1532		2417
Single-Storm R				
Little Rock, AR	1174	1957	2682	3591
Indianapolis, IN	698	1021	1277	1532
Devils Lake, ND	460	664	834	1004

Source: Wischmeier and Smith (1978).

If R values or K values from other sources are employed, compatible units must be used. The topographic factors, L (McCool et al., 1989) and S (McCool et al., 1987), adjust the predicted erosion rates to give greater erosion rates on longer and/or steeper slopes, when compared with a USLE "standard" slope of 9 percent and slope length of 22 m. The differences are attributed to increasing rill erosion rates, as more runoff accumulates with longer slopes, and greater erosive forces occurring with steeper gradients. These factors can be calculated from the equations

$$L = \left(\frac{l}{22} \right)^m \qquad (5.9a)$$

where L = slope length factor,
l = slope length in m,
m = dimensionless exponent.

McCool et al. (1989) recommended that for conditions where rill erosion and interrill erosion were about equal on a 9 percent, 22-m-long slope, then m could be found from the equation

$$m = \frac{\sin \theta}{\sin \theta + 0.269 \, (\sin \theta)^{0.8} + 0.05} \qquad (5.9b)$$

where θ = field slope steepness in degrees = $\tan^{-1}\left(\dfrac{s}{100} \right)$,

and where s = field slope in percent.

For conditions where rill erosion is greater than interrill erosion (like soils with a large silt or fine sand content), m should be increased up to 75 percent; where rill erosion is less than interrill erosion (on short slopes or high-clay-content soils), m should be decreased down to 50 percent.

McCool et al. (1987) presented a set of S factors based on the slope steepness. For slopes shorter than 4 m,

$$S = 3.0 \, (\sin \theta)^{0.8} + 0.56 \qquad (5.10a)$$

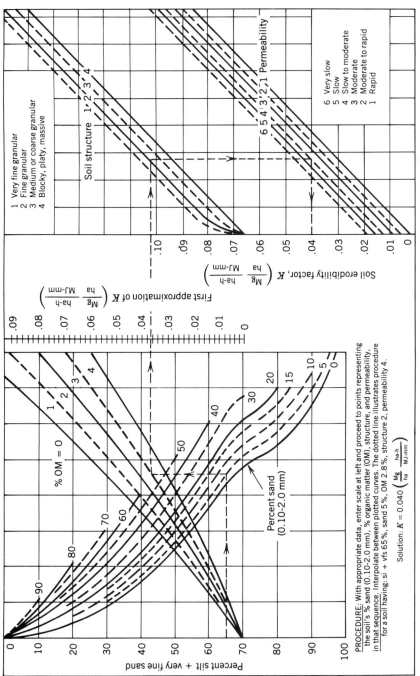

Fig. 5.6   Soil erodibility K factor nomograph in SI units. (*Source:* Foster et al., 1981.)

For slopes longer than 4 m and $s < 9$ percent,

$$S = 10.8 \sin \theta + 0.03 \qquad (5.10b)$$

For slopes longer than 4 m and $s \geq 9$ percent,

$$S = 16.8 \sin \theta - 0.50 \qquad (5.10c)$$

The slope length is measured from the point where surface flow originates (usually the top of the ridge) to the outlet channel or a point downslope where deposition begins. RUSLE considers nonuniform concave or convex slopes (Renard et al., 1991), as do many process-based erosion prediction programs.

The cover-management factor C includes the effects of cover, crop sequence, productivity level, length of growing season, tillage practices, residue management, and expected time distribution of erosive events. Table 5.2 gives the cropping and residue management variables as percentages of the soil loss for crops to that for continuous fallow. The annual distribution of the rainfall erosivity index will vary with geographic location. Examples are given in Fig. 5.7. The C factor for a given crop rotation is found by first multiplying the soil loss ratios for each

Table 5.2 Percentage of Soil Loss from Crops to Corresponding Loss from Continuous Fallow

Cover, Sequence, and Management[a]	Spring Planting		Soil Loss Ratio (%)[d] for Crop Stage Period and Canopy Cover[e]							
	Residue[b] (kg)	Cover[c] (%)	F	SB	1	2	3:80	3:90	3:96	4L
Continuous										
Corn, RdL, sprg, TP	5000	—	36	60	52	41	—	24	20	30
Small grain	5000	60	—	16	14	12	7	4	2	—
Meadow							1			
Rotation										
Rowcrop after meadow			12	27	23	20	—	14	12	21
Corn after beans, sprg,			47	78	65	51	—	30	25	37
Beans after corn, sprg, TP			39	64	56	41	—	21	18	
Conservation Tillage										
Beans or corn after corn	5000	60	—	13	11	10	—	10	8	20
Corn after beans		30	—	33	29	25	22	18	14	33
Small grain after corn	5000	60	—	16	14	13	7	4	2	

[a]RdL, crop residue left in field; sprg, spring tillage; TP, plowed with moldboard.
[b]Dry mass per hectare, after winter loss and reductions by grazing or partial removal, 5000 kg/ha represents a yield of 6 to 8 Mg/ha.
[c]Percentage of soil surface covered by plant residue mulch after crop seeding. The difference between spring residue and that on the surface after crop seeding is reflected in the soil loss ratios as residues mixed with the topsoil.
[d]The soil loss ratios assume that the indicated crop sequence and practices are followed consistently. One-year deviations from normal practices do not have the effect of a permanent change.
[e]Crop stage periods: F, rough fallow; SB, seedbed until 10% canopy cover; 1, establishment until 50% canopy cover; 2, development until 75% canopy cover; 3, maturing until harvest for three different levels of canopy cover; 4L, residue or stubble.
Source: Wischmeier and Smith (1978).

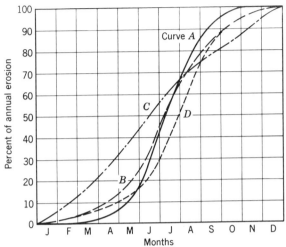

Fig. 5.7 Monthly distribution of the rainfall and runoff erosivity index. Curve A: north-western Iowa, northern Nebraska, and southeastern South Dakota. Curve B: northern Missouri, central Illinois, Indiana, and Ohio. Curve C: Louisiana, Mississippi, western Tennessee, and eastern Arkansas. Curve D: Atlantic Coastal Plains of Georgia and the Carolinas. (Redrawn from Smith and Wischmeier, 1962.)

growth period in a crop rotation (Table 5.2) by the percentage of annual erosion during each respective period (Fig. 5.7). These products are then summed and expressed as a decimal value to give the C factor (Example 5.3). Many state agencies have developed C factor tables for their respective climate pattern and cropping practices to aid field staff in working with the USLE.

The conservation practice P can be found from the equation

$$P = P_c \times P_s \times P_t \qquad (5.11)$$

where P_c = contouring factor based on slope (Table 5.3),

P_s = strip cropping factor for crop strip widths recommended in Table 5.3 (1.0 for contouring only or for alternating strips of corn and small grain, 0.75 for 4-year rotation with 2 years of row crop, and 0.50 with 1 year of row crop),

P_t = terrace sedimentation factor (1.0 for no terraces, 0.2 for terraces with graded channel sod outlets, and 0.1 for terraces with underground outlets).

In humid regions, strip cropping and terracing include contouring which re-quires engineering design; the other erosion factors are more agronomic in nature. As will be discussed in Chapter 8, terracing affects the slope length so that the L factor in a terracing system is altered to determine the amount of sediment delivered to the terrace. The P_t factor will predict the amount of sediment actually delivered from a given terrace. It is possible to predict either sediment detached from the cropping area, and not include the terrace factor, or sediment leaving the field, and include the terrace factor (see Example 5.2).

Table 5.3 Contouring Factor P_c, Maximum Slope Lengths, and Maximum Strip Crop Widths for Different Slopes[a]

Land Slope (%)	P_c value	Maximum Slope Length[b] (m)	Maximum Strip-Crop Width (m)
1–2	0.6	120	40
3–5	0.5	90	30
6–8	0.5	60	30
9–12	0.6	36	24
13–16	0.7	24	24
17–20	0.8	18	18
21–25	0.9	15	15

[a]Factor for farming upslope and downslope is 1.0.
[b]Maximum slope length for strip cropping can be twice that for contouring only.
Source: Wischmeier and Smith (1978).

Some of the process-based erosion computer programs predict the erosion rates in the upslope areas and then route both the flow and the sediment transport along the terrace and through the drainage system, noting areas of sediment detachment and/or deposition.

☐ Example 5.1

Determine the soil loss for the following conditions: Location, Memphis, Tennessee. $K = 0.01$ Mg/ha-year, $l = 120$ m, $s = 10$ percent (5.7°), $C = 0.18$ (approximate for corn–corn–oats–meadow rotation with good management), and the field is to be contoured.

Solution. From Fig. 5.5, read $R = 5000$ as the erosion index. From Eq. 5.9b, assuming a moderate relationship between rill and interrill erosion, calculate m.

$$m = \frac{\sin 5.7}{\sin 5.7 + 0.269 \, (\sin 5.7)^{0.8} + 0.05} = 0.52$$

From Eq. 5.9a, calculate L:

$$L = \left(\frac{120}{22}\right)^{0.52} = 2.41$$

From Eq. 5.10c, calculate S:

$$S = 16.8 \sin 5.7 - 0.5 = 1.17$$

From Table 5.3 note that $P_c = 0.6$; from Eq. 5.11, note that $P_s = 0.75$ and $P_t = 1$; calculate P:

$$P = 0.6 \times 0.75 \times 1 = 0.45$$

Substitute in Eq. 5.7.

$$A = 5000 \times 0.01 \times 2.41 \times 1.17 \times 0.18 \times 0.45 = 11.4 \text{ Mg/ha-year}$$

Using the RUSLE computer program, the calculated coefficients were $R = 5160$, $LS = 2.81$, $C = 0.151$, and $P = 0.39$, and the annual erosion was estimated as 8.5 Mg/ha-year. The C factor was calculated for ridge tillage system. □

□ Example 5.2

If the soil loss tolerance for the conditions in Example 5.1 is 5 Mg/ha, what practices could be adopted to accomplish this level?

Solution. Since C, P, and L are the only variables that can be changed, the following combinations are possible by substituting the appropriate factors in Eq. 5.7.

Conservation Practice	L (factor)	C (factor)	P (factor)	Soil Loss (Mg/ha)	Remarks
(1) None	2.41	0.18	0.6	15	Soil loss too high
(2) Strip cropping	2.41	0.18	0.45	11	Soil loss too high
(3) Strip cropping and reduced C	2.41	0.12	0.30	5.1	Satisfactory
(4) Conservation tillage, no strip crop	2.41	0.05	0.6	4.2	Satisfactory
(5) Terrace at 24-m spacing	1.05	0.18	0.45	5.0	Loss to channel
			0.1	1.1	Loss from field
(6) Terrace with conservation tillage	1.05	0.05	0.45	1.4	Loss to channel
			0.1	0.3	Loss from field

Note that strip cropping alone is not adequate, unless the C factor is reduced by changing the rotation or other practices to reduce C to about 0.12. Conservation tillage alone will meet the tolerable loss level, as will terracing. If conservation tillage and terracing are both used, then a considerably higher C factor could be tolerated. Note that the addition of terraces changes not only the P factor but also the L factor. □

□ Example 5.3

If strip cropping is selected as the most desirable conservation practice in Example 5.2, what recommendations could be made for the cover-management practice?

Solution. From line 3 in Example 5.2, the maximum C factor to keep soil loss to 5 Mg/ha is 0.12. Assuming a corn–oats–meadow–meadow rotation, spring plowing, and corn residue plowed under, the following calculations were made using values from Table 5.2 and Fig. 5.7 (curve C), respectively.

Since $0.104 < 0.12$, the trial rotation is satisfactory. The RUSLE program esti-

Crop	Months	Soil Loss in Percentage of Continuous Fallow (Table 5.2)	Percentage of Annual Erosion (Fig. 5.7)	C Factor
Corn, first year	Jan.–Apr.	1	32	0.0032
	May	60	12	0.0720
	June	52	12	0.0624
	July	41	12	0.0492
	Aug.–Sept.	24	22	0.0528
	Oct.–Dec.	30	10	0.0300
Small grain	Jan.–Mar.	30	22	0.0660
with meadow	Apr.	16	10	0.0160
seeding	May	14	12	0.0168
	June	13	12	0.0156
	July–Aug.	4	20	0.0080
	Sept.–Dec.	1	24	0.0024
Meadow	Jan.–Dec.	1	100	0.0100
Meadow	Jan.–Dec.	1	100	0.0100
			Total	0.4144
			Average C for four years	0.104

mated a C factor of 0.132 for the trial rotation and specified a 5 percent soil loss for the meadow crop. □

By evaluation of the factors in Eq. 5.7, the predicted soil loss can be determined for a given set of conditions. If the soil loss is higher than the minimum required to maintain productivity, it may be reduced by changing the cover-management practices or the conservation practices.

Both physical and economic factors, as well as social aspects, need to be considered in establishing soil loss tolerances, sometimes called T values. These values vary with topsoil depth and subsoil properties from 4.5 to 11 Mg/ha. They are the maximum rates of soil erosion that will permit a high level of crop productivity to be sustained economically and indefinitely. Criteria for control of sediment pollution may dictate lower tolerance values.

□ Example 5.4

Determine the soil loss from a single 10-year return period storm for continuous corn, for the topography given in Example 5.1. The storm came during the second corn year in the rotation and during the first month (Period 1) after planting. All residue from the previous crop was left and incorporated by plowing.

Solution. From Table 5.1, read R = 2682 for nearest station at Little Rock, Arkansas, and from Table 5.2, read from line 1, C = 0.52. Substitute in Eq. 5.7.

$$A = 2682 \times 0.01 \times 2.41 \times 1.17 \times 0.52 \times 0.45 = 18 \text{ Mg/ha} \qquad \square$$

Sediment delivery downstream in a watershed may be estimated from the USLE and a sediment delivery ratio. The soil loss equation estimates gross sheet and rill erosion, but does not account for sediment deposited en route to the place of measurement nor for gully or channel erosion downstream. The sediment delivery ratio is defined as the ratio of sediment delivered at a location in the stream system to the gross erosion from the drainage area above that point. This ratio varies widely with size of area, steepness, density of drainage network, and many other factors. The sediment delivery ratio varies roughly as the inverse of the drainage area to the power of 0.2. For watersheds of 2.6, 130, 1300, and 26000 ha, ratios of 0.65, 0.33, 0.22, and 0.10, respectively, were suggested as average values by Roehl (1962).

EROSION CONTROL PRACTICES

In this chapter, the primary emphasis is on describing and estimating erosion by water from cultivated farmland and other nonpoint sources. Vegetated waterways and terraces are important control measures, but they will be discussed in later chapters, since specialized design procedures are required.

5.10 Contouring

This practice is the performing of field operations, such as plowing, planting, cultivating, and harvesting, approximately on the contour. It reduces surface runoff by impounding water in small depressions, and decreases the development of rills. The distribution of contouring in the United States is shown in Fig. 5.8. The greatest concentration is in the eastern wheat belt, where the benefits include both erosion reduction and conservation of water.

The relative effectiveness of contouring for controlling erosion on various slopes is shown by the conservation practice factor P calculated from Eq. 5.11. These

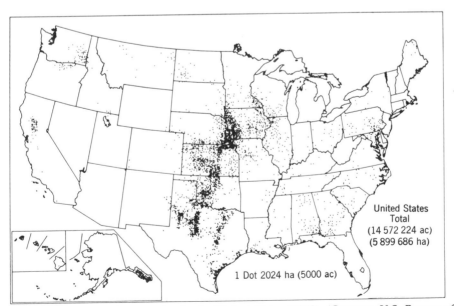

United States
Total
(14 572 224 ac)
(5 899 686 ha)

1 Dot 2024 ha (5000 ac)

Fig. 5.8 Distribution of contouring in the United States (1969). (Courtesy U.S. Bureau of the Census.)

values are applicable to fields relatively free from gullies and depressions other than grassed waterways and for slope lengths shown in Table 5.3. If ridge tillage is practiced, the storage capacity of the furrows is materially increased and the conservation practice factor is reduced to about half that shown for contouring. Contouring on steep slopes or under conditions of high rainfall intensity and soil erodibility will increase gullying because row breaks may release the stored water. Breakovers cause cumulative damage as the volume of water increases with each succeeding row.

The effectiveness of contouring is also impaired by changes in the infiltration capacity of the soil due to surface sealing. Depression storage is reduced after tillage operations cease and consolidation takes place. Studies by Harrold (1947) showed that contour cultivation together with good sod waterways reduced watershed runoff by 75 to 80 percent at the beginning of the season. This reduction dropped to as low as 20 percent at the end of the year, leaving an annual average reduction in runoff resulting from contouring of 66 percent.

Because of nonuniform slopes in most fields, all crop rows cannot be on the true contour. To establish row directions, a guideline (true contour) is laid out at one or more elevations in the field. On small fields of uniform slope, one guideline may be sufficient. Another guideline should be established if the slope along the row direction exceeds 1 to 2 percent in any row laid out parallel to the guideline. A small slope along the row is desirable to prevent runoff from a large storm breaking over the ridges. Where practical, field boundaries for contour farming should be relocated on the contour or moved, to eliminate odd-shaped fields that would result in short, variable-length rows called point rows.

5.11 Strip Cropping

Strip cropping is the practice of growing alternate strips of different crops in the same field. For controlling water erosion, the strips are on the contour, but in dry regions, strips are placed normal to the prevailing wind direction for wind erosion control. The distribution of strip cropping in the United States is shown in Fig. 5.9. The greatest concentration is in Montana and North Dakota, with alternate strips of small grain and fallow where wind erosion is prevalent.

The three general types of strip cropping are shown in Fig. 5.10. In contour strip cropping, layout and tillage are held closely to the contour and the crops follow a definite rotational sequence. With field strip cropping, strips of uniform width are placed across the general slope. When used with adequate grassed waterways, the strips may be placed where the topography is too irregular to make contour strip cropping practical. Field strip cropping may also be used for wind erosion control. Buffer strip cropping has strips of a grass or legume crop between contour strips of crops in the regular rotations. Buffers may be even or irregular in width or placed on critical slope areas of the field. Their main purpose is to give protection from erosion or allow for areas of deposition. The type used depends on cropping system, topography, and types of erosion hazard.

Rotations that provide strips of close-growing perennial grasses and legumes alternating with grain and row crops are the most effective for strip cropping. Their effectiveness in reducing runoff and soil loss is illustrated in Table 5.4 in which a 4-year rotation is compared with continuous cotton.

The three general methods of laying out contour strip cropping are (1) both

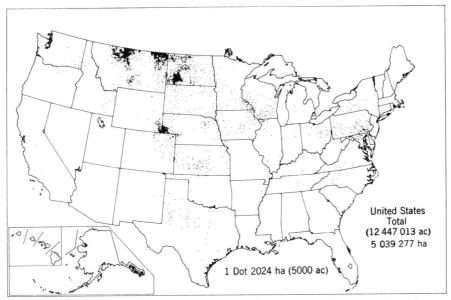

Fig. 5.9 Distribution of strip cropping in the United States for both water and wind erosion control (1969). (Courtesy U.S. Bureau of the Census.)

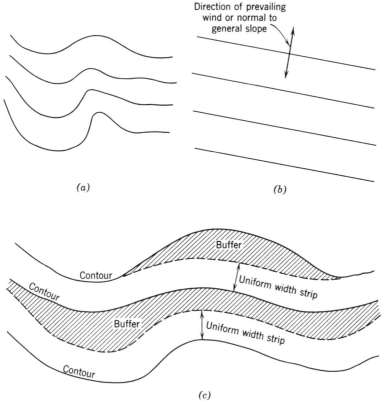

Fig. 5.10 Three types of strip cropping: (a) contour, (b) field, and (c) buffer.

Table 5.4   Runoff and Soil Losses on Class III Land with 7 Percent Slope at Watkinsville, Georgia

Rotation	Runoff (mm)	Soil Loss (Mg/ha)
First-year fescue	104	2.2
Second-year fescue	15	0.0
Corn	48	2.0
Cotton	99	7.4
Rotation average	66	2.9
Continuous cotton	254	30.9

Source: USDA–ARS (1963).

edges of the strips on the contour; (2) one or more strips of uniform width laid out from a key or base contour line; and (3) alternate uniform-width and variable-width correction or buffer strips.

The recommended maximum widths for contour strip cropping, shown in Table 5.3, reduce the computed soil loss as indicated by the corresponding conservation practice P_c. Strip width should be convenient for multiple-row equipment operation.

In some areas, insect damage resulting from the extended exposed crop borders has been a serious disadvantage of strip cropping. Establishment of rotations that give a minimum of protective harbor to insects, spray programs, and other approved insect-control measures reduce this problem.

A number of secondary factors should be taken into account when considering contour systems of farming. Studies showed that although power requirements are about equal for contour versus upgrade and downgrade operations, savings in time and fuel consumption averaged 13 to 19 percent, respectively, in favor of contour operations (Barger, 1938). Upgrade and downgrade operations contribute to other inefficiencies such as wheel slippage, stops for gear changes, and tractor and implement wear caused by starts and stops. Operation of implements such as combines and balers is simplified when implement speed is uniform.

5.12   Tillage Practices

The essential basis for tillage is the preparation of a seedbed, but the role of tillage has become increasingly important as a conservation tool. Its primary purpose is to provide an adequate soil and water environment for the plant; its role as a means of weed control has diminished with increased use of herbicides and improved timing of operations.

The effect of tillage on erosion is a function of its effect on such factors as surface residue, aggregation, surface sealing, infiltration, and resistance to wind and water movement. Excessive tillage destroys structure, increasing the susceptibility of the soil to erosion.

One of the major benefits of minimum tillage is the increased residue left on the surface. Such residue is extremely effective in reducing erosion. One of many similar studies showed that ridge tillage reduced runoff by two thirds and erosion from 12 to 1 Mg/ha compared with conventional tillage (Kramer and Hjelmfelt,

1989). Other methods for reducing runoff, such as putting checks in furrows or imprinting, can also reduce erosion.

5.13 Soil and Water Conservation Districts

The purpose of the soil and water conservation district in the United States is to provide a local group organization for the conservation of soil, water, and related resources and to promote better land use. As a condition for receiving benefits under the Soil Conservation and Domestic Allotment Act, passed by Congress in 1935, the states were required to enact suitable laws providing for the establishment of soil conservation districts. This legislation has been patterned largely after the standard state soil conservation districts law prepared by the SCS (1936). Most districts are called soil and water conservation districts. After such districts are established in local communities, they may request technical assistance from such agencies as the SCS, State Cooperative Extension Service, and county officials for carrying out erosion control and other land-use management activities.

Provided they are not in contradiction to other state laws, a soil and water conservation district may have the power (1) to conduct surveys, investigations, and research relating to soil erosion control programs; (2) to conduct demonstration projects; (3) to carry out preventive and control measures on the land; (4) to cooperate and make agreements with farmers and to furnish technical and financial aid; (5) to make available to land occupiers, through sale or rent, machinery, equipment, fertilizers, and so on; (6) to develop conservation plans for farms; (7) to take over erosion control projects, either state or federal; (8) to lease, purchase, or acquire property to carry out objectives of the program; and (9) to sue or to be sued in the name of the district. In some states, the district has the power of eminent domain. In others, the principal results of the district setup have been to assist in the development of conservation farm plans and provide technical assistance in adopting conservation practices.

REFERENCES

Aneke, D. O. (1985). "The Effect of Changes in Catchment Characteristics on Soil Erosion in Developing Countries (Nigeria)." *Agr. Eng.* **66**, 131–135.

Barger, E. L. (1938). "Power, Fuel, and Time Requirements of Contour Farming." *Agr. Eng.* **19**, 153–157.

Beer, C. E., and H. P. Johnson (1963). "Factors in Gully Growth in the Deep Loess Area of Western Iowa." *ASAE. Trans.* **6**, 237–240.

Bradford, J. M., D. A. Farrell, and W. E. Larson (1973). "Mathematical Evaluation of Factors Affecting Gully Stability." *Soil Sci. Soc. Am. Proc.* **37**, 103–107.

Foster, G. R. (1988). *User Requirements, USDA–Water Erosion Prediction Project (WEPP).* USDA–ARS National Soil Erosion Lab., West Lafayette, IN.

Foster, G. R., D. K. McCool, K. G. Renard, and W. C. Moldenhauer (1981). "Conversion of the Universal Soil Loss Equation to SI Metric Units." *J. Soil Water Cons.* **36**, 355–359.

Foster, G. R., and L. D. Meyer (1972). "A Closed-Form Soil Erosion Equation for Upland Areas." In *Sedimentation: Symposium to Honor Professor H. A. Einstein*, H. W. Shen (ed.), pp. 12.1–12.19. H. W. Shen, Fort Collins, CO.

Harrold, L. L. (1947). "Land-Use Practices on Runoff and Erosion from Agricultural Watersheds." *Agr. Eng.* **28**, 536–566.

—— (1949). "Soil Loss as Determined by Watershed Measurements." *Agr. Eng.* **30**, 137–140.

Hudson, N. (1985). *Soil Conservation.* Batsford Academic and Educational, London.

Knisel, W. G. (1980). *CREAMS: A Field-Scale Model for Chemicals, Runoff, and Erosion from Agricultural Management Systems.* Conservation Research Report No. 26. USDA, Washington, DC.

Kohnke, H., and A. R. Bertrand (1959). *Soil Conservation.* McGraw-Hill, New York.

Kramer, L. A., and A. T. Hjelmfelt, Jr. (1989). *Watershed Erosion from Ridge-Till and Conventional-Till Corn.* Paper No. 89-2511. ASAE, St. Joseph, MI.

Lane, L. J., G. R. Foster, and A. D. Nicks (1987). *Use of Fundamental Erosion Mechanics in Erosion Prediction.* Paper No. 87-2540. ASAE, St. Joseph, MI.

Lane, L. J., and M. A. Nearing. (1989). *USDA-Water Erosion Prediction Project: Hillslope Profile Model Documentation.* NSERL Report No. 2. USDA–ARS National Soil Erosion Research Laboratory, West Lafayette, IN.

Liebenow, A. M., W. J. Elliot, J. M. Laflen, and K. D. Kohl (1990). "Interrill Erodibility: Collection and Analysis of Data from Cropland Soils." *ASAE Trans.* **33**, 1882–1888.

McCool, D. K., L. C. Brown, G. R. Foster, C. K. Mutchler, and L. D. Meyer (1987). "Revised Slope Steepness Factor for the Universal Soil Loss Equation." *ASAE Trans.* **30**, 1387–1396.

McCool, D. K., G. R. Foster, C. K. Mutchler, and L. D. Meyer (1989). "Revised Slope Length Factor for the Universal Soil Loss Equation." *ASAE Trans.* **32**, 1571–1576.

Mihara, Y. (1952). "Effect of Raindrops and Grass on Soil Erosion." In *Proceedings, 6th International Grassland Congress*, pp. 987–990.

Renard, K. G., G. R. Foster, G. A. Weesies, and J. P. Porter (1991). "RUSLE Revised Universal Soil Loss Equation." *J. Soil Water Cons. Soc.* **46**, 30–33.

Roehl, J. N. (1962). *Sediment Source Areas, Delivery Ratios and Influencing Morphological Factors.* Publ. No. 59. Int. Assoc. of Scientific Hydrology, Commission of Land Erosion.

Schertz, D. L., W. C. Moldenhauer, S. J. Livingston, G. A. Weesies, and E. A. Hintz (1989). "Effect of Past Soil Erosion on Crop Productivity in Indiana" *J. Soil Water Cons. Soc.* **44**, 604–608.

Smith, D. D., and W. H. Wischmeier (1962). "Rainfall Erosion." *Adv. Agron.* **14**, 109–148.

U.S. Department of Agriculture (USDA) (1989). *The Second RCA Appraisal, Soil, Water, and Related Resources on Nonfederal Land in the United States, Analysis of Condition and Trends.* GPO, Washington, DC.

U.S. Department of Agriculture–Agricultural Research Service (USDA–ARS) (1963). *Conservation Methods for Soils in the Southern Piedmont.* Agr. Inform. Bull. 269. Washington, DC.

U.S. Soil Conservation Service (SCS) (1936). *A Standard State Soil Conservation Districts Law.* GPO, Washington, DC.

Vanoni, V. A. (1975). *Sedimentation Engineering.* ASCE, New York.

Watson, D. A., and J. M. Laflen (1986). "Soil Strength, Slope, and Rainfall Intensity Effects on Interrill Erosion." *ASAE Trans.* **29**, 98–102.

Wischmeier, W. H., and D. D. Smith (1958). "Rainfall Energy and its Relation to Soil Loss." *Am. Geophys. Union Trans.* 39, 285–291.

—— (1978). *Predicting Rainfall Erosion Losses-A Guide to Conservation Planning.* USDA Handbook 537. GPO, Washington, DC.

PROBLEMS

5.1 If the soil loss at Memphis, Tennessee, for a given set of conditions is 11.2 Mg/ha, what is the expected soil loss at your present location if all factors are the same except the rainfall factor?

5.2 If the degree of slope is increased from 2 to 10 percent, on a 60-m-long slope, what is the relative increase in erosion caused by water, assuming other factors are constant?

5.3 If the soil loss for a given set of conditions is 4.5 Mg/ha for a 60-m length of slope, what soil loss could be expected for a 240-m slope length if the slope is 5 percent?

5.4 Determine the soil loss for a field at your present location if $K = 0.015$, $l = 91$ m, $s = 10$ percent, $C = 0.2$, and upslope and downslope farming is practiced. What conservation practice should be adopted if the soil loss is to be reduced to 7 Mg/ha?

5.5 If the soil loss for upslope and downslope farming at your present location is 80 Mg/ha from a field having a slope of 6 percent and a slope length of 120 m, what will be the soil loss if the field is terraced with a horizontal spacing of 24 m? Assume that the cover-management conditions remain unchanged.

5.6 Compute the average annual soil loss from a terraced field at your present location assuming a slope of 6 percent, $C = 0.2$, $K = 0.03$, and with terraces at 24-m intervals. Determine the soil lost (1) at the terrace outlet and (2) to the terrace channel (use contouring factor).

5.7 If the average annual soil loss in Indianapolis, Indiana, is 10 Mg/ha, compute the annual erosion for a return period of 10 years, assuming all other soil loss factors are the same. Compute the soil loss for a single 10-year return period storm at this location.

5.8 Determine the contour strip width to the nearest complete equipment round for six 0.75-m row width equipment, where the land slope is 6 percent.

5.9 If the soil loss from a field with a 5 percent slope is 45 Mg/ha for upslope and downslope farming, and the cover-management factor is 0.25, what is the estimated soil loss if the field is contoured and the C factor is changed to 0.15? What would be the soil loss if strip cropping was substituted for contouring?

5.10 A rill channel is observed at two locations during a storm, 20 and 60 m from the top of the hill. The slopes are 5 and 4 percent, the rill widths are 150 and 220 mm, and the hydraulic radii are estimated to be 0.01 and 0.04 m, respectively. The sediment transport rate is measured and found to be 200 and 800 kg/m-s. The rill erodibility (K_r) is estimated to be 0.004 s/m, critical shear (τ_c) 2.5 Pa, and transport coefficient (B) 100. Calculate the total sediment transport capacity and erosion rate for each position.

5.11 Calculate the interrill erosion rate during a storm with an intensity of 32 mm/h, an interrill erodibility of 2 000 000 kg-s/m⁴, and a slope of 12 percent. (Convert intensity to units of m/s.)

CHAPTER 6

Wind Erosion and Control Practices

In the arid and semiarid regions of the United States, large areas are affected by wind erosion. The Great Plains region, an area especially subject to soil movement by wind, represents about 20 percent of the total land area in the United States. Wind erosion not only removes soil, but also damages crops, fences, buildings, and highways. Fine soil particles are lost along with nutrients, which can result in reduced crop yields. Figure 6.1 shows the distribution of wind erosion hazard in the states west of the Mississippi River. Figure 6.2 shows some of the crop damage that can result from severe wind erosion.

Many humid regions are also damaged by wind erosion. The areas most subject to damage are the sandy soils along streams, lakes, and coastal plains and the organic soils. Peats and mucks constitute about 10 million ha located in 34 states. Although the extent of sandy areas is not available, it probably is as much or more than the acreage of peats and mucks.

6.1 Wind Distribution with Height

Air movement of the free atmosphere is retarded near the surface of the soil. In immediate contact with the soil surface, the air is nearly at rest because of the drag forces between the air and the soil surface. Internal shear of the air allows an increased velocity with height above the soil until all effect of the soil surface and/or vegetation is dissipated.

Wind velocities are normally measured at a given height at major airports and first-order weather stations. For other heights, such as required for predicting evapotranspiration, velocities can be estimated since they vary approximately as the logarithm of the height.

A mean wind velocity-profile equation over stable surfaces is

$$u_z = \frac{u^*}{k} \ln\left(\frac{z - d}{z_0}\right)$$

(6.1)

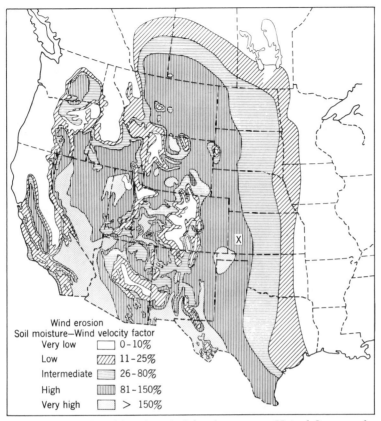

Fig. 6.1 Relative potential soil loss by wind for the western United States and southern Canada as a percentage of that in the vicinity of Garden City, Kansas, marked by X. (*Source:* Chepil et al., 1962.)

where u_z = wind velocity at z height (L/T),

u_* = friction velocity (L/T) = $\left(\dfrac{\tau_0}{\rho}\right)^{1/2}$,

τ_0 = shear stress at the boundary (F/L^2),

ρ = air density (M/L^3),

k = von Karman's constant, usually taken as 0.4,

z = height above a reference surface (L),

d = an effective surface roughness height (L),

z_0 = a roughness parameter (L).

The friction velocity u_* is a characteristic velocity in a turbulent boundary layer. The equation is valid only in the first few meters of height above the surface under neutral temperature conditions. These conditions exist when no heat is added or subtracted from the surface. Over short crops and smooth surfaces, the effective roughness height d is small or nearly zero. Both d and the roughness parameter z_0 are subject to considerable variation as crops bend and weave with the wind. For a wide range of crops, Stanhill (1969) found the equation

Fig. 6.2 Example of severe wind erosion in California. (Courtesy SCS.)

$$d = 0.7h \qquad (6.2)$$

applies where d is the effective roughness height and h is the crop height. The roughness parameter z_0, as defined in Eq. 6.1, is the height above d where the velocity profile extrapolates to zero. It can be estimated (Tanner and Pelton, 1960) from

$$z_0 = 0.13h \qquad (6.3)$$

Wind velocities with height are shown in Fig. 6.3 for three surfaces. Approximate values of d, z_0, and h are shown for wheat. For grass and snow cover, the plotted points were computed from Eq. 6.1. Appropriate values of d and z_0 were selected and u_*, the friction velocity, was computed from the velocities at 4.9-m height. Velocities at other heights were determined using the computed u_* value. As shown, the equation fits the curves reasonably well.

WIND EROSION

6.2 Types of Soil Movement

Saltation, suspension, and surface creep are the three types of soil movement. Saltation is the process where fine particles (0.1 to 0.5 mm in diameter) are lifted from the surface and follow distinct trajectories under the influence of air resistance and gravity (Fig. 6.4). When the particles return to the surface, they may rebound or become embedded when impacting the surface, but in either case, they

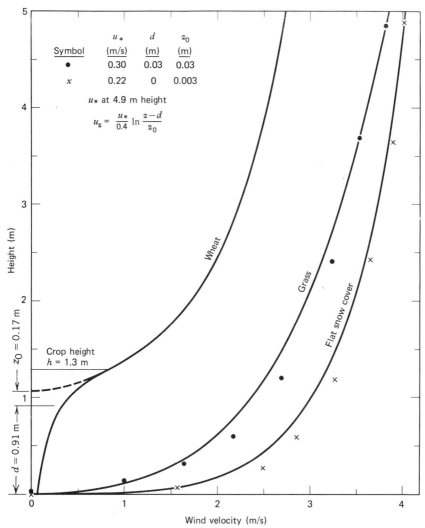

Fig. 6.3 Wind velocity distribution with height over various surfaces. (Curves are at different free wind velocities.)

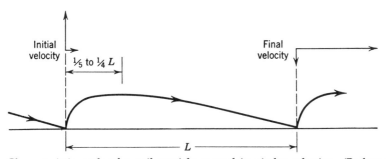

Fig. 6.4 Characteristic path of a soil particle moved in air by saltation. (Redrawn from Bagnold, 1941.)

initiate movement of other particles to create an "avalanching" effect of additional soil movement. Most saltation occurs within 0.3 m of the surface.

Suspended finer particles (0.02 to 0.1 mm in diameter) dislodged by saltating particles may remain aloft for an extended period. Suspended particles are often abraded by saltating particles and represent 3 to 10 percent of eroding particles. Sand-sized particles or aggregates (0.5 to 2 mm in diameter) are set in motion by the impact of saltating particles, and tend to roll or creep along the surface. Creep has been shown to account for 7 to 25 percent of soil movement (Chepil, 1945).

6.3   Mechanics of Wind Erosion

For a precise understanding of the mechanics of wind erosion, an analysis must be made of the nature and magnitude of the forces as they react on soil particles.

The wind erosion process may be divided into the three simple but distinct phases: (1) initiation of movement, (2) transportation, and (3) deposition.

Initiation of Movement.   Soil movement is initiated as a result of turbulence and velocity of the wind. The fluid threshold velocity is defined as the minimum velocity required to produce soil movement by direct action of the wind, whereas the impact threshold velocity is the minimum velocity required to initiate movement from the impact of soil particles carried in saltation. Except very near the surface and at low velocities (less than about 1 m/s), the surface wind is always turbulent. Wind speeds of 5 m/s or less at 0.3-m height are usually considered nonerosive for mineral soils.

Transportation.   The quantity of soil moved is influenced by the particle size, gradation of particles, wind velocity, and distance across the eroding area. Winds, being variable in velocity and direction, produce gusts with eddies and cross currents that lift and transport soil. The quantity of soil moved varies as the cube of the excess wind velocity above the constant threshold velocity, the square root of the particle diameter, and the gradation of the soil.

The rate of soil movement increases with the distance from the windward edge of the field or eroded area. Fine particles drift and accumulate on the leeward side of the area or pile up in dunes. Increased rates of soil movement with distance from the windward edge of the area subject to erosion are the result of increasing amounts of erosive particles, thus causing greater abrasion and a gradual decrease in surface roughness.

The atmosphere has a tremendous capacity for transporting soil, particularly those soil fractions less than 0.1 mm in diameter. It is estimated that the potential carrying capacity for 1 km³ of the atmosphere is many gigagrams of soil, depending on the wind velocity. As much as 224 kg/ha was deposited in Iowa in 1937 from a dust storm originating in the Texas Panhandle.

Deposition.   Deposition of sediment occurs when the gravitational force is greater than the forces holding the particles in the air. The process generally occurs when there is a decrease in wind velocity caused by vegetation or other physical barriers, such as ditches, vegetation, and snow fences. Raindrops may also take dust out of the air.

6.4   Estimating Wind Erosion

The wind erosion equation is not a simple product of erodibility parameters, but is a set of complex relationships among those parameters that affect erosion. Currently nomograph solutions are available and widely used. Computer models are being developed that will eventually replace the graphical solutions (Skidmore et al., 1970; Hagen, 1991). The present wind erosion model is

$$E = f(I,K,C,L,V) \tag{6.4}$$

where E = the estimated average annual soil loss (Mg/ha-year),
I = the soil erodibility index (Mg/ha-year),
K = the ridge roughness factor,
C = climate factor,
L = unsheltered length of eroding field in meters,
V = vegetative cover factor.

Erodibility Index I.   The above factors are not independent, but must be combined to estimate wind erosion. The wind erodibility I is a function of the soil aggregates greater than 0.84 mm in diameter. The following regression equation was developed from estimates of I given in Woodruff and Siddoway (1965),

$$I = 525 \times (2.718)^{(-0.04F)} \tag{6.5}$$

where I is the wind erodibility, and F is the percentage of dry soil fraction greater than 0.84 mm. The fraction of dry soil can vary during the season and can also be altered with changes in soil water content and organic matter. Table 6.1 summarizes soil erodibility values for different textures of soil. Increased wind erosion has also been observed on knolls, and Table 6.2 records adjustment factors that should

Table 6.1   Wind Erodibility Indices for Different Soil Textures

Predominant Soil Texture and Soil Erodibility Index	Erodibility Group	Soil Erodibility Index I^a (Mg/ha-year)
Loamy sands and sapric organic material	1	360–700
Loamy sands	2	300
Sandy loams	3	200
Clays and clay loams	4	200
Calcareous loams	4L	200
Noncalcareous loams, silt loam <20 percent clay, and hemic organic soils	5	125
Noncalcareous and silt loams >20 percent clay	6	100
Silt, noncalcareous silty clay loam and fibric organic soils	7	85
Wet or rocky soils not susceptible to erosion	8	—

[a]The I factors for Group 1 vary from 360 for coarse sands to 700 for very fine sands. Use 500 for an average.
Source: SCS (1988).

Table 6.2   Knoll Erodibility Adjustment Factors

Slope Change in Prevailing Wind Erosion Direction (%)	Knoll Adjustment to I (factor)	Increase at Crest Area Where Erosion is Most Severe (factor)
3	1.3	1.5
4	1.6	1.9
5	1.9	2.5
6	2.3	3.2
8	3.0	4.8
10	3.6	6.8

Source: SCS (1988).

be multiplied by I to account for the increased erosion on windward sides and tops of knolls.

Surface crusting caused by wetting and drying will reduce erosion for most soils, but should not be considered when making annual estimates. With computer prediction models, it will be possible to better account for crusting effects and the interaction between time of crusting and occurrence of wind erosion events (Hagen, 1991).

Erodibility can be decreased by increasing the amount of clods on the soil surface. This can be accomplished on some soils by timely tillage. Although widely practiced as an emergency method to reduce wind erosion, current methods of estimating erosion do not allow for inclusion of clod forming in reducing the predicted erosion. Computer models will have the ability to include tillage effects on soil roughness and erodibility (Hagen, 1991).

Roughness Factor K.   The roughness factor K is a measure of the effect of ridges made by tillage and planting implements on erosion rate. Ridges absorb and deflect wind energy, and trap moving soil particles. Too much roughness, however, causes turbulence which may accelerate particle movement. Ridge roughness can be estimated from the equation

$$K_r = 4 \frac{h^2}{d} \tag{6.6}$$

where K_r = ridge roughness in mm,
h = ridge height in mm,
d = ridge spacing in mm.

From the ridge roughness K_r, the roughness factor K can be calculated by the regression relationship derived from Woodruff and Siddoway (1965):

$$K = 0.34 + \frac{12}{(K_r + 18)} + 6.2 \times 10^{-6} \, K_r^2 \tag{6.7}$$

If there is a dominant wind direction, and the ridges are normal to that direction, then K is assumed to equal 1.00 regardless of the soil roughness.

Climate Factor C.   The climate factor is an index of climatic erosivity, which includes the wind speed and the soil surface moisture. It is expressed as a percentage of the C factor for Garden City, Kansas. Figure 6.1 shows the distribution of C factors for the United States. C can be calculated for a given climate by methods presented in Woodruff and Siddoway (1965).

Unsheltered Distance L.   The L factor represents the unsheltered distance in meters along the prevailing wind erosion direction for the field or area to be evaluated. This distance is the length from a sheltered edge of a field, parallel to the direction of the prevailing wind, to the end of the unsheltered field.

Vegetative cover factor V.   The effect of vegetative cover in the wind erosion equation is expressed by relating the kind, amount, and orientation of vegetative material to its equivalent of small grain residue. The small grain equivalent (Lyles and Allison, 1981) can be calculated from the relationship.

$$V = aR_w{}^b \tag{6.8}$$

where V = vegetative cover factor expressed as the small grain equivalent in kg/ha,
   a,b = crop constants from Table 6.3,
   R_w = quantity of residue to be converted to small grain equivalent in kg/ha.

Calculating Erosion.   To calculate an estimated annual erosion, the soil erodibility index is determined from Eq. 6.5 or Table 6.1. The effect of knolls may also be included by multiplying I by the appropriate factor from Table 6.2. The estimated annual wind erosion can then be calculated by one of the following equations which have been derived from statistical analyses of nomographs published

Table 6.3   Crop Residue Coefficients for Predicting Vegetative Cover Factor V

Crop	Orientation	Height (mm)	Length (mm)	Row Space (mm)	Orientation to Flow	a	b
Cotton	Flat-random		250			0.077	1.17
Cotton	Standing	340		750	Normal	0.188	1.15
Forage sorghum	Standing	150		750	Normal	0.353	1.12
Rape	Flat-random		250			0.064	1.29
Rape	Standing	250		250	Normal	0.103	1.4
Silage corn	Standing	150		750	Normal	0.229	1.14
Soybeans	1/10 standing	60		750		0.016	1.55
Soybeans	Flat-random		250			0.167	1.17
Sunflowers	Flat-random		430			0.011	1.37
Sunflowers	Standing	430		750	Normal	0.021	1.34
Winter wheat	Standing	250		250	Normal	4.31	0.97
Winter wheat	Flat-random		250			7.28	0.78

Source: Lyles and Allison (1981).

in Woodruff and Siddoway (1965). If the product of I, K, C, and L is less than 5.5×10^6, then

$$E = 0.0015 \times 2.718^{(-V/4500)} \times \left(I^{1.87} \times K^2 \times \left(\frac{C}{100} \right)^{1.3} \times L^{0.3} \right) \qquad (6.9a)$$

If the product of I, K, C, and L is equal to or greater than 5.5×10^6, then

$$E = 2.718^{(-V/4500)} \times \left(I \times K \times \frac{C}{100} \right) \qquad (6.9b)$$

□ *Example 6.1*

A field 800 m long, north and south, near Ford County, Kansas ($C = 80$), contains 25 percent nonerodible clods (>0.84 mm). Several knolls with 3 percent slopes are in the field. A crop of forage sorghum was produced in 750-mm rows, and 500 kg/ha of 150-mm-tall stubble remains standing in the field. The ridge roughness is 100 mm.

Solution.   The calculated soil loss is as follows.
(1)   Calculate I from the clod content (Eq. 6.5).

$$I = 525 \, (2.718)^{(-0.04 \times 25)} = 193 \text{ Mg/ha-year}$$

(2)   Calculate the effect of the 3 percent knolls from Table 6.2.

$$I = 193 \times 1.3 = 251 \text{ Mg/ha-year}$$

(3)   Calculate the roughness factor K for a ridge roughness of 100 mm (Eq. 6.7).

$$K = 0.34 + \frac{12}{100 + 18} + 6.2 \times 10^{-6} \, (100)^2 = 0.5$$

(4)   Calculate the vegetative cover factor V for a cover of 500 kg/ha (Eq. 6.8). Assume that for a height of 150 mm and a spacing of 750 mm, $a = 0.353$ and $b = 1.12$ (Table 6.3).

$$V = 0.353 \times (500)^{1.12} = 372 \text{ kg/ha equivalent}$$

(5)   Check the $IKCL$ product.

$$IKCL = 251 \times 0.5 \times 80 \times 800 = 8 \times 10^6 \, (>5.5 \times 10^6)$$

Complete the calculation by substituting the above values into Eq. 6.9b.

$$E = 2.718^{(-372/4500)} \times \left(251 \times 0.5 \times \frac{80}{100} \right) = 92 \text{ Mg/ha-year}$$

If the above loss is unacceptable, examining Eq. 6.9b shows that the loss can be reduced by reducing the length of the field with respect to the prevailing wind direction or by increasing the residue cover.   □

CONTROL PRACTICES

6.5 Cultivated Crops

In general, close-growing crops are more effective for erosion control than are intertilled crops. The effectiveness of crops is dependent on stage of growth, density of cover, row direction, width of rows, kind of crop, and climatic conditions. Pasture or meadow tends to accumulate soil if there is a good growth of vegetation resulting from deposition of soil eroded from neighboring cultivated fields. Good management practices such as rotation grazing are important.

Intertilled crops such as corn, cotton, and vegetables offer some protection. Experience has shown that close-growing crops gained soil, but intertilled crops lost soil by wind erosion. The best practice is to seed the crop normal to prevailing winds. In the Great Plains region the rows are usually in an east–west direction. A good crop rotation that will maintain soil structure and conserve moisture should be followed. Crops adapted to soil and climatic conditions and providing as much protection against blowing as practical are recommended. For instance, in the Great Plains region forage sorghum and Sudan grass are quite resistant to drought and are effective in preventing wind erosion. Stubble mulch farming and cover crops between intertilled crops in more humid regions aid in controlling blowing until the plants become established. In some dry regions emergency crops with low water requirements are often planted on summer fallow land before seasons of high-intensity winds. In muck soils where vegetable crops are grown, miniature windbreaks consisting of rows of small grain are sometimes planted. Stabilization of sand dunes with vegetation may be accomplished by establishing grasses and then reforesting. Vegetation used should have the ability to grow on sandy soil, the ability to grow in the open, firmness against the wind, and long life. It should also provide a dense cover during critical seasons, provide as uniform an obstruction to the wind as possible, reduce the surface wind velocity, and form an abundance of crop residue.

6.6 Shrubs and Trees

The SCS (1969) estimated that over 3 240 000 ha of land was in need of protection by shelterbelts and windbreaks. Although a windbreak is defined as any type of barrier for protection from winds, it is more commonly associated with mechanical or vegetative barriers for buildings, gardens, orchards, and feed lots. A shelterbelt is a longer barrier than a windbreak, usually consisting of shrubs and trees, and is intended for the conservation of soil and water and for the protection of field crops. About 450 000 km of windbreaks and shelterbelts have been planted in the United States since the middle of the past century. Not only are windbreaks and shelterbelts valuable for wind erosion control, but they also save fuel, increase livestock gains, reduce evaporation, prevent firing of crops from hot winds, catch snow during the winter months, provide better fruiting in orchards, and make spraying of trees for insect control more effective, as well as provide farm woodlots and wildlife refuge.

The relative wind velocity at a height of 0.4 m above the soil surface near a windbreak is shown in Fig. 6.5. The data were obtained for a slat fence barrier 19 times longer than its height and having an average density of 50 percent but more

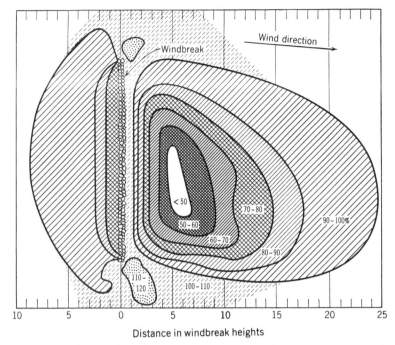

Fig. 6.5 Percentage of normal wind velocity near a windbreak having an average density of 50 percent. (Redrawn from Bates, 1944.)

open in the lower than in the upper half. Similar results could be expected from a tree shelterbelt with an equivalent density.

Shelterbelts should be moderately dense from ground level to treetops if they are to be effective in filtering the wind and lifting it from the surface. Since the wind velocity at the ends of the belt as given in Fig. 6.5 is as much as 20 percent greater than velocities in the open, it is evident that long shelterbelts are more effective than short ones. An opening or break in an otherwise continuous belt results in a similar increase in velocity and will reduce the area protected. Roads through shelterbelts should therefore be avoided, and, when essential, they should cross the belt at an angle or should be curved. In establishing the direction of shelterbelts, records of wind direction and velocity, particularly during vulnerable seasons, should be considered, and the barrier should be oriented as nearly as possible at right angles to the prevailing direction of winds.

From wind-tunnel tests Woodruff and Zingg (1952) found that the distance of full protection from a windbreak is

$$d = 17h \ (v_m/v)\cos \theta \qquad (6.10)$$

where d = distance of full protection (L),

h = height of the barrier in the same units as d (L),

v_m = minimum wind velocity at 15-m height required to move the most erodible soil fraction (L/T),

v = actual wind velocity at 15-m height (L/T),

θ = the angle of deviation of prevailing wind direction from the perpendicular to the windbreak.

Chepil (1959) reported that v_m for a smooth bare surface after erosion has been initiated and, before wetting by rainfall and subsequent surface crusting, was about 9.6 m/s. Equation 6.10 is valid only for wind velocities below 18 m/s. It may also be adapted for estimating the width of strips by using the crop height in the adjoining strip in the equation.

Typical shelterbelts for the Great Plains are shown in Fig. 6.6. A tight row of shrubs on the windward side is desirable, and, when combined with conifers and low, medium, and tall deciduous trees, the shelterbelt provides a compact and rather dense barrier. Such an extensive shelterbelt may not always be required. Single-row belts are preferred in many areas because fewer trees and less land are needed, and the shelterbelt is easier to cultivate and maintain. Studies in North Dakota indicated that shelterbelts with a density of much less than 50 percent were effective in trapping snow for distances up to 90 m with a 5-m single row of trees. Local recommendations should be followed for varieties, spacings, and other practices.

Shelterbelts are designed principally to control erosion, but crop production may be increased even though some land is occupied by the trees. In most situations in the Great Plains, Stoeckeler (1965) reported that where shelterbelts occupy 1 to 5 percent of the gross land area, optimum crop gains are most likely to be obtained. Among the crops most responsive to shelterbelt protection are garden crops, flower bulbs, citrus, and fruit trees. Those with the least response are drought-hardy crops.

6.7 Field and Contour Strip Cropping

Field and contour strip cropping consists of growing alternate strips of clean-culti-vated and close-growing crops in the same field. Field strip cropping is laid out parallel to a field boundary or other guideline. In some of the plains states, strips of fallow and grain crops are alternated. The chief advantages of strip cropping are (1) physical protection against blowing, provided by the vegetation; (2) soil erosion limited for a distance equal to the width of strip; (3) greater conservation of water, particularly from snowfall; and (4) the possibility of earlier harvest. The chief disadvantages are machine problems in farming narrow strips and greater number of edges to protect in case of insect infestation.

The strips should be of sufficient width to be convenient to farm, yet not so wide as to permit excessive erosion. The width of strips depends on the intensity of wind, row direction, crops, and erodibility of the soil and can be determined from Eq. 6.10.

The required width of field strips as determined by Eq. 6.10 can be increased to the extent that each strip is fully protected from wind by a standing crop stubble or some other barrier on its windward side as determined from Eq. 6.10. The zone of protection is about 10 times the barrier height, so for a crop stubble it is essentially negligible.

6.8 Primary and Secondary Tillage

The objective of primary and secondary tillage for wind erosion control is to produce a rough, cloddy surface with some plant residue exposed on the surface. To obtain maximum roughness the land normally should be cultivated as soon after a rain as possible. Large clods as well as a high percentage of large aggregates are desirable.

Fig. 6.6 Typical shelterbelts. (*a*) A half-mile-long windbreak of evergreens, deciduous trees, and shrubs protects a Nebraska farmstead and fields. (*b*) A three-row evergreen and broadleaf tree windbreak. (*c*) A three-row broadleaf windbreak in winter. (Courtesy SCS.)

Small ridges normal to the direction of prevailing winds are effective in wind erosion control. The effect of such ridges on wind movement is shown in Fig. 6.7. Very low velocities were observed between the ridges and approximately 20 mm below the crest. In the furrow the direction of movement was opposite to that of the wind. The decrease in wind velocity and change in direction between the ridges cause soil deposition. At a height of about 150 mm above the ridges there is an area of considerable turbulence.

Tillage may be quite effective as an emergency control measure. Soil blowing usually starts in a small area where the soil is less stable or is more exposed than in other parts of the field. Where the entire field starts to blow, the surface should be put in a rough and cloddy condition as soon as practicable. This tillage should begin at the windward side of the field and continue by making widely spaced trips across the field. When the field has been stripped, the areas between the strips may then be cultivated.

Crop residues exposed on the surface are an effective means of control, especially when combined with a rough soil surface. This practice is usually called stubble mulch tillage. The effectiveness of various soil treatments on wind velocity and soil erosion are shown in Fig. 6.8. These experiments were carried out in a wind tunnel. All surfaces were exposed to the same wind velocity as measured at a height of 300 mm. The reduction in wind velocity was greatest for the ridges plus straw. The effect of the straw mulch in reducing soil erosion was greater than for ridges alone. The combined effect of straw and ridges was appreciably greater than

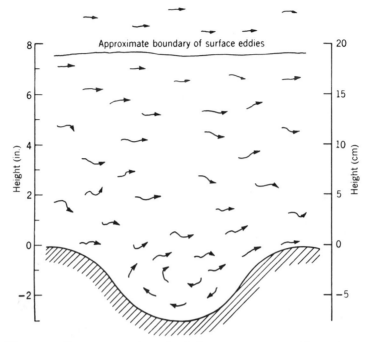

Fig. 6.7 Diagrammatic representation of wind structure across ridges. (Redrawn from Chepil and Milne, 1941.)

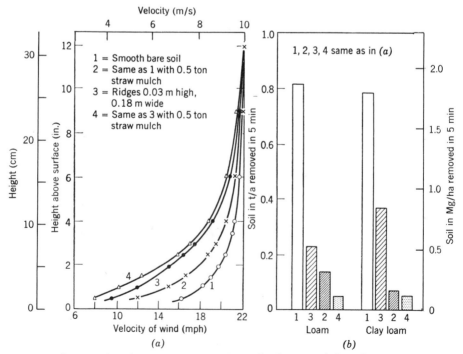

Fig. 6.8 Influence of surface treatment on (*a*) wind velocity and (*b*) soil erosion. (Redrawn from Chepil, 1944.)

the effect of either straw or ridges. As compared with bare soil, 1120 kg/ha of wheat straw reduced the amount of erosion by 83 and 88 percent for loam and clay loam soils, respectively.

Crop residues act in two ways: they reduce wind velocity and trap eroding soil. Short stubble is generally less effective than long stubble. A mixture of straw and stubble on the surface provides more protection against erosion than equivalent amounts of straw or stubble alone. The higher the wind velocity, the greater the quantity of crop residue required.

The effect of various quantities of crop residue on soil loss for the soil aggregate fraction less than 0.42 mm in diameter is shown in Fig. 6.9. The data show that the soil loss varies inversely as the 0.8 power of the weight of plant residue. With 60 percent of the soil fractions less than 0.42 mm, the soil loss may be reduced from 12 to 2 Mg/ha by increasing the residue from 280 to 2240 kg/ha, respectively.

6.9 Mechanical Methods

Mechanical barriers such as windbreaks are of little importance for field crops, but they are frequently employed for the protection of farmsteads and small areas. Some of the mechanical methods of control include slat or brush fences, board walls, and vertical burlap, plastic, or paper strips, as well as the surface protection, such as brush matting, rock, and gravel. These barriers may be classed as semipermeable or impermeable. Semipermeable windbreaks are usually more effective than impermeable structures because of diffusion and eddying effects on the leeward side of the barrier. A slat snow fence is a good example of this action. Slat

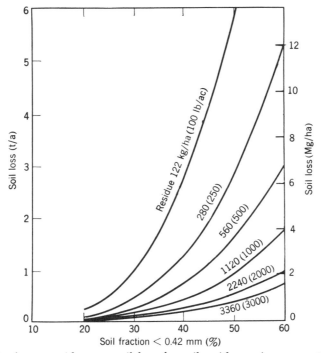

Fig. 6.9 Effect of crop residues on soil loss for soils with varying amounts of erodible fractions. (Redrawn from Englehorn et al., 1952.)

fences, picket fences, and vertical burlap, plastic, or paper strips are sometimes used for the protection of vegetable crops in organic soils. Brush matting, debris, rock, and gravel may be suitable in stabilizing sand dune areas.

Terraces have some effect on wind erosion. In the Texas Panhandle, terraces lost less soil than the interterraced area and, in some instances, gained soil. Most of the soil that was lost from the interterraced area was collected in the terraces. Where vegetation was growing on the ridge and in the channel, terraces were still more effective.

6.10 Conserving Soil Water

The conservation of soil water, particularly in arid and semiarid regions, is of utmost importance for wind erosion control as well as for crop production. The means of conserving water fall into three categories: increasing infiltration, reducing evaporation, and preventing unnecessary plant growth. In practice these can be accomplished by such practices as level terracing, contouring, mulching, and selection of suitable crops.

The greater the amount of mulch on the surface, the greater the quantity of soil water conserved. The effect of crop residue and tillage practices on the conservation of soil water is shown in Fig. 6.10. These data show the percentages of rainfall conserved from April to September. Where the straw was disked or plowed in, some of the residue was left on the surface, and where the soil was plowed and disked, no straw was present. Although there was no runoff for the basin-listed area, the quantity of soil water conserved was about half that for the treatment with 4.5 Mg/ha of straw on the surface, probably owing to greater evaporation. For the same reason disking and plowing were less effective than any of the mulch treatments.

The time of seedbed preparation in some regions has considerable effect on the conservation of soil water. This effect is shown in Fig. 6.11 for winter wheat in Hays, Kansas. Late plowing (September–October) resulted in a decrease in available moisture as compared with early plowing (July) or summer fallow. Wheat

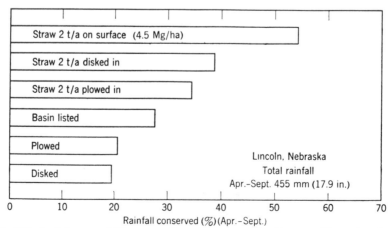

Fig. 6.10 Effect of straw and tillage practices on storage of soil water. (From data by Duley and Russel, 1939.)

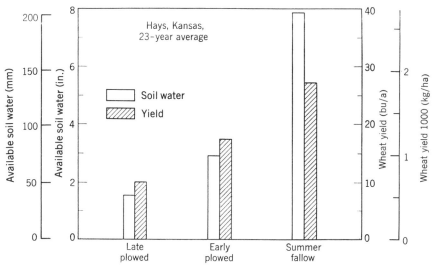

Fig. 6.11   Effect of time of seedbed preparation on available soil water at seeding time and on wheat yields. (From data by Hallsted and Mathews, 1936.)

yields for late and early plowing were 37 and 65 percent, respectively, of that for summer fallow.

Organic soils do not blow appreciably if the soil is moist. Where the subsoil is wet, rolling the soil with a heavy roller will increase capillary movement and moisten the surface layer. Controlled drainage where the water level is maintained at a specified depth may also reduce blowing. In irrigated areas the surface is often wet down by overhead sprinkling at critical periods.

Terracing is a good water conservation practice where level or conservation bench terraces are suitable and where the slopes are gentle enough so that the water can be spread over a relatively large area. Such practices as contouring, strip cropping, and mulching are effective in increasing the total infiltration and thereby the total soil water available to crops. Field strip cropping generally does not conserve as much water as contour strip cropping, but it is somewhat more effective in reducing surface wind velocities.

6.11   Conditioning Topsoil

Since wind erosion is influenced to a large extent by the size and apparent density of aggregates, an effective method of conditioning the soil against wind erosion is adopting practices that produce nonerosive aggregates (greater than 1 mm in diameter) or large clods. During the periods of the year when the soil is bare or has a limited amount of crop residue, control of erosion may depend on the degree and stability of soil aggregation.

Tillage may or may not be beneficial to soil structure, depending on the water content of the soil, type of tillage, and number of operations. For optimum resistance to wind erosion in semiarid regions it is desirable to perform primary tillage as soon as practical after a rain. The number of operations should be a minimum because tillage has a tendency to reduce soil aggregate size and to pulverize the soil. Secondary tillage for seedbed preparation should be delayed as long as practical.

Good crop and soil management practices are necessary to maintain a desired soil structure. In the Great Plains a typical recommended rotation may include a row crop, fallow, and winter wheat. The organic matter in the soil should be maintained at a high level and lime and fertilizers applied where necessary.

Soil structure is affected by the climatic influences of the season, rainfall, and temperature. Freezing and thawing generally have a beneficial effect in improving soil structure where sufficient water is present; however, in dry regions the soil is more susceptible to erosion because of rapid breakdown of the clods into smaller aggregates.

REFERENCES

Bagnold, R. A. (1941). *The Physics of Blown Sand and Desert Dunes*. Methuen, London.

Bates, C. G. (1944). *The Windbreak as a Farm Asset*. USDA Farmers Bull. 1045 (rev.). GPO, Washington, DC.

Chepil, W. S. (1944). "Utilization of Crop Residues for Wind Erosion Control." *Sci. Agr.* **24**, 307–319.

———(1945). "Dynamics of Wind Erosion. I. Nature of Movement of Soil by Wind." *Soil Sci.* **60**, 305–320.

———(1959). "Wind Erodibility of Farm Fields." *J. Soil Water Cons.* **14**(5), 214–219.

Chepil, W. S., and R. A. Milne (1941). "Wind Erosion of Soil in Relation to Roughness of Surface." *Soil Sci.* **52**, 417–431.

Chepil, W. S., F. H. Siddoway, and D. V. Armbrust (1962). "Climate Factor for Estimating Wind Erodibility of Farm Fields." *J. Soil Water Cons.* **17**(4), 162–165.

Duley, F. L., and J. C. Russel (1939). "The Use of Crop Residues for Soil and Moisture Conservation." *Am. Soc. Agron. J.* **31**, 703–709.

Engelhorn, C. L., A. W. Zingg, and N. P. Woodruff (1952). "The Effects of Plant Residue Cover and Clod Structure on Soil Losses by Wind." *Soil Sci. Soc. Am. Proc.* **16**, 29–33.

Hagen, L. J. (1991) "A Wind Erosion Prediction System to Meet User Needs." *J. Soil Water Cons.* **46**(2), 106–111.

Hallsted, A. L., and O. R. Mathews (1936). *Soil Moisture and Winter Wheat with Suggestions on Abandonment*. Kansas Agr. Expt. Sta. Bull. **273**. Kansas State University, Manhattan, KS.

Lyles, L., and B. E. Allison (1981). "Equivalent Wind-Erosion Protection from Selected Crop Residues." *ASAE Trans.* **24**, 405–408.

Skidmore, E. L., P. S. Fisher, and N. P. Woodruff (1970). "Wind Erosion Equation: Computer Solution and Application." *Soil Sci. Soc. Am. Proc.* **34**(5), 931–935.

Stanhill, G. (1969). "A Simple Instrument for Field Measurement of Turbulent Diffusion Flux." *J. Appl. Meteorol.* **8**, 509–513.

Stoeckeler, J. J. (1965). "The Design of Shelterbelts in Relation to Crop and Yield Improvement." In *World Crops*, pp. 3–8, Grampian Press Ltd., England.

Tanner, C. B., and W. L. Pelton (1960). "Potential Evapotranspiration Estimates by the Approximate Energy Balance Method of Penman." *J. Geophys. Res.* **65**, 3391–3413.

U.S. Soil Conservation Service (SCS) (1969). "Soil and Water Conservation Needs Inventory." Preliminary Report, Base Year 1967. *Soil Cons.* **35**(5), 99–109.

———(1988). "Wind Erosion." In *National Agronomy Manual*, Chap. 5, pp. 502-1 to 502-159. Am. Soc. Agron., Madison, WI.

Woodruff, N. P., and F. H. Siddoway (1965). "A Wind Erosion Equation." *J. Soil Sci. Soc. Am. Proc.* **29**, 602–608.

Woodruff, N. P., and A. W. Zingg (1952). *Wind-Tunnel Studies of Fundamental Problems Related to Windbreaks*. SCS TP-112. SCS, Washington, DC.

PROBLEMS

6.1 Calculate the estimated annual erosion for a 500-m (1500-ft)-long field located in north Central Kansas that has a soil with 30 percent nonerodible clods (>0.84 mm), several knolls with 4 percent slopes in the field, a crop of flat wheat residue estimated to be 300 kg/ha, and a ridge roughness of 50 mm.

6.2 Calculate the percentage change in predicted erosion rates if a field that initially has a flat wheat residue cover of 500 kg/ha and a roughness of 30 mm is cultivated to give a cover of 100 kg/ha but a ridge roughness of 125 mm using both Eqs. 6.9a and 6.9b. Comment on the effect of cultivation on erosion.

6.3 Determine the spacing between windbreaks that are 15 m (50 ft) high if the 5-year return period wind velocity at a 15-m (50-ft) height is 16 m/s (35 mph) and the wind direction deviates 10 degrees from the perpendicular to the field strip. Assume a smooth, bare soil surface and a fully protected field.

6.4 Determine the full protection strip width for field strip cropping if the crop in the adjacent strip is wheat 0.9 m (3 ft) tall and the wind velocity at 15-m (50-ft) height is 9 m/s (20 mph) at 90 degrees with the field strip.

CHAPTER 7

Vegetated Waterways

The design of vegetated waterways is more complex than the design of channels lined with concrete or other stable material because of variation in the roughness coefficient with depth of flow, stage of vegetal growth, hydraulic radius, and velocity.

7.1 Uses of Vegetated Waterways

Runoff must flow in a controlled manner that will not result in channel erosion or gully formation. Flow may be concentrated by the natural topography or by contour furrows, terraces, or other human works. In any event considerable amounts of energy are dissipated as flow proceeds down a slope. A flow of 1.4 m³/s for 30 m down a 5 percent slope releases energy at the rate of over 20 kW. If this energy acts on bare soil, considerable quantities of soil particles will be detached and transported by the water. The resultant gullies may separate a field into several parts, making farming difficult and decreasing the farm value. Roads, bridges, buildings, and fences frequently are jeopardized by gully development. Soil carried from eroded areas contributes to costly downstream sedimentation damage and pollution.

The basic approach to the control of such erosion involves (1) reduction of peak flow rates through the channel by full utilization of field protection practices, and (2) provision of a stable channel that can handle the flow that remains. This stabilization is best accomplished by providing vegetal protection for the channel together with modifying the cross section and grade of the channel so as to limit the flow velocities to a level that the vegetation can withstand.

Providing properly proportioned channels protected by vegetation is frequently a complete solution to the problem of gully formation. For large runoff volumes or steep channels, it may be necessary to supplement the vegetated watercourse by permanent gully control structures. Vegetated waterways should be used to handle natural concentrations of runoff or to carry the discharge from terrace systems, contour furrows, diversion channels, or emergency spillways for farm ponds or other structures. Vegetated waterways should not be used for continuous flows, such as may discharge from tile drains, as prolonged wetness in the waterway will result in poor vegetal protection. Special measures must be taken in the planning and execution of control measures on large gullies (see Chapter 9).

DESIGN

7.2 Determination of Runoff

In the design of a vegetated watercourse, the functional requirements should be determined and then the channel proportioned to meet these requirements. The capacity of the waterway should be based on the estimated runoff from the contributing drainage area. The 10-year return period storm is a sound basis for vegetated waterway design, except in flood spillways for dams. For exceptionally long watercourses it may be desirable to estimate the flow for each of several reaches of the channel to account for changing drainage area. For short channels the estimated flow at the waterway outlet is the practical design value.

7.3 Shape of Waterway

The cross-sectional shape of the channel as it is constructed may be parabolic, trapezoidal, or triangular. The parabolic cross section approximates that of natural channels. Under the normal action of channel flow, deposition, and bank erosion, the trapezoidal and triangular sections tend to become parabolic. In some channels no earthwork is necessary; the natural drainageway or meadow outlet is adequate, and only the width and location of the waterway need be defined.

A number of factors influence the choice of the shape of cross section. Channels built with a blade-type machine may be trapezoidal if the bottom width of the channel is greater than the minimum width of the cut. Triangular channels may also be readily constructed with blade equipment. Trapezoidal channels having bottom widths less than a mower swath are difficult to mow. Flat triangular or parabolic channels with side slopes of 4 : 1 (4 horizontal to 1 vertical) or flatter may be easily maintained by mowing. Side slopes of 4 : 1 or flatter are also desirable to facilitate crossing with farm equipment.

Broad-bottom trapezoidal channels require less depth of excavation than do parabolic or triangular shapes. During low flow periods, sediment may be deposited in trapezoidal channels with wide, flat bottoms. Uneven sediment deposition may result in meandering of higher flows and development of turbulence and eddies that will cause local damage to vegetation. Triangular channels reduce meandering, but high velocities may damage the bottom of the waterway.

Parabolic cross sections should usually be selected for natural waterways. A trapezoidal section with a slight V bottom is most easily constructed where the waterway is artificially located as in a terrace outlet along a fence line.

The geometric characteristics of the three shapes of cross section are given in Fig. 7.1. This figure defines the three types of cross section and gives formulas necessary for computing the hydraulic characteristics of each.

7.4 Selection of Suitable Vegetation

Soil and climatic conditions are primary factors in the selection of vegetation. Vegetation recommended for various regions of the United States is indicated in

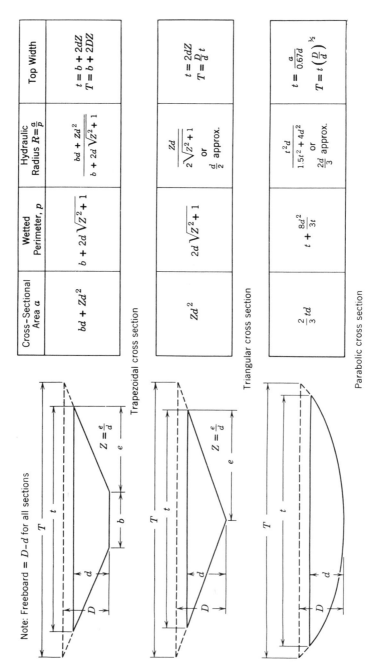

Cross-Sectional Area a	Wetted Perimeter, p	Hydraulic Radius $R=\frac{a}{p}$	Top Width
$bd + Zd^2$	$b + 2d\sqrt{Z^2+1}$	$\dfrac{bd + Zd^2}{b + 2d\sqrt{Z^2+1}}$	$t = b + 2dZ$ $T = b + 2DZ$

Trapezoidal cross section

Cross-Sectional Area a	Wetted Perimeter, p	Hydraulic Radius $R=\frac{a}{p}$	Top Width
Zd^2	$2d\sqrt{Z^2+1}$	$\dfrac{Zd}{2\sqrt{Z^2+1}}$ or $\dfrac{d}{2}$ approx.	$t = 2dZ$ $T = \dfrac{D}{d}\, t$

Triangular cross section

Cross-Sectional Area a	Wetted Perimeter, p	Hydraulic Radius $R=\frac{a}{p}$	Top Width
$\dfrac{2}{3}\,td$	$t + \dfrac{8d^2}{3t}$	$\dfrac{t^2 d}{1.5t^2 + 4d^2}$ or $\dfrac{2d}{3}$ approx.	$t = \dfrac{a}{0.67d}$ $T = t\left(\dfrac{D}{d}\right)^{\frac{1}{2}}$

Parabolic cross section

Note: Freeboard $= D - d$ for all sections

Fig. 7.1 Channel cross section, wetted perimeter, hydraulic radius, and top width formulas.

Table 7.1 Vegetation Recommended for Grassed Waterways

Geographical Area of United States	Vegetation[a]
Northeastern	Kentucky bluegrass, red top, tall fescue, white clover
Southeastern	Kentucky bluegrass, tall fescue, Bermuda, brome, Reed canary
Upper Mississippi	Brome, Reed canary, tall fescue, Kentucky bluegrass
Western Gulf	Bermuda, King Ranch bluestem, native grass mixture, tall fescue
Southwestern	Intermediate wheatgrass, western and tall wheatgrass, smooth brome
Northern Great Plains	Smooth brome, western wheatgrass, red top, switchgrass, native bluestem mixture

[a]Recommended vegetation does not necessarily apply to all areas in the region.

Table 7.1. Other factors to be considered are duration, quantity, and velocity of runoff, ease of establishment of vegetation, time required to develop a good protective cover, suitability to the farmer with respect to utilization of the vegetation as a seed or hay crop, spreading of vegetation to adjoining fields, cost and availability of seed, and retardance to shallow flows in relation to sedimentation. If herbicides are used on the crops, it may be necessary to establish herbicide-tolerant vegetation.

7.5 Design Velocity

The ability of vegetation to resist erosion is limited. The permissible velocity in the channel is dependent on the type, condition, and density of vegetation and the erosive characteristics of the soil. Uniformity of cover is very important, as the stability of the most sparsely vegetated areas controls the stability of the channel. Permissible velocities for bunch grasses or other nonuniform covers are lower than those for sod-forming grasses. Bunch grasses produce nonuniform flow with highly localized erosion. Their open roots do not bind the soil firmly against erosion.

Permissible velocities are also influenced by bed slope. Suggested design values for velocity are given in Table 7.2. It should be recognized that the design velocity is an average velocity rather than the actual velocity in contact with the vegetation or with the channel bed. Figure 7.2 shows the velocity distribution in a grass-lined channel and illustrates this point. Though the average velocity in the cross section is about 0.8 m/s, the velocity in contact with the vegetation and bed is less than 0.3 m/s. Design of vegetated waterways is based on the Manning formula (Eq. 7.1) for velocity.

7.6 Roughness Coefficient

Slope and hydraulic radius are calculated readily from the geometry of the channel; however, the roughness coefficient is more difficult to evaluate. Extensive tests by Ree (1949, 1958, 1976, 1977) and others have provided techniques and data to determine roughness coefficients for various types of vegetation. Figure 7.3

Table 7.2   Permissible Velocities for Vegetated Channels

Cover	Slope Range[a] (%)	Permissible Velocity (m/s)[b] Erosion-Resistant Soils	Easily Eroded Soils
Bermuda grass	0–5	2.4	1.8
	5–10	2.1	1.5
	>10	1.8	1.2
Bahia Buffalo grass Kentucky bluegrass Smooth brome Blue grama Tall fescue	0–5 5–10 >10	2.1 1.8 1.5	1.5 1.2 0.9
Grass mixtures Reed canary grass	0–5 5–10	1.5 1.2	1.2 0.9
Lespedeza sericea Weeping lovegrass Yellow bluestem Redtop Alfalfa Red fescue	0–5[c]	1.0	0.8
Common lespedeza[d] Sudan grass[d]	0–5	1.0	0.8

[a]Do not use on slopes steeper than 10 percent except for vegetated side slopes in combination with a stone, concrete, or highly resistant vegetative center section.

[b]Use velocities exceeding 1.5 m/s only where good covers and proper maintenance can be obtained.

[c]Do not use on slopes steeper than 5 percent except for vegetated side slopes in combination with a stone, concrete, or highly resistant vegetative center section.

[d]Annuals—use on mild slopes or as temporary protection until permanent covers are established.

Source: SCS (1975).

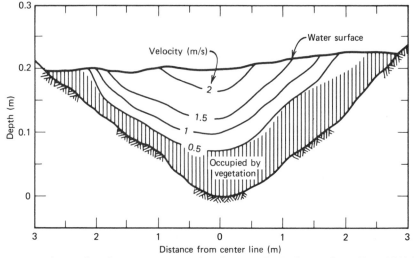

Fig. 7.2   Velocity distribution in a grass-lined channel. (Redrawn from Ree, 1949.)

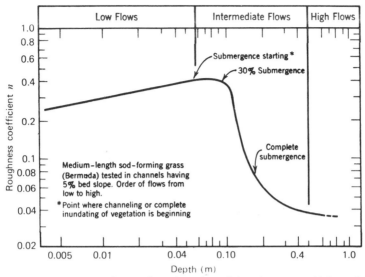

Fig. 7.3 Hydraulic behavior of a medium-length sod-forming grass. (Adapted from Ree, 1949.)

illustrates the complexity of the problem. The roughness coefficient varies with the depth of flow. Shallow flows encounter a maximum resistance because the vegetation is upright in the flow. The slight increase in resistance in the low flow is due to the greater bulk of vegetation encountered with increasing depth. Intermediate flows bend over and submerge some of the vegetation, and resistance drops off sharply as more and more vegetation is submerged.

Resistance to flow is also influenced by the gradient of the channel. Decreasing resistance results from higher velocities on steeper slopes with an accompanying greater flattening of the vegetation. Type and condition of vegetation have a great influence on the retardance. Newly mown grass offers less resistance than rank growth. Long plants, stems, and leaves tend to whip and vibrate in the flow, introducing and maintaining considerable turbulence. The cross-sectional shape of the channel has only minor influence on the roughness coefficient in the range of cross sections commonly used.

The product vR, velocity multiplied by hydraulic radius, has been found to be a satisfactory index of channel retardance for design purposes. Vegetation has been grouped into five retardance categories designated A through E. Table 7.3 gives a portion of this classification of vegetation by degree of retardance, and Fig. 7.4 shows the $n-vR$ curves for five retardance categories. For small channels it has been common practice to use $n = 0.04$ for vegetated waterways. In many channels this may be satisfactory, but careful consideration should be given to the vegetation and flow conditions, especially for long, large channels where refinements in design may result in lower construction costs.

7.7 Channel Capacity

The channel must be proportioned to carry the design runoff at average velocities less than or equal to the permissible velocity. This is accomplished by application of the Manning formula,

Table 7.3 Classification of Vegetal Cover According to Retardance

Retardance	Cover	Condition	Height (m)
A	Reed canary	Excellent stand, tall	0.9
	Yellow bluestem *Ischaemum*		
	Smooth brome	Good stand, mowed	0.3–0.4
	Bermuda	Good stand, tall	0.3
	Native grass mixture	Good stand, unmowed	
	Tall fescue	Good stand, unmowed	0.5
B	*Lespedeza sericea*	Good stand, not woody, tall	0.5
	Grass–legume mixture	Good stand, uncut	0.5
	Reed canary	Good stand, mowed	0.3–0.4
	Tall fescue with bird's foot trefoil or ladino	Good stand, uncut	0.5
	Blue grama	Good stand, uncut	0.3
	Bahia	Good stand, uncut	0.2
	Bermuda	Good stand, mowed	0.15
C	Redtop	Good stand, headed	0.4–0.5
	Grass–legume mix — summer	Good stand, uncut	0.2
	Centipede grass	Very dense cover	0.15
	Kentucky bluegrass	Good stand, headed	0.2–0.3
	Bermuda	Good stand, cut	0.1
	Red fescue	Good stand, headed	0.3–0.5
D	Buffalo grass	Good stand, uncut	0.1–0.2
	Grass–legume mixture — fall	Good stand, uncut	0.2
	Lespedeza sericea	After cutting	0.1
E	Bermuda grass	Good stand	0.1
	Burned stubble		

Source: SCS (1975).

$$v = \frac{R^{2/3} \, s^{1/2}}{n}$$

(7.1)

where v = average velocity of flow in m/s,
  n = roughness coefficient of the channel,
  $R = a/p$, the cross-sectional area divided by the wetted perimeter in m,
  s = hydraulic gradient (slope of the channel) in m/m.

The dimensions of the channel must be so selected to satisfy the continuity equation

$$q = av$$

(7.2)

where q = flow rate to be carried by the channel in m³/s.

Nomographs in Figs. 7.5, 7.6, and 7.7 give solutions to the Manning equation for retardance classes B, C, and D, respectively. For trapezoidal channels with 4 : 1 side slopes, the depth of flow can be determined for the required cross-sectional area, bottom width, and hydraulic radius using Fig. 7.8 if the top width is greater than 10 times the depth. These approximations greatly simplify the calculations,

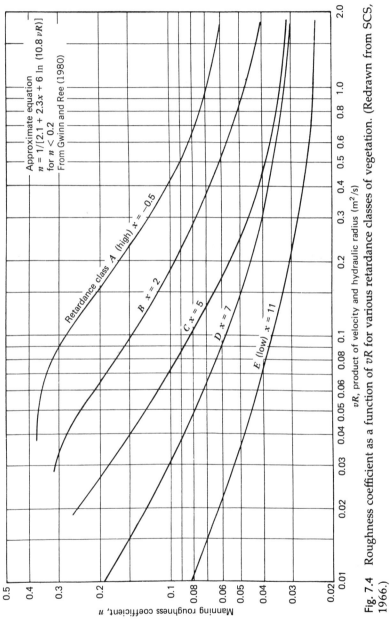

Fig. 7.4 Roughness coefficient as a function of vR for various retardance classes of vegetation. (Redrawn from SCS, 1966.)

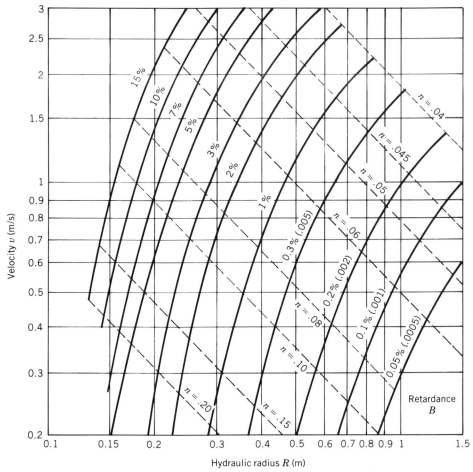

Fig. 7.5 Nomograph of the Manning equation for vegetated channels of retardance Class B (high vegetal retardance). (*Source:* SCS, 1966.)

but when many solutions are required the problem can more easily be solved on a computer (see Appendix I). The channel should be designed to carry the runoff at a permissible velocity under conditions of minimum retardance (short grass) that may be encountered during the runoff season. This condition establishes the basic proportions of the channel, for example, the bottom width of a trapezoidal channel. Additional depth should then be added to the channel to provide adequate capacity under conditions of maximum retardance (long grass). A freeboard of 0.1 to 0.15 m should be added to the design depth. Examples 7.1 and 7.2 illustrate design procedure.

☐ *Example 7.1*

Design a trapezoidal grassed waterway to carry 5 m³/s (177 cfs) on a 3 percent slope on erosion-resistant soil. The vegetation is to be Bermuda grass, and the channel should have 4:1 side slopes.

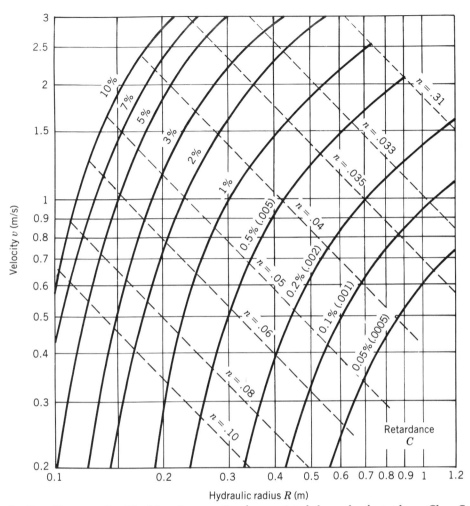

Fig. 7.6 Nomograph of the Manning equation for vegetated channels of retardance Class C (moderate vegetal retardance). (*Source:* SCS, 1966.)

Solution. From Table 7.2 the permissible velocity is 2.4 m/s (8 ft/s). Table 7.3 shows Bermuda grass in retardance Class D when mowed and in Class B when long. To design for stability against erosion, the mowed condition is the more critical. Entering Fig. 7.7 for Class D retardance with $v = 2.4$ m/s (8 ft/s) and a slope of 3 percent, $R = 0.31$ m (1.02 ft). The trapezoidal cross-sectional area must be $5/2.4 = 2.08$ m^2 (22 ft^2). From Fig. 7.8 for $R = 0.31$ m and $a = 2.08$ m^2, read $b = 4$ m and a depth of flow $d = 0.4$ m (1.3 ft). These dimensions will provide a stable channel with $v = 2.4$ m/s (8 ft/s).

The design depth must now be increased when the grass is long with retardance Class B because the velocity is reduced. The previous bottom width of 4 m must be retained. By trial and error, select a depth of 0.5 m (1.5 ft) that will have $a = 3$ m^2 (32.3 ft^2) and $R = 0.37$ m (1.2 ft). Entering Fig. 7.5 with $R = 0.37$ m and a slope of 3 percent gives $v = 1.7$ m/s (5.6 ft/s). At the 0.5-m depth, $q = 3.0 \times 1.7 = 5.1$ m^3/s (177 cfs), which is within 10 percent of the design value. An alternate solution is to select the approximate additional depth increment, 0.16 m (0.13 +

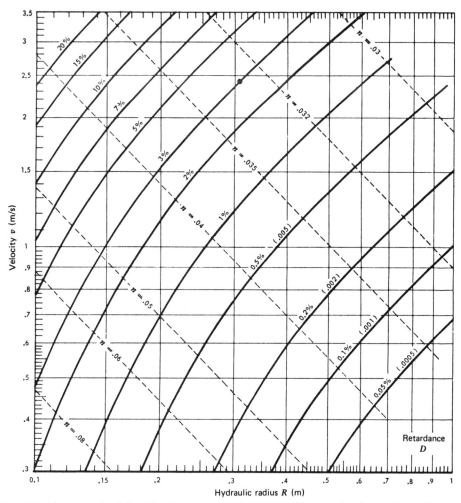

Fig. 7.7 Nomograph of the Manning equation for vegetated channels of retardance Class D (low vegetal retardance). (*Source:* SCS, 1966.)

0.03) from Table 7.4 for the increased roughness. The depth of flow is 0.41 + 0.16 = 0.57 m (1.9 ft). An SCS computer design program (see Appendix I) calculated a bottom width of 4.3 m, a top width of 8.2 m, and depth of 0.49 m for the long grass condition.

An additional freeboard or safety factor of 0.1 m (0.3 ft) should be included. This would increase the depth to 0.67 m (2.2 ft) making the constructed top width $T = 3.9 + (2 \times 0.67 \times 4) = 9.3$ m (30.5 ft). □

The example shows that the bottom width is determined by the need to not exceed the permissible velocity under the mowed condition of minimum retardance, and that the depth is determined by the need to provide capacity under conditions of maximum retardance and to allow for freeboard.

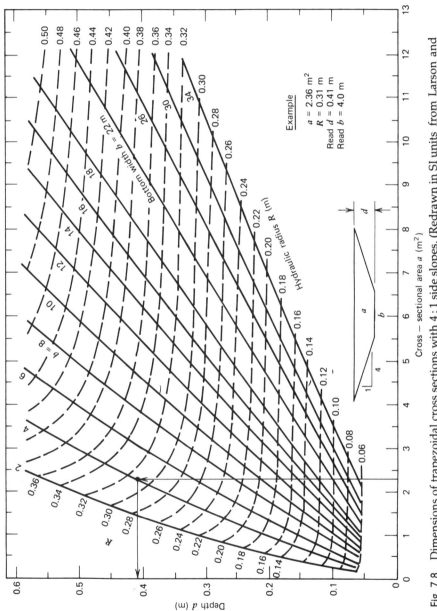

Fig. 7.8 Dimensions of trapezoidal cross sections with 4 : 1 side slopes. (Redrawn in SI units from Larson and Manbeck, 1960.)

Table 7.4 Approximate Channel Depth Increment for Grassed Waterways with Retardance Increase from Class C (Short) to Class B (Long)

Slope Range (%)	Channel Cross Section and Width		
	Triangular or Parabolic $t = 2-6 \, m$ *Trapezoidal* $b = 2-6 \, m$	*Parabolic* $t = 6-27 \, m$ *Trapezoidal* $b = 6-27 \, m$	*Trapezoidal* $b \geq 27 \, m$
1–2	0.17 m	0.15 m	0.12 m
2–5	0.13	0.12	0.09
≥5	0.10	0.09	0.06

Note: For change from retardance D to B add 0.03 m to the above values for all slopes and cross sections. For change from retardance D to C use 0.03 m as the depth increment for all slopes and cross sections.

Source: Larson and Manbeck (1960) as computed by C. L. Larson.

☐ Example 7.2

Design a parabolic grassed waterway for the same conditions as in Example 7.1.

Solution. From Example 7.1 the hydraulic requirements for a stable channel for Class D retardance are $a = 2.08 \, m^2$ (22 ft^2) and $R = 0.31 \, m$ (1.02 ft). Using the approximate relationship for parabolic channels of $d = 1.5R$, $d = 1.5 \times 3.1 = 0.46 \, m$ (1.52 ft). From Fig. 7.1,

$$t = 3a/2d = (3 \times 2.08)/(2 \times 0.46) = 6.78 \, m \ (22.2 \, ft)$$

These dimensions provide for a stable channel.

The additional increase in depth when in Class B retardance (long grass) is 0.12 for D to C and 0.03 for C to B, to give 0.15 m (0.5 ft) from Table 7.4. The design depth of flow is

$$d = 0.46 + 0.15 = 0.61 \, m \ (2 \, ft)$$

and the constructed depth including 0.1 m freeboard is

$$D = 0.61 + 0.1 = 0.71 \, m \ (2.3 \, ft)$$

The top width is

$$T = 6.78(0.71/0.46)^{1/2} = 8.42 \, m \ (27.6 \, ft)$$

A more accurate method for obtaining the flow depth with long grass is to assume a new depth of 0.57 m (trial and error), which results in

$$R = 0.57/1.5 = 0.38 \, m \ (1.25 \, ft)$$

From Fig. 7.5 read $v = 1.75 \, m/s$ (5.6 ft/s). Since the top width of a parabolic cross section by definition varies as the square root of the depth, the new top width is

$$t = 6.78 \ (0.57/0.46)^{1/2} = 7.54 \text{ m } (25.8 \text{ ft})$$

Then the calculated flow rate is

$$q = (2/3) \ 7.54 \times 0.57 \times 1.75 = 5.01 \text{ m}^3/\text{s } (177 \text{ cfs})$$

which is adequate. An SCS program (Appendix I) required a top width of 7.6 m before freeboard to ensure a velocity less than 2.4 m/s. Both solutions are similar to the dimensions of the trapezoidal cross section in Example 7.1.   □

Waterways are often designed using computer programs, tables, or charts developed for a particular area. An engineer or technician working in a given region gains confidence in methods particularly adapted to local conditions.

Immediately after construction, the channel may be called on to carry runoff under conditions of little or no vegetation. It is not practical to design for this extreme condition. In many channels, it may be practical and desirable to divert flow from the channel until vegetation is established. In others, the possibility that high runoff will occur before vegetation is established is accepted as a calculated risk.

7.8 Drainage

Waterways that are located in seepy draws or below seeps, springs, or pipe outlets will be wet for long periods. The wet condition will inhibit the development and maintenance of a good vegetal cover and will cause the soil to be in a weak, erosive condition. Subsurface drainage or diversion of such flow is essential to the success of the waterway.

A small concrete or asphalt channel of about 0.2-m² cross section is sometimes placed in the bottom of a waterway to carry prolonged low flows. Seepage along the sides or upper end of the waterway may be intercepted by subsurface drains. Drains should be placed on one side of the center of the waterway to prevent erosion leading to exposure of the drain in case of failure of the waterway.

WATERWAY CONSTRUCTION

7.9 Shaping Waterways

The procedure and earthwork involved in shaping a waterway depend on the topographic situation and the equipment available. If the watercourse is to be located in a natural waterway or meadow outlet where there is little gullying, only smoothing and normal seedbed preparation are required. Some improvement in alignment of the channel may be desired to remove sharp bends. This will improve the hydraulic characteristics, facilitate farming operations, and reduce channel maintenance. If the waterway is to reclaim an established gully, considerable earthwork is required. The gully must be filled and the waterway cross section properly shaped.

Small waterways may be easily shaped with regular farm equipment. Large gullies, however, should be constructed with a bulldozer or other heavy earth-moving equipment.

ESTABLISHMENT OF VEGETATION

7.10 Seedbed Preparation

Soil in the waterway should be finely tilled, and brought to a high fertility level in accordance with the soil and plant requirements. Manure incorporated into the seedbed furnishes both plant nutrients and organic material which will increase the soil's resistance to erosion.

7.11 Seeding

Waterway seeding mixtures should include some quick-growing annual for temporary control as well as a mixture of hardy perennials for permanent protection. Seed should be either broadcast or drilled nonparallel to the direction of flow. Mulching after seeding helps to secure a good stand. When a channel must carry high flows before seedlings can become established, sodding may be justified.

Special materials can be applied to control infiltration and runoff as a means of facilitating revegetation. Such materials include chemical soil stabilizers, plastic, fiber, or other mesh or net covers (Fig. 7.9), asphalt mulches, and plastic or other surface covers. The primary purpose of such materials is to stabilize the soil surface during the establishment of grass seedlings. The materials reduce rate of drying and crusting and absorb the energy of overland flow, raindrops, and/or wind, thus protecting the soil.

Fig. 7.9   Paper net being installed to secure straw mulch on a newly seeded sod waterway. (Courtesy SCS.)

WATERWAY MAINTENANCE

7.12 Causes of Failure

Failure of vegetated waterways may result from insufficient capacity, excessive velocity, or inadequate vegetal cover. The first two of these are largely a matter of design. The condition of the vegetation, however, is influenced not only by the initial preparation of the waterway but also by the subsequent management. Use of a waterway, especially in wet weather, as a lane, stock trail, or pasture injures the vegetation and often results in failure. Terraces that empty into a waterway at too steep a grade may erode into the terrace channel, damaging both terrace and waterway. Careless handling of machinery in crossing a waterway may injure the sod. When land adjacent to the waterway is being plowed, the ends of furrows abutting against the vegetated strip should be staggered to prevent flow concentration down the edges of the watercourse.

7.13 Controlling Vegetation

Waterways should be mowed and raked several times a season to stimulate new growth and control weeds. A rotary-type mower cuts the grass fine enough to make raking unnecessary. Annual application of manure or fertilizer maintains a dense sod. Any breaks in the sod should be repaired. Rodents that are damaging waterways should be controlled.

7.14 Sediment Accumulation

Good conservation practice on the watershed is the most effective means of controlling sedimentation. Accumulated sediments smother vegetation and restrict the capacity of the waterway. Extending vegetal cover well up the side slopes of the waterway and into the outlets of terrace channels helps to prevent sediment from being deposited in the watercourse. Control of vegetation to prevent a rank, matted growth reduces the accumulation of sediment. High allowable design velocities also decrease sedimentation.

VEGETATION OF LARGE GULLIES

The discussion to this point has been applicable mainly to waterways and to the stabilization of small gullies. Larger gullies may be controlled by reduction of the surface inflow, by shaping and intensive natural or artificial revegetation, or by the installation of control structures (see Chapter 9).

7.15 Control of Inflow

Large gullies generally have a contributing watershed area of considerable size or one with a high runoff potential. To facilitate the vegetation and control of the gully, the normal conservation practices that will protect small gullies must be replaced with more effective methods. Diversion terraces, constructed above the gully head and carefully laid out on a grade that will resist channel erosion, can be used to intercept runoff from the watershed area above the gully and then convey it to a safely stabilized outlet area.

7.16 Sloping Gully Banks

Bank sloping should be done only to the extent required for establishment of vegetation or for facilitating tillage operations. Where trees and shrubs are to be established, rough sloping of the banks to about 1:1 should be sufficient. Where gullies are to be reclaimed as grassed waterways, sloping of banks to 4:1 or flatter usually is desired.

7.17 Natural Revegetation

If the runoff that has caused the gully is diverted, and livestock are fenced from the gullied area, plants will begin to establish naturally. A gradual succession of plant species eventually will protect the gullied area with grasses, vines, shrubs, or trees native to the area. In some gullied areas, the development of vegetation may be stimulated by fertilizing and by spreading a mulch to conserve water and protect young volunteer plants. Vertical gully banks may be roughly sloped to prevent collapse and provide improved conditions for natural seeding. The opportunity to provide protective cover by natural revegetation frequently is overlooked, and unnecessary expenditures are made for structures and plantings.

7.18 Artificial Revegetation

Selection of vegetation to be established artificially in a reclaimed gully should be governed by the use intended for the planted area. Grasses and legumes may be planted if the vegetation is to be used for a hay or pasture crop. Where gullies are reclaimed as drainageways in cultivated fields, sod-forming vegetation should be selected to permit crossing of the drainageway with farm machines.

In some areas trees and shrubs are easier to establish in gullies than are grasses, particularly if the gully is not to be shaped to permit operation of farm implements. Shrubs, such as dogwoods and lespedezas, are desirable for establishing the gullied area as a wildlife refuge. Trees and shrubs should be planted in accordance with local recommendations for control of erosion and establishment of wildlife refuges.

REFERENCES

Barfield, B. J., R. C. Warner, and C. T. Haan (1983). *Applied Hydrology and Sedimentology for Disturbed Areas*. Oklahoma Technical Press, Stillwater, OK.

Chow, V. T. (1959). *Open Channel Hydraulics*. McGraw-Hill, New York.

Gwinn, W. R., and W. O. Ree (1980). "Maintenance Effects on the Hydraulic Properties of a Vegetation-Lined Channel." *ASAE Trans.* **23**, 636–642.

Larson, C. L., and D. M. Manbeck (1960). "Improved Procedures in Grassed Waterway Design." *Agr. Eng.* **41**, 694–696.

Ree, W. O. (1949). "Hydraulic Characteristics of Vegetation for Vegetated Waterways." *Agr. Eng.* **30**, 184–187, 189.

————(1958). "Retardation Coefficients for Row Crops in Diversion Terraces." *ASAE Trans.* **1**, 78–80.

————(1976). *Effect of Seepage Flow on Reed Canarygrass and Its Ability to Protect Waterways.* ARS-S-154. USDA-ARS, Washington, DC.

————(1977). *Performance Characteristics of a Grassed-Waterway Transition.* ARS-S-158. USDA-ARS, Washington, DC.

Ree, W. O., and V. J. Palmer (1949). *Flow of Water in Channels Protected by Vegetative Linings.* USDA Tech. Bull. 967.

U.S. Soil Conservation Service (SCS) (1966). *Handbook of Channel Design for Soil and Water Conservation.* SCS-TP-61. Washington, DC.

————(1979). *Engineering Field Manual for Conservation Practices.* SCS, Washington, DC.

PROBLEMS

7.1 Determine the velocity of flow in a parabolic-, a triangular-, and a trapezoidal-shaped waterway, all having a cross-sectional area of 2 m² (21.5 ft²), a depth of flow of 0.3 m (1.0 ft), a channel slope of 4 percent, and a roughness coefficient of 0.04. Assume 4:1 side slopes for the trapezoidal cross section.

7.2 Design a parabolic-shaped grassed waterway to carry 1.5 m³/s (53 cfs). The soil is easily eroded; the channel has a slope of 4 percent; and a good stand of Bermuda grass, cut to 60 mm (2½ in), is to be maintained in the waterway.

7.3 Design a trapezoidal-shaped waterway with 4:1 side slopes to carry 0.6 m³/s (21 cfs) where the soil is resistant to erosion and the channel has a slope of 5 percent. Brome grass in the channel may be either mowed or long when maximum flow is expected.

7.4 Design a parabolic-shaped waterway to carry 1.8 m³/s (64 cfs) from a terraced field where the soil is resistant to erosion and the channel slope is 6 percent. Tall fescue is normally mowed, but the channel should have adequate capacity when the grass is long.

7.5 Design a trapezoidal-shaped emergency spillway for a farm pond to carry 1.8 m³/s (64 cfs) with retardance Class C condition. The maximum velocity in the spillway is 1.5 m/s (4.9 fps). Around the end of the dam the channel slope is to be 2 percent, but it is increased to 7 percent from this section to the stream channel below. The depth of flow should not exceed 0.3 m (1.0 ft). Determine the bottom width, design depth, and recommend design slopes for both the 2 and 7 percent sections.

7.6 If the velocity of flow is 2.2 m/s (7.2 fps) for complete submergence ($n = 0.035$) of Bermuda grass described in Fig. 7.3, what is the velocity at 30 percent submergence considering only the change in roughness coefficient? If the hydraulic radius was also reduced from 0.24 to 0.12 m (0.8 to 0.4 ft), what is the expected velocity?

7.7 At low flow in a channel the velocity is 0.6 m/s (2 fps) and the hydraulic radius is 0.15 m (0.5 ft) for a short grass with retardance Class D. From Fig. 7.4 determine the relative change in the roughness coefficient for long grass with retardance Class B for the same conditions. If the velocity is increased to 3 m/s (10 fps) and the hydraulic radius to 0.3 m (1.0 ft) for the long grass (Class B), what is the change in the roughness coefficient?

7.8 A diversion terrace is needed for a flow of 0.85 m³/s (30 cfs). Assume a poor stand of grass and use an allowable velocity of 0.9 m/s (3.0 fps). Determine d, D, and the required slope using a 2.5-m (8.2-ft) bottom width. Allow 0.1 m (0.3 ft) of freeboard with long grass (class D). Use a trapezoidal cross section with 4 : 1 side slopes.

CHAPTER 8

Terracing

Terracing is a method of erosion control accomplished by constructing broad channels across the slope of rolling land. It has been estimated that over 36 million ha (90 million ac) of cropland in the United States could be more effectively protected from runoff and erosion damage through the use of well-designed and well-maintained terrace systems. The use of diversions, a form of terrace, to protect bottomlands, buildings, and special areas from damaging overland flow and erosion from hillsides is becoming increasingly important. The first terraces consisted of large steps or level benches. For several thousand years, bench terraces have been widely adopted over the world. In the United States, ditches that functioned as terraces were constructed across the slopes of cultivated fields by farmers in the southern states during the latter part of the 18th century.

As technology has advanced, terrace design has been adapted to the hydrologic and erosion control needs of the treated areas. Channel cross sections have been modified to become more compatible with modern mechanization. During the 1940s and 1950s broadbase terraces with vegetated outlets were commonly installed. Since the 1970s, parallel terraces with steep grassed backslopes have become popular. As shown in Fig. 8.1, most terraces are found in the central and south central states as well as in the Southeast.

8.1 Functions of Terraces

The functions of terraces in humid areas are to decrease the length of the hillside slope, thereby reducing rill erosion; to prevent the formation of gullies; and to allow sediment to settle from runoff water, thereby improving the quality of surface water leaving the field. Crop rows are usually parallel to the terrace channel, so terracing includes contouring as a conservation practice. Since terracing requires additional investment and causes some inconvenience in farming, it should be considered only where other cropping and soil management practices, singly or in combination, will not provide adequate erosion control or water management.

In drier areas, terraces serve to retain runoff and increase the amount of water available for crop production or for recharging of shallow aquifers. Such retention

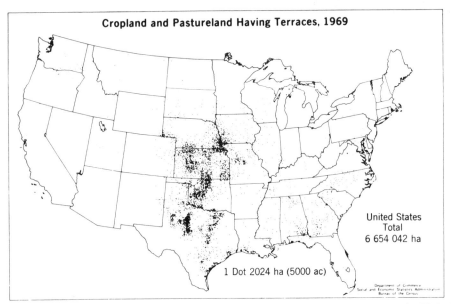

Fig. 8.1 Geographic distribution of cropland and pastureland having terraces in the United States. (Courtesy U.S. Bureau of the Census, 1969.)

of water also reduces the risk of wind erosion. Terracing can also be used as an aid in surface irrigation, particularly in rice-growing areas.

TERRACE CLASSIFICATION

Terraces are classified by alignment, cross section, grade, and outlet (ASAE, 1989).

8.2 Classification by Alignment

Terrace alignment can be either nonparallel or parallel. Nonparallel terraces follow the contour of the land regardless of alignment. Some minor adjustments frequently made to eliminate sharp turns and short rows are installation of additional outlets, use of variable grade, and installation of vegetated turning strips. Parallel terraces aid in farming operations and should be installed wherever possible. Parallel terraces require more cut and fill volumes during construction than nonparallel systems.

8.3 Classification by Cross Section

There are numerous shapes of terrace cross sections (Fig. 8.2*a–c*). The bench terrace (Fig. 8.2*c*) has improved farmability under very steep (20 to 30 percent) conditions where labor is cheap or land is in short supply. The bench terrace provides for efficient distribution of water under both irrigated and dryland production. Some bench terraces are not suited to mechanized farming systems.

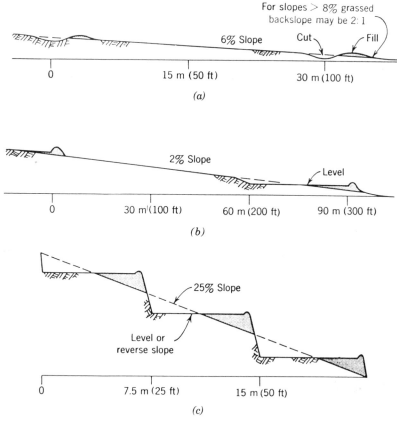

Fig. 8.2 Classification of terrace shapes. (*a*) Three-segment section broadbase. (*b*) Conservation bench broadbase. (*c*) Bench.

Bench-type systems can be constructed on nearly flat lands to improve irrigation efficiency (see Chapter 19).

The broadbase shape includes the three-segment section, the conservation bench, and the grassed backslope terrace (Figs. 8.2*a–b* and 8.3). The three-segment section terrace (Fig. 8.2*a*) is more common for mechanized farming systems on moderate slopes (6 to 8 percent). All slopes on the three-segment section broadbase (Figs. 8.2*a* and 8.3) are sufficiently flat for the operation of farm machinery. Lengths of each side slope are designed to match the width of equipment that operates on those slopes.

The conservation bench variation (Fig. 8.2*b*) incorporates a wide, flat channel uphill of the embankment to provide a maximum area for infiltration of runoff water. A comparison of the infiltration abilities between the conservation bench and the three-segment section is shown in Fig. 8.4. The grassed backslope terrace (Fig. 8.3*c*) is constructed with a 2 : 1 backslope that is usually seeded to permanent grass. This terrace reduces land slopes between terraces and improves farmability. When field slopes are uniform, a constant terrace cross section is recommended. With nonuniform slopes, some sections will be predominantly fill and some predominantly cut (Fig. 8.5).

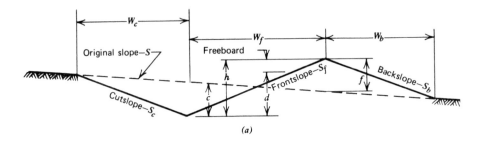

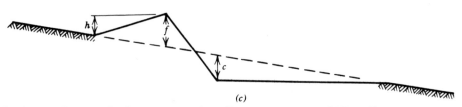

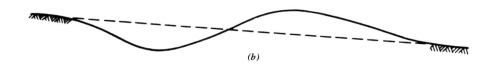

(c)

Fig. 8.3 (a) Design of a three-segment broadbase cross section. (b) Broadbase cross section after 10 years of farming. (c) Grassed backslope cross section.

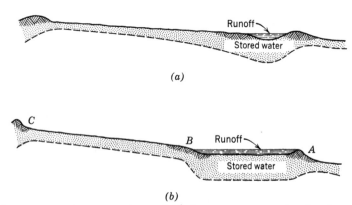

Fig. 8.4 Comparison of cross sections of (a) three-segment section level terrace and (b) conservation bench terrace showing available soil water. Ratio of storage area AB to runoff area BC may be varied to suit soil, cover, and topographic conditions. (After Hauser et al., 1962.)

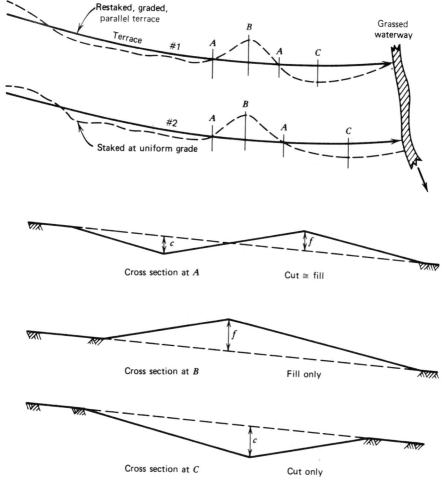

Fig. 8.5 Layout of parallel terraces with varying cuts and fills.

8.4 Classification by Grade

The channels in terraces can be graded toward an outlet or level. Graded or channel-type terraces are designed to remove excess water in such a way as to minimize erosion. Terraces control erosion by reducing the slope length of overland flow and then by conducting the intercepted runoff to a safe outlet at a nonerosive velocity. The reduced flow velocities in the channel also allow for deposition of eroded sediment. Water conservation may be a secondary benefit. Because of the importance of constructing and maintaining a satisfactory channel, this type of terrace should not be built on soils that are too stony or steep or have topsoil too shallow to permit adequate construction.

Level terraces are constructed to conserve water and control erosion. In low to moderate rainfall regions they trap and hold rainfall for infiltration into the soil profile. They may be suitable for this same purpose on permeable soils in high rainfall areas. Frequently, it is necessary to excavate soil from both sides of the embankment to achieve sufficient height to store the design runoff without overtopping or piping through the embankment by the entrapped water. The channel is level and is sometimes closed at both ends to ensure maximum water retention. On slopes over 2 percent, water in the channel is spread over a relatively small area, thus limiting the effect on crop yield. The conservation bench cross section (Figs. 8.2b and 8.4b) was designed to overcome this deficiency.

8.5 Classification by Outlet

Terraces may be classified as blocked outlets (all water infiltrates the terrace channel), permanently vegetated outlets (grassed waterway or a vegetated area), or underground outlets (water is removed through subsurface drains). Combinations of outlets may be employed to meet specific problems. The selection of outlets is discussed more fully in the next section.

PLANNING THE TERRACE SYSTEM

The terrace system should be coordinated with the water management system for the farm, giving adequate consideration for proper land use. Terrace systems should be planned by watershed areas and should include all terraces that may be constructed at a later date. Where practicable, adjacent farms having fields in the same drainage area may have joint terracing systems. Factors such as fence and road locations must be considered.

8.6 Selection of Outlets

One of the first steps in planning is the selection of outlets or disposal areas. Outlet types include natural drainage ways, constructed channels, sod flumes, permanent pasture or meadow, road ditches, wasteland, concrete or stabilized channels, pipe drains, and stabilized gullies.

Natural drainage ways, where properly vegetated, provide a desirable and economical outlet. Where these channels are inadequate, constructed waterways along field boundaries or pipe outlets may be considered. The design, construction, and maintenance of vegetated outlets and watercourses as discussed in Chapter 7 are applicable for terrace outlets. Terrace outlets onto pastureland should be staggered by increasing the length of each terrace a few meters, starting with the lowest terrace. Sod flumes and concrete channels are to be avoided because of excessive cost and maintenance problems. Road ditches and active gullies may scour or enlarge if terrace runoff is added. Outletting into road ditches may not be allowed by some highway agencies.

8.7 Terrace Location

After a suitable outlet is located, the next step is the location of the terraces. Factors that influence terrace location include (1) land slope, (2) soil conditions, such as degree and extent of erosion, (3) proposed land use, (4) boulders, trees, gullies, and other impediments to cultivation, (5) farm roads, (6) fences, (7) row layout, (8) type of terrace, and (9) outlet. Minimum maintenance, ease of farming, and adequate control of erosion are the criteria for good terrace location. Better alignment of terraces can usually be obtained by placing the terrace ridge just above eroded spots, gullies, and abrupt changes in slope. Satisfactory locations for roads and fences are on the ridge, on the contour, or on the spoil beside the outlet.

Unless there are obvious reasons for doing otherwise, the top terrace is laid out first, starting from the outlet end. It is important that the top terrace be properly located so that it will not overtop and cause failure of other terraces below. General rules for the location of the top terrace are: (1) the drainage area above the top terrace ordinarily should not exceed 1 ha. (2) if the top of the hill comes to a point, the spacing may be increased to one and one half times the regular spacing. (3) on long ridges, where the terrace approximately parallels the ridge, the regular spacing should be specified. (4) if short, abrupt changes in slope occur, the terrace should be placed just above the break.

Obstructions or topographic features below the top terrace, or the need to make terraces parallel, may necessitate locating a terrace at some other point downslope. This terrace is called the key terrace because terraces above and/or below the key terrace are set out at normal spacing from it.

A typical terrace layout is shown in Fig. 8.6. The top terrace 1a and 1b is a diversion that intercepts runoff from the pasture and prevents overflow to the cultivated land below. Since the slope below terrace 1a is uniform, terraces 2a and 3a are laid out parallel to it. Because terraces 2bc and 3bc are each longer than 500 m, outlets are provided at each end.

Level terraces are located in much the same manner as graded terraces. On flat slopes in the Great Plains, level terraces are sometimes constructed so that runoff is allowed to flow from one terrace to the next by opening alternate ends of the terraces.

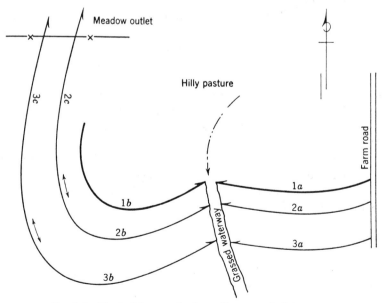

Fig. 8.6 Typical layout for broadbase graded terraces.

TERRACE DESIGN

The design of a terrace system involves specifying the proper spacing and location of terraces, the design of a channel with adequate capacity, and development of a farmable cross section. For the graded terrace, runoff must be removed at nonerosive velocities in both the channel and the outlet. Soil characteristics, cropping and soil management practices, and climatic conditions are the most important considerations in terrace design.

8.8 Terrace Spacing

Spacing is expressed as the vertical distance between the channels of successive terraces. For the top terrace, the spacing is the vertical distance from the top of the hill to the bottom of the channel. This vertical distance is commonly known as the vertical interval or V.I. The horizontal interval H.I. is found by dividing V.I. by the slope (m/m). The H.I. for parallel terraces is usually selected as an even number of rounds for row-crop equipment. The vertical interval is more convenient than the horizontal interval for terrace layout and construction with surveying equipment.

Graded. Graded terrace spacing is often expressed as a function of land slope by the empirical formula

$$\text{V.I.} = Xs + Y \tag{8.1}$$

where V.I. = vertical interval between corresponding points on consecutive terraces or from the top of the slope to the bottom of the first terrace in m,

X = constant for geographical location as given in Fig. 8.7,

Y = constant for soil erodibility and cover conditions during critical erosion periods

= 0.3, 0.6, 0.9, or 1.2, with the low value for highly erodible soils with no surface residue and the high value for erosion-resistant soils with conservation tillage (ASAE, 1989),

s = average land slope above the terrace in percent.

Spacings thus computed may be varied as much as 25 percent to allow for soil, climatic, and tillage conditions. Terraces are seldom recommended on slopes over 20 percent, and in many regions slopes from 10 to 12 percent are considered the maximum.

Where soil loss data are available, spacing should be based on slope lengths using contouring and the appropriate cover-management factor. This will result in soil losses within the tolerable loss as outlined in Chapter 5. Estimation of terrace spacing is illustrated in the following examples.

□ *Example 8.1*

If the soil loss was 16 Mg/ha (7.1 t/a) at Memphis, Tennessee, for $K = 0.1$, $l = 120$ m (394 ft), $s = 8$ percent, $C = 0.2$, and $P = 0.6$ (contouring) in the USLE, what maximum slope length and corresponding terrace spacing are needed to reduce the soil loss to the terrace channel to 7 Mg/ha (3.1 t/a)?

Solution.   From Eqs. 5.9,

$$\theta = \tan^{-1}\left(\frac{8}{100}\right) = 4.57°$$

$$m = \frac{\sin 4.57}{\sin 4.57 + 0.269\ (\sin 4.57)^{0.8} + 0.05} = 0.48$$

$$L = \left(\frac{120}{22}\right)^{0.48} = 2.26$$

The maximum L to reduce loss to 7 Mg/ha is

$$L = 2.26 \times \frac{7}{16} = 0.99$$

Calculate l to achieve the above L factor value.

$$l = 0.99^{1/0.48} \times 22 = 21.5 \text{ m (70.5 ft)}$$

Calculate the vertical interval.

$$\text{V.I.} = \frac{8}{100} \times 21.5 = 1.7 \text{ m (5.6 ft)} \qquad \square$$

☐ Example 8.2

Compute the terrace spacing for Example 8.1 from Eq. 8.1 assuming that the soil has a low intake rate and that good cover conditions exist.

Solution.   From Fig. 8.7, read $X = 0.15$ and assume $Y = 0.6$. Substitute in Eq. 8.1.

$$\text{V.I.} = 0.15 \times 8 + 0.6 = 1.8 \text{ m (5.9 ft)}$$

Since the permissible spacing based on soil loss takes into account more of the erosion variables, a vertical interval of 1.7 m (5.6 ft) as computed in Example 8.1 is preferred. ☐

Level.   The spacing for level terraces is a function of channel infiltration and runoff; however, in more humid areas, where erosion control is important, the slope length may limit the spacing. The storage capacity of the terrace should be adequate to prevent overtopping from upslope runoff, and the infiltration rate in the channel should be sufficiently high to prevent serious damage to crops. Spacings vary widely in different parts of the country, and the SCS or other agencies should be consulted to determine local practices.

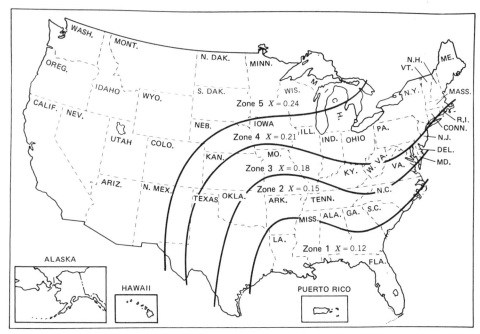

Fig. 8.7   Values of X in the terrace spacing equation (Eq. 8.1). (*Source:* ASAE, 1989.)

8.9   Terrace Grades

The gradient in the channel must be sufficient to provide adequate drainage while removing the runoff at nonerosive velocities. The minimum slope is desirable from the standpoint of soil loss. Grades may be uniform or variable.

In the uniform-graded terrace, the slope remains constant throughout its entire length. A grade of 0.4 percent is common in many regions; however, grades may range from 0.1 to 0.6 percent, depending on soil and climatic factors. Generally, the steeper grades are recommended for impervious soils and short terraces.

ASAE (1989) recommends maximum velocities of 0.5 m/s for extremely erosive soils and 0.6 m/s for most other soils, when the roughness coefficient in the Manning formula is taken as 0.035. Recommended minimum and maximum grades are given in Table 8.1.

The variable-graded terrace is more effective because the capacity increases toward the outlet with a corresponding increase in runoff. The grade may vary from a minimum at the upper portion to a maximum at the outlet end to reduce the velocity in the upper reaches. This reduction provides for greater absorption of runoff and more deposition of sediment. Variable gradient makes flexibility in design possible. For instance, either constant velocity or constant capacity could be provided by varying the grade in the channel. Such designs are sometimes required, particularly in large diversion terraces.

8.10   Terrace Length

Size and shape of the field, outlet possibilities, rate of runoff as affected by rainfall and soil infiltration, and channel capacity are factors that influence terrace length.

Table 8.1   Maximum and Minimum Terrace Grades

Terrace Length or Length from Upper End of Long Terraces (m)	Maximum Slope (%)
≤30	2.0
31–60	1.2
61–150	0.5
151–365	0.35
≥366	0.3
	Minimum Slope (%)
Soils with slow internal drainage	0.2
Soils with good internal drainage	0.0

Source: Values for the maximum slope are from Beasley (1963); those for minimum slope are from ASAE (1989).

The number of outlets should be the minimum consistent with good layout and design. Extremely long graded terraces are to be avoided; however, long lengths may be reduced in some terraces by dividing the flow midway in the terrace length and draining the runoff to outlets at both ends of the terrace (Fig. 8.6). The length should be such that erosive velocities and large cross sections are not required. On permeable soils longer terraces may be permitted than on impermeable soils. The maximum length for graded terraces generally ranges from about 300 to 500 m, depending on local conditions. The maximum applies only to that portion of the terrace that drains toward one of the outlets.

There is no maximum length for level terraces, particularly where blocks or dams are placed in the channel about every 150 m. These dams prevent total loss of water from the entire terrace and reduce gully damage should a break occur. The ends of the level terrace may be left partially or completely open to prevent overtopping in case of excessive runoff.

8.11   Terrace Cross Section

The terrace cross section should provide adequate capacity, have broad farmable side slopes, and be economical to construct with available equipment. For design purposes, the cross section of a broadbase terrace can be considered a triangular channel as shown in Fig. 8.3a. The flow depth d is the height to the top of the ridge h less a freeboard of about 0.08 m. After smoothing, the ridge and bottom widths will be about 1 m, which will give a cross section that approximates the shape of a terrace after 10 years of farming (Fig. 8.3b).

In designing the cross section, the frontslope width W_f is specified to be equal to the machinery width ordinarily used for row-crop operations. The depth of flow is determined from the runoff rate for a 10-year return period storm or for the required runoff volume for storage-type terraces. When the side slope widths are equal ($W_c = W_f = W_b = W$), cuts and fills from the geometry are

$$c + f = h + sW \qquad (8.2)$$

where c = cut (L),
f = fill (L),
h = depth of channel including freeboard (L),
s = original land slope (L/L),
W = width of side slope (L).

For a balanced cross section, cut and fill are equal. Larson (1969) has developed other similar equations for grassed backslope terraces shown in Fig. 8.3c. These terraces have backslopes too steep to farm and are planted to permanent grass to stabilize the slope. Fill soil may be obtained from the lower side of the terrace, which tends to reduce the land slope between terraces.

□ *Example 8.3*

For a channel depth h of 0.3 m, and for eight-row (0.75-m row width) equipment 6 m wide on 7 percent land slope, compute the cut and fill heights and the slope ratios for the frontslope and the backslope assuming a balanced cross section.

Solution. From Eq. 8.2, letting the cut equal the fill,

$$c = f = (h + sW)/2$$
$$= (0.33 + 0.07 \times 6)/2 = 0.38 \text{ m } (1.3 \text{ ft})$$

By geometry for the frontslope,

$$S_f = 6/0.33 = 18.2 \text{ or round to } 18:1 \text{ slope ratio}$$

Similarly, for the backslope or cutslope,

$$S_b = 6/(0.33 + 0.07 \times 6) = 8.0 \text{ or } 8:1 \text{ sideslope ratio} \qquad □$$

For practical reasons, a terrace is usually constructed with a uniform cross section from the outlet to the upper end, although this construction results in the upper portion of the channel being overdesigned. On the conservation bench terrace, the ridge is generally built up to provide a settled height of 0.3 m above the level of the bench. The ends of the bench are blocked to retain 0.15 m of water on each bench before they overtop into a vegetated waterway. The terrace is generally constructed with a 5:1 backslope and planted with grass.

8.12 Pipe Outlets

Pipe outlet terraces shown in Fig. 8.8 were known in the early 1900s, but the present version was developed in Iowa in the 1960s. These outlets eliminate the need for grassed waterways. By straightening the terrace at natural channels with an earth fill, it is easier to make adjacent terraces parallel. The pipe outlet as shown in Fig. 8.8b has an orifice plate to restrict the outflow. This restriction ensures that subsurface drains are not overloaded, and that sediment in the runoff has time to settle in the terrace channel, improving the quality of the runoff water. To provide for storage in the terrace, the top of the terrace ridge may be constructed at the

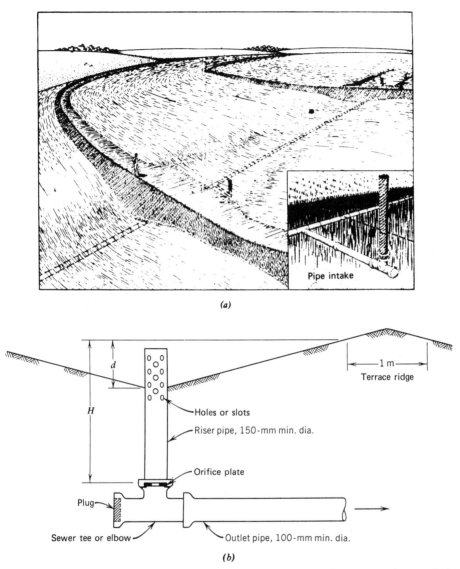

Fig. 8.8 (a) Grassed backslope and pipe outlet graded terraces. (b) Details of controlled-flow pipe intake. (Redrawn from SCS, 1979.)

same elevation along its length even though the bottom of the channel may have a slope to the pipe inlet. These terraces may be constructed with a grassed backslope as shown in Fig. 8.8a.

A variation of the pipe outlet terrace is a sediment and water control basin frequently used in the Midwest (Nolte and VanVliet, 1984). These structures can be less expensive and better suited to greater variations in topography than terrace systems. They are constructed across waterways to prevent or reclaim gullies, reduce the sediment leaving the farm, or conserve water. Slopes may be farmable or have a grass-covered, steep backslope.

8.13 Terrace Channel Capacity

With graded terraces, the rate of runoff is more important than total runoff, whereas both rate and total runoff influence the design of level, pipe outlet, and conservation bench terraces. Graded terraces are designed as drainage channels or waterways, and level terraces function as storage reservoirs. The terrace channel acts as a temporary storage reservoir subjected to unequal rates of inflow and outflow. The Manning velocity equation given in Chapter 7 is suitable for design. A large roughness coefficient of 0.06 is recommended (ASAE, 1989) to ensure that the channel will carry the design runoff under the most severe channel conditions without overtopping. The maximum design velocity will vary with the erosiveness of the soil but should rarely exceed 0.6 m/s for soil devoid of vegetation. The channel depth should permit a freeboard of about 20 percent of the total depth after allowing for settlement of the fill.

For graded terraces, the design peak runoff rate should be based on a storm return period of 10 years (see Chapter 4) and an appropriate duration. The runoff volume for level, pipe outlet, and conservation bench terraces should be based on a 10-year 24-h-duration storm. The orifice for pipe outlet and the pipe are usually selected so that the 24-h design storm will be removed in a 48-h period. Orifice flow is discussed in Chapters 9 and 17, and subsurface pipes in Chapter 14.

8.14 Layout Procedure

A tripod level and the application of surveying techniques along with field experience are sufficient for terrace layout. When available, topographic maps are especially helpful in planning terrace systems (Wittmuss, 1988).

If maps are not available, planning and layout can be completed on the site. The first step is to measure the predominant slope above the terrace and to determine a suitable vertical interval (Eq. 8.1). Stakes are normally set along the proposed terrace every 15 m, although intervals are shorter if turns in the line are sharp. The grade in the channel is provided by placing the stakes on the desired grade, allowance being made at the outlet to compensate for the difference in the elevations of the constructed terrace channel and the stake line. Additional terraces are staked in the same manner. Terraces may be made parallel by adjusting the location of the stakes set for uniform graded terraces as shown in Fig. 8.5. Realignment of these stakes should be limited to provide a cut of not more than 0.3 m below the bottom of the normal channel or a fill height not in excess of 0.9 m. For parallel pipe outlet terraces these heights are more variable. Beasley (1963), Larson (1969), and SCS (1979) describe other procedures for locating parallel, variable-cut terraces, for balancing the cut and fill volumes, and for computing water storage requirements. The procedures involve analysis of short segments and are suited to computer solutions (see Appendix I).

The staking procedure for graded terraces that drain into an established grass waterway is illustrated in Fig. 8.9. The difference in elevations of the stake line and the bottom of the constructed channel will depend on the land slope, shape of the cross section, and locations of the staked lines in relation to the terrace channel. Normally, the center line of the ridge is staked as shown. On slopes of more than about 6 percent, the staked line may be lower than the bottom of the channel. In Fig. 8.9, the vertical scale has been exaggerated to show y more clearly. As shown, the rod reading at 0 + 15 is 1.86 m, assuming $y = 0.15$ m and a 0.06-m slope in

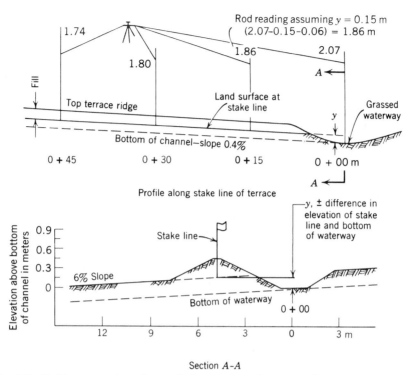

Fig. 8.9 Staking procedure for graded terraces with a grassed waterway outlet.

15 m. Note that the remaining rod readings decrease by 0.06 m for each 15-m station to establish the desired grade of 0.4 percent.

Use of wide farm equipment makes the elimination of sharp turns and point rows important. In many instances, with a relatively small amount of land forming, the topography of a field can be sufficiently changed to permit parallel terraces. Figure 8.10 illustrates the layout adjustment and point row elimination that can be achieved by parallel terrace construction. Pipe outlet terraces facilitate making terraces parallel without land forming.

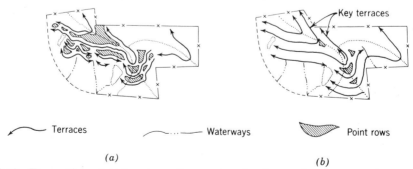

Fig. 8.10 Terrace layout to improve alignment and reduce point rows. (a) Conventional layout with many point rows. (b) Parallel terraces with few point rows. (Redrawn from Jacobson, 1961.)

TERRACE CONSTRUCTION

8.15   Construction Equipment

A variety of equipment is available for terrace construction. Terracing machines include the bulldozer, pan or elevating scraper, motor patrol or blade grader, and elevating grader. Smaller equipment, such as moldboard and disk plows, are suitable for slopes of less than about 8 percent, but the rate of construction is much less than with heavier machines. Soil and crop conditions are likely to be most suitable for construction in the spring and fall.

8.16   Settlement of Terrace Ridges

The amount of settlement in a newly constructed terrace ridge depends largely on soil and water conditions, type of equipment, construction procedure, and amount of vegetation or crop residue. The percentage of settlement based on unsettled height will vary as follows: moldboard plow or bulldozer, 10 to 20 percent; elevating grader, 10 to 25 percent; blade grader, 0 to 5 percent. These data are applicable for soils in good tillable condition with little or no vegetation or residue and for normal construction procedures. In general, those machines that compact the loose fill during construction result in less settling than those that carry the soil to the ridge, such as the elevating grader.

TERRACE MAINTENANCE

Proper maintenance is as important as the original construction of the terrace; however, it need not be expensive since normal farming operations will usually suffice. The terrace should be watched carefully during the first year after construction.

8.17   Tillage Practices

In a terraced field, all farming operations should be carried out as nearly parallel to the terrace as possible. The most evident effect of tillage operations after several years is the increase in the base width of the terrace.

The effect of one-way plowing, in which the furrow slice is moved up the slope and the dead furrow placed in the channel, is shown in Fig. 8.11. Except for a soil loss of about 7 percent from the channel, the soil has been transferred from the interterraced slope to the frontslope.

DIVERSIONS

Diversions or diversion terraces effectively protect bottomland from hillside runoff and divert water away from buildings, cropped fields, and other special-purpose areas. A typical application is shown in Fig. 8.12. The most effective control of gullies may often be accomplished by diverting runoff from above the gully and causing it to flow in a controlled manner to some suitably protected outlet.

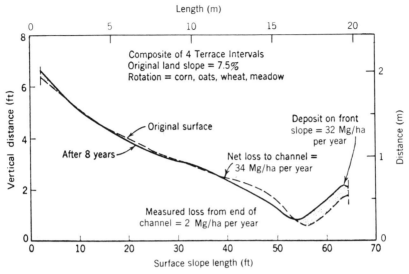

Fig. 8.11 Effect of one-way plowing for 8 years on soil movement between terraces at Bethany, Missouri. (Redrawn from Zingg, 1942.)

8.18 Design of Diversions

A diversion is a channel constructed around the slope and given a slight gradient to cause water to flow to a suitable outlet. The capacity of diversion channels should be based on estimates of peak runoff for the 10-year return period if they are to empty into a vegetated waterway. If the diversion is to outlet into a

Fig. 8.12 Typical diversion to protect cultivated bottomland from damaging hillside runoff. (*Source:* SCS, 1973.)

permanent structure, the design should be the same as for the structure. The design procedures for diversions are the same as for vegetated waterways discussed in Chapter 7.

Cross-section design may vary to suit soil, land slope, and maintenance needs. Side slopes of 4 : 1 with bottom widths that permit mowing are frequently used. Since sediment deposition is often a problem in diversions, the designed flow velocity should be kept as high as the channel protection will permit. In the event that the channel cross section has been designed to permit cultivation, the velocity of flow must be based on bare soil conditions, that is, a maximum of about 0.5 m/s.

REFERENCES

American Society of Agricultural Engineers (ASAE) (1989). *Design, Layout, Construction and Maintenance of Terrace Systems.* ASAE Standard S268.3. St. Joseph, MI.

Beasley, R. P. (1963). *A New Method of Terracing.* Missouri Agr. Expt. Sta. Bull. **699** (rev. July). University of Missouri, Columbia, MO.

Hauser, V. L., and M. B. Cox (1962). "A Comparison of Level and Graded Terraces in the Southern High Plains." *ASAE Trans.* **5**, 75–77.

Hauser, V. L., and M. B. Cox (1962). "Evaluation of Zingg Conservation Bench Terrace." *Agr. Eng.* **43**, 462–464, 467.

Jacobson, P. (1961). "A Field Method for Staking Cut and Fill Terraces." *Agr. Eng.* **42**, 684–687.

Jacobson, P. (1962). "A New Method for Bench Terracing Steep Slopes." *ASAE Trans.* **6**, 257–258.

Larson, C. L. (1969). "Geometry of Broad-Based and Grassed-Backslope Terrace Cross Sections." *ASAE Trans.* **12**, 509–511.

Nolte, B. H., and C. B. VanVliet (1984). *Mini-terraces.* Cooperative Extension Service, Ohio State University, Columbus, OH.

U.S. Soil Conservation Service (SCS) (1973). *Drainage of Agricultural Lands.* Water Information Center, Port Washington, NY.

———(1979). *Engineering Field Manual for Conservation Practices* (litho.). SCS, Washington, DC.

Wittmuss, H. (1988). "A Study of Time Required for Planning, Staking, and Designing Parallel Terrace Systems." *ASAE Trans.* **30**, 1076–1081.

Zingg, A. W. (1942). "Movement Within the Surface Profile of Terraced Lands." *Agr. Eng.* **23**, 93–94.

PROBLEMS

8.1 On one graph, plot two curves with the slope in percent (0 to 10) as the abscissa, and the vertical interval and the horizontal spacing for graded terraces as ordinates. Follow recommendations specified for your area and for resistant soil with good cover.

8.2 Determine the time required for the flow to travel a distance of 100 m (328 ft) in a terrace channel having a slope of 0.3 percent, $n = 0.04$, depth of flow of 0.3 m (1 ft), and side slopes for $W_c = W_f = 4.3$ m (14 ft).

8.3 Determine rod readings for the first four 15-m (50-ft) stations for a uni-form-graded terrace having a slope of 0.3 percent if the rod reading at the outlet is 1.7 m (5.6 ft). If the stake line is 0.15 m (0.5 ft) higher than the bottom of the finished terrace channel, what should be the rod readings be at these four stations?

8.4 If the rod reading at the outlet of the top terrace is 1.46 m (4.8 ft) and the vertical interval (V.I.) for the second terrace is 1.38 m (4.5 ft), what should be the rod reading at the outlet of the second terrace assuming the survey instrument has not been moved?

8.5 Determine the slopes for three sections of a 300-m (984-ft)-long terrace to give a constant velocity of about 0.6 m/s (2 fps). Assume that 0.085 m³/s (3 cfs) of surface runoff enters the channel at the upper end of each 100-m (300-ft) section, $n = 0.04$, and the cutslope and frontslope are 8:1.

8.6 Compute the cut volume and the fill volume for a 200-m (656-ft) terrace on a 4 percent slope with a depth of flow of 0.3 m. Assume the three slope widths are 4.3 m (14 ft), a freeboard of 0.08 m, and a balanced cross section (cut = fill).

8.7 Determine the vertical interval from Eq. 8.1 for graded terraces on re-sistant soil and good cover with a slope of 6 percent at your present location. What is the peak runoff from the second terrace if the terrace length is 366 m (1200 ft), $C = 0.5$, and $T_c = 10$ min? Use the rational method for a return period of 10 years (see Chapter 4).

8.8 The runoff from a 460-m (1500-ft)-long graded terrace with a channel grade of 0.4 percent is 0.3 m³/s (11 cfs). Assuming a roughness coefficient of 0.03, and side slope widths are 4.3 m (14 ft), compute the depth of flow using the Manning formula and determine the total depth of the channel allowing a free-board of 0.1 m.

8.9 A system of graded terraces for your present location is to be made parallel and to be farmed with four-row equipment 0.76-m (30-in.) rows. To provide the nearest complete rounds of travel for this equipment, what slope distance between terraces is needed if the soil is highly erodible and poor cover and the land slope is 2 percent?

8.10 If a pipe outlet terrace is to store 50 mm (2 in.) of runoff from a 10-year return period 24-h-duration storm, determine the opening size in mm (in.) for the orifice to remove this volume in 48 h. Assume an average pressure head of 1.5 m (4.9 ft) on the orifice, an orifice discharge coefficient of 0.6, and a runoff area above the terrace of 1 ha (2.47 ac).

8.11 For an implement width of 5 m (16 ft), a depth of flow of 0.25 m (0.8 ft), and a 4 percent slope, what are the slope ratios of the cutslope, frontslope, and backslope of a three-section broadbase terrace?

8.12 What would be the slope ratios for Problem 8.11 if the land slope is 10 percent?

CHAPTER 9

Conservation Structures

Major channels, whether they are designed to convey irrigation water, drainage flow, or flood runoff, often require stabilization by structures of concrete, metal, or wood. This chapter discusses hydraulic principles in relation to the design of major channel stabilization structures and some uniquely related to drainage or irrigation.

Provision of a stable channel frequently involves reducing the gradient of the channel to maintain velocities below an erosive level. Much of the fall in the channel is taken up by structures that are designed to dissipate the energy of the falling water. The gradient of the channel reaches between structures is designed to maintain nonsilting and nonscouring velocities.

9.1 Temporary and Permanent Structures

Temporary Structures. Temporary structures can be recommended only in situations where cheap labor and materials can be used. Increasing mechanization and high labor costs have resulted in a decline in the popularity of temporary channel stabilization structures. For gully control, shaping the channel and establishing vegetation in accordance with the principles discussed in Chapter 7 provide more efficient and effective control. Temporary structures may be constructed of creosoted planks, rocks, logs, brush, woven wire, sod, or earth.

Smith (1952) reported on the performance of 50 temporary structures that had been used on the SCS experimental farm at Bethany, Missouri; only 5 percent of the structures were found to have functioned as intended. It was concluded that vegetal protection was established as easily without temporary structures as with them.

Permanent Structures. Structures constructed of permanent materials may be required to control the overfall at the head of a large gully, to drop the discharge from a vegetated waterway into a drainage ditch, to take up the fall at various points in any channel, or to provide for discharge through earth fills. Figure 9.1 shows the profile of a gully that has been reclaimed by methods involving the use

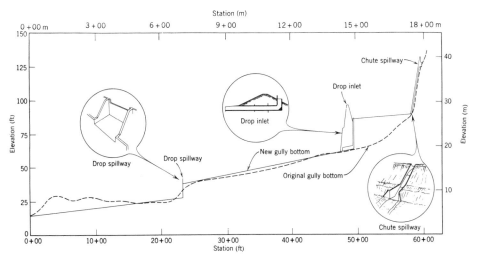

Fig. 9.1 Profile of a gully showing the application of three types of permanent structures.

of several types of permanent structures. Standard designs are available from SCS (1979) and USBR (1987).

9.2 Functional Requirements of Control Structures

Not only must a control structure have sufficient capacity to pass the design discharge, but the kinetic energy of the discharge must be dissipated within the confines of the structure in a manner and to a degree that will protect both the structure and the downstream channel from damage or erosion. The two primary causes of failure of permanent control structures are (1) insufficient hydraulic capacity and (2) insufficient provision for energy dissipation.

9.3 Design Features

The basic components of a hydraulic structure are the inlet, the conduit, and the outlet. Structures are classified and named in accordance with the form of these three components. Figure 9.2 identifies the various types of inlets, conduits, and outlets that are commonly used. In addition to these hydraulic features, the structure must include suitable wing walls, side walls, head wall extensions, and toe walls to prevent seepage under or around the structure and to prevent damage from local erosion. These structural components are identified in Fig. 9.3 for one common type of structure. It is important that a firm foundation be secured for permanent structures. Wet foundations should be avoided or provided with adequate artificial drainage. Surface soil and organic material should be removed from the site to allow a good bond between the structure and the foundation material.

Models. The design criteria for conservation structures have been developed from intensive observation of the behavior of small-scale laboratory models. The results of such laboratory studies have been summarized in empirical formulas, graphs, or tables that relate certain critical dimensions of the structure to charac-

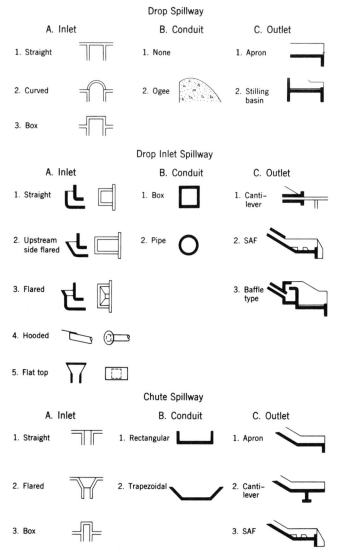

Fig. 9.2 Classification of components of hydraulic structures. (Adapted from SCS, 1979.)

teristics of the flow. Additional information on models may be found in Murphy (1950).

Critical Depth. A given quantity of water in an open conduit may flow at two depths having the same energy head. When these depths coincide, the energy head is a minimum and the corresponding depth is termed the critical depth (Fig. 9.4). Most control structures include a section at which flow at critical depth occurs. Thus, design equations for certain components of structures are often expressed as functions of critical depth.

The expression for critical depth at a rectangular section may be developed as follows. The specific energy head at a section with reference to the channel bed is

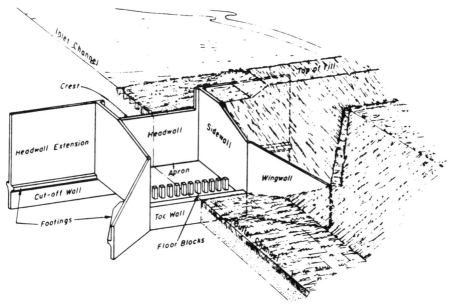

Fig. 9.3 Straight drop spillway showing structural components. (*Source:* SCS, 1979.)

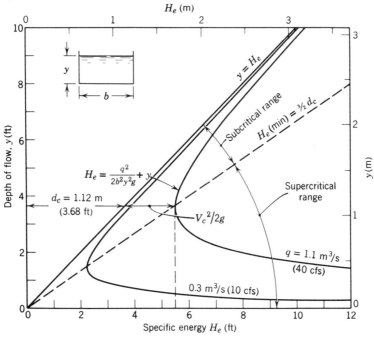

Fig. 9.4 Depth of flow and specific energy for two flow rates in a rectangular channel. (Adapted from USBR, 1987.)

$$H_e = y + \frac{v^2}{2g} = y + \frac{q^2}{2a^2g} = y + \frac{q^2}{2b^2y^2g} \tag{9.1a}$$

where H_e = specific energy (L),
y = depth of flow (L),
v = velocity of flow (L/T),
g = acceleration of gravity (L/T^2),
q = flow rate (L^3/T),
a = cross-sectional area of flow (L^2),
b = width of flow (L).

Differentiating with respect to y,

$$\frac{dH_e}{dy} = 1 - \frac{q^2}{b^2y^3g} \tag{9.1b}$$

Setting $dH_e/dy = 0$ to determine y when H_e is a minimum, and letting this value of y be d_c, the critical depth,

$$1 - \frac{q^2}{b^2y^3g} = 0$$

$$d_c = \sqrt[3]{\frac{q^2}{b^2g}} \tag{9.2}$$

Equation 9.1a may also be solved to determine the depth at which maximum discharge will occur for a given energy, H_e. Solving for q in Eq. 9.1a yields

$$q^2 = 2y^2b^2g(H_e - y) \tag{9.3}$$

Assuming H_e is constant and setting $dq/dy = 0$,

$$\frac{dq}{dy} = \frac{2b^2g(2yH_e - 3y^2)}{2[2y^2b^2g(H_e - y)]^{1/2}} = 0 \tag{9.4}$$

If $y = d_c$, then by solving Eq. 9.4,

$$d_c = \frac{2}{3}H_e \tag{9.5}$$

Thus, for a given H_e, maximum flow occurs at the critical depth or when $H_e = (3/2)y$. Equation 9.3 is plotted in Fig. 9.5 for $H_e = 1.68$ m showing a maximum q of 1.13 m^3/s at a depth of 1.12 m, which is $(2/3)(1.68)$. The curve shows that the discharge would be less at any other depth either above or below the critical depth of 1.12 m.

Hydraulic Jump.   When a given flow changes from a flow depth y less than critical depth to a flow depth greater than critical depth, the phenomenon is referred to as a hydraulic jump. The profile and depth–energy relationships of a

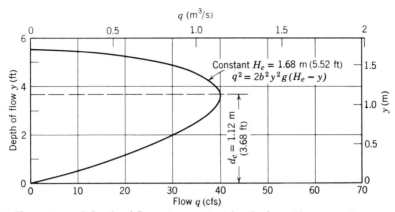

Fig. 9.5 Flow rate and depth of flow at a constant level of specific energy in a rectangular channel.

hydraulic jump are shown in Fig. 9.6. Energy resulting from velocity is converted to energy of elevation and some energy is lost to friction through turbulence in the process. Control structures are often designed so that a hydraulic jump forms within the downstream portion of the structure and velocity is reduced to a nonerosive level in the subcritical range. When the inflow is at critical depth, the two depths, d_1 and d_2, in Fig. 9.6 are equal; thus, no jump would occur.

DROP SPILLWAYS

A typical drop spillway is shown in Fig. 9.3. Drop spillways may have a straight, arched, or box-type inlet. The energy dissipator may be a straight apron or some type of stilling basin.

9.4   Function and Limitations

Drop spillways are installed in channels to establish permanent control elevations below which an eroding stream cannot lower the channel floor. These structures

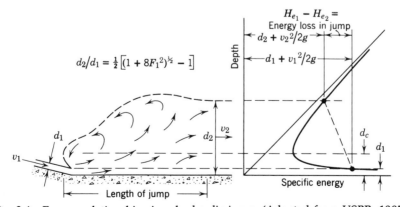

Fig. 9.6 Energy relationships in a hydraulic jump. (Adapted from USBR, 1987.)

control the stream grade not only at the spillway crest itself but also through the ponded reach upstream. Drop structures placed at intervals along a channel can stabilize it by changing its profile from a continuous steep gradient to a series of more gently sloping reaches. Where relatively large volumes of water must flow through a narrow structure at low head, the box-type inlet is preferred. The curved inlet serves a similar purpose and also gives the advantage of arch strength where masonry construction is used. Drop spillways are usually limited to drops of 3 m, flumes or drop-inlet pipe spillways being used for greater drops.

9.5 Design Features

Capacity. The free flow (no submergence) capacity for drop spillways is given by the weir formula

$$q = 0.55CLh^{3/2} \tag{9.6}$$

where q = the discharge in m^3/s,
C = weir coefficient (for English units),
L = weir length in m,
h = depth of flow over crest in m.

The length L is the sum of the lengths of the three inflow sides of a box inlet, the circumference of an arch inlet, or the crest length of a straight inlet. The value of C varies considerably with entrance conditions. Blaisdell and Donnelly (1951) have prepared correction charts to modify C for a wide range of conditions of entrance and crest geometry for box inlets. Where the ratio of head to box width is 0.2 or greater, the ratio of the width of the approach channel to the total length L is greater than 1.5, and no dikes or other obstacles are within $3h$ of the crest, a value of $C = 3.2$ may be used with an accuracy of ±20 percent. Discharge characteristics of a typical box-inlet drop spillway are given in Fig. 9.7. A value of $C = 3.2$ will

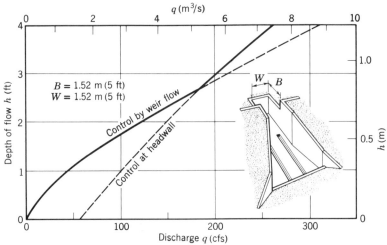

Fig. 9.7 Discharge-head relationships of a box-inlet drop spillway. (*Source:* Blaisdell and Donnelly, 1951.)

also give satisfactory results for the straight inlet or the control section of a flood spillway (see Chapter 10). The inlet should have a freeboard of 0.15 m above h, the height of the water surface.

Whenever the tailwater is nearly up to or above the crest of the inlet section, submergence decreases the capacity of the structure. When such conditions occur, other design equations apply (beyond the scope of this text).

Apron Protection.   The kinetic energy gained by the water as it falls from the crest must be dissipated and/or converted to potential energy before the flow is discharged from the structure. For straight-inlet drop structures the dissipation and conversion of energy are accomplished in either a straight apron or a Morris and Johnson (1942) stilling basin. Dimensions for the Morris and Johnson stilling basin are given in Fig. 9.8. For larger structures the Morris and Johnson outlet is preferred, as it results in a shorter apron and the transverse sill induces a hydraulic jump at the toe of the structure. The longitudinal sills serve to straighten the flow and prevent transverse components of velocity from eroding the side slopes of the downstream channel. The flow pattern through a Morris and Johnson stilling basin is shown in dimensionless form in Fig. 9.9. A stilling basin for the box-inlet drop spillway is shown in Fig. 9.10.

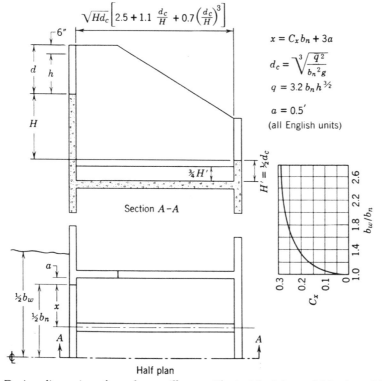

Fig. 9.8 Design dimensions for a drop spillway with straight inlet and Morris and Johnson outlet. (*Source:* Morris and Johnson, 1942.) (Another design is shown in Fig. 9.3.)

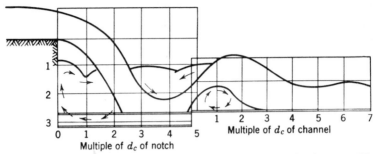

Fig. 9.9 Flow pattern through a drop spillway with a Morris and Johnson stilling basin. (*Source:* Morris and Johnson, 1942.)

CHUTES

Chutes are designed to carry flow down steep slopes through a concrete-lined channel rather than by dropping the water in a free overfall.

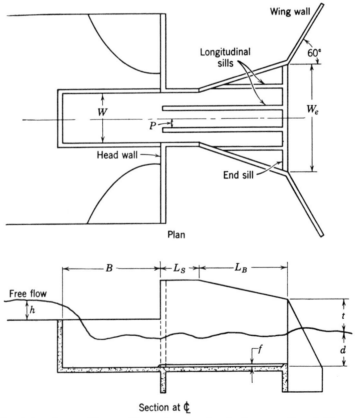

Section at ₵

Fig. 9.10 Box-inlet drop spillway. (See reference for design equations.) (*Source:* Blaisdell and Donnelly, 1951.)

9.6   Function and Limitations

Chutes may be used for the control of elevation changes up to 6 m. They usually require less concrete than do drop-inlet structures of the same capacity and drop; however, there is considerable danger of undermining of the structure by rodents, and, in poorly drained locations, seepage may threaten foundations. Where there is no opportunity to provide temporary storage above the structure, the chute with its inherent high capacity is preferred over the drop-inlet pipe spillway. The capacity of a chute is not decreased by sedimentation at the outlet.

9.7   Design Features

Capacity.   Chute capacity normally is controlled by the inlet section. Inlets may be similar to those for straight-inlet or box-inlet drop spillways, and in such inlets capacity formulas already discussed will apply. Blaisdell and Huff (1948) have investigated the performance of other types of flume entrances.

Outlet Protection.   The cantilever-type outlet should be used where the channel grade below the structure is unstable. In other situations, either the straight-apron or Saint Anthony Falls (SAF) outlet is suitable. The straight apron is applicable to small structures. Figure 9.11 shows dimensions of this type of outlet protection.

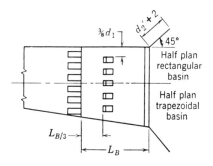

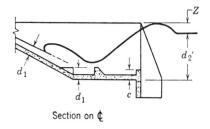

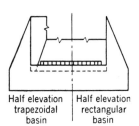

Fig. 9.11   The Saint Anthony Falls (SAF) stilling basin. (See reference for design equations.) (*Source:* Blaisdell, 1948.)

FORMLESS FLUME

9.8 Function and Limitations

This structure has the advantage of low-cost construction. It may replace drop spillways where the fall does not exceed 2 m and the width of notch required does not exceed 7 m. The flume is constructed by shaping the soil to conform to the shape of the flume and applying a 0.13-m layer of concrete reinforced with wire mesh. No forms are needed; thus, the construction is simple and inexpensive. The formless flume should not be used where water is impounded upstream (danger of undermining the structure by seepage) or where freezing occurs at great depth.

9.9 Design Features

Figure 9.12 shows the design features and dimensions of the formless flume. The capacity is given by Eq. 9.6 using a C value of 3.9. This weir coefficient accounts for the increased cross-sectional area because the sides of the weir slope outward rather than vertically and the entrance is rounded. The depth of the notch, D, is h plus a freeboard of 0.15 m.

PIPE SPILLWAYS

Pipe spillways may take the form of a simple conduit under a fill (Fig. 9.13*b*), or they may have a riser on the inlet end and some type of structure for outlet protection (Fig. 9.13*a*). The pipe in Fig. 9.13*c*, called an inverted siphon, is often

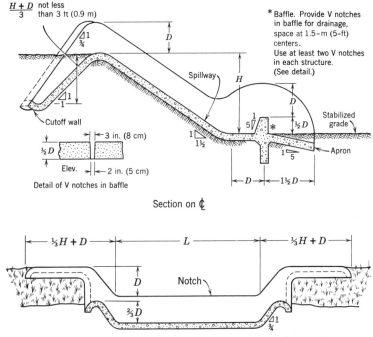

Fig. 9.12 A formless flume. (Based on design by Wooley et al., 1941.)

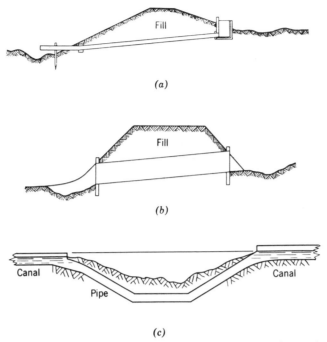

(a)

(b)

(c)

Fig. 9.13 Types of pipe spillways. (*a*) Drop-inlet pipe spillway with cantilever outlet. (*b*) Simple culvert. (*c*) Inverted siphon.

used when water in an irrigation canal must be conveyed under a natural or artificial drainage channel. Inverted siphons must withstand hydraulic pressures much higher than those encountered in other pipe spillways and therefore require special attention to structural design.

9.10 Function and Limitations

The pipe spillway used as a culvert has the simple function of providing for passage of water under an embankment. When combined with a riser or drop inlet, the pipe spillway serves to lower water through considerable drop in elevation and to dissipate the energy of the falling water. Drop-inlet pipe spillways are thus frequently used as gully control structures. This application is usually made where water may pond behind the inlet to provide temporary storage. The hydraulic capacity of pipe spillways is related to the square root of the head; hence they are relatively low-capacity structures. This characteristic is used to advantage where discharge from the structure is to be restricted.

9.11 Design Features

Culverts. Culvert capacity may be controlled either by the inlet section or by the conduit. The headwater elevation may be above or below the top of the inlet section. Several possible flow conditions are represented in Fig. 9.14. Solution of a culvert problem is primarily the determination of the type of flow that will occur under given headwater and tailwater conditions. Consider a culvert as shown in Figs. 9.14*a* and 9.14*b*. Pipe flow (conduit controlling capacity) will occur under

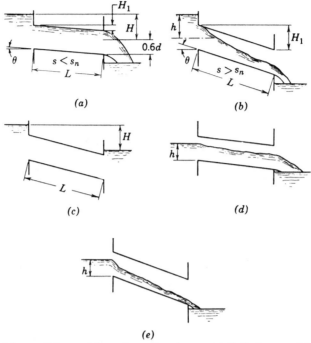

Fig. 9.14 Possible conditions of flow through culverts. (*a*) Full: free outfall, pipe flow. (*b*) Part full: free outfall, orifice flow. (*c*) Full: outfall submerged, pipe flow. (*d*) Inlet not submerged: conduit controls, open channel flow. (*e*) Inlet not submerged: inlet controls, weir flow. (Modified from Mavis, 1943.)

most conditions when the slope of the culvert is less than the neutral slope and entrance capacity is not limiting. The neutral slope s_n for small angles of θ is

$$s_n = \tan \theta = \sin \theta = \frac{H_f}{L} = K_c \frac{v^2}{2g} \tag{9.7}$$

where θ = slope angle of conduit in degrees,
H_f = friction loss in conduit length L (L),
L = length of conduit (L),
K_c = friction loss coefficient (1/L),
v = velocity of flow (L/T),
g = gravitational constant (L/T²).

In some situations inlet losses may be so great that pipe flow will not occur even though the culvert is on less than neutral slope. The capacity of a culvert under conditions of full pipe flow is given by

$$q = \frac{a\sqrt{2gH}}{\sqrt{1 + K_e + K_b + K_cL}} \tag{9.8}$$

where q = flow capacity (L³/T),
a = conduit cross-sectional area (L²),

H = head causing flow (L),
K_e = entrance loss coefficient,
K_b = loss coefficient for bends in the culvert.

Values of K_b, K_c, and K_e are given in Appendix C. Since most culverts do not have bends, K_b can often be omitted.

If the conduit is at greater than neutral slope and the outlet is not submerged, the flow will be controlled by the inlet section on short-length culverts, and orifice flow will prevail. Capacity is then given by

$$q = aC\sqrt{2gh} \qquad (9.9)$$

where a = cross-sectional area (L²),
h = head to the center of the orifice (L) (either SI or English units).

The coefficient C for a sharp-edged orifice is 0.6. For more detailed values and for other orifices, consult Brater and King (1976) or other hydraulics books. Examples 9.1 and 9.2 serve to clarify the above discussions.

In situations where the headwater elevation does not reach the elevation of the top of the inlet section, there is again the possibility of control of flow by either the conduit or the inlet section. In Figs. 9.14d and 9.14e, the conduit controls if the slope of the conduit is too flat to carry the maximum possible inlet flow at the required depth. This depth is equal to the headwater depth above the inlet invert minus the static head loss resulting from entrance losses and acceleration. Conditions of control at the entrance section occur when the slope of the conduit is greater than that required to move the possible flow through the inlet. For conditions of control by the entrance section, solution for circular culverts may be made from Fig. 9.15. This figure will also apply when the inlet is submerged. Examples 9.3 and 9.4 clarify the solution for conditions of an unsubmerged inlet. Values of the roughness coefficient n for conduits may be found in Appendix B.

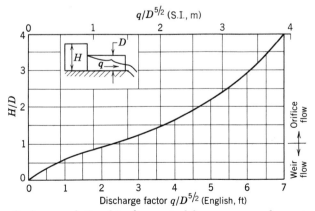

Fig. 9.15 Stage-discharge relationship for control by a square-edge entrance inlet to a circular pipe. (Redrawn from Mavis, 1943.)

☐ Example 9.1

Determine the capacity of a 762-mm (30-in.) diameter corrugated culvert 18.29 m (60 ft) long with a square-edged entrance. Elevation of the inlet invert is 127.92 m (419.7 ft), and the elevation of the outlet invert is 127.71 m (419.0 ft). Headwater elevation is 129.54 m (425.0 ft), and tailwater elevation is 126.80 m (416.0 ft).

Solution. Assume pipe flow prevails, for which Eq. 9.8 applies. From Appendix C, find $K_e = 0.5$ and $K_c = 0.112$. Head H is from the headwater elevation to an elevation of $0.6d$ above the elevation of the outlet invert, which is $129.54 - (127.71 + 0.6 \times 0.762) = 1.37$ m.

$$q = \frac{3.14 \times 0.762^2}{4} \left[\frac{2 \times 9.8 \times 1.37}{1 + 0.5 + 0.112 \times 18.29} \right]^{1/2}$$

$$= 1.26 \text{ m}^3/\text{s (44.4 cfs)}$$

From Eq. 9.7 determine the neutral slope of the culvert for a discharge of 1.26 m³/s.

$$s_n = \frac{0.112 \times 1.26^2}{2 \times 9.8 \times 0.45^2} = 0.043 = 4.3 \text{ percent}$$

Actual slope of the culvert is $(127.92 - 127.71)/18.29$, or 0.0115. Since the culvert slope is less than the neutral slope, pipe flow prevails and 1.26 m³/s is the discharge. Checking for orifice control at the entrance ($h = 1.24$ m) gives a discharge of 1.34 m³/s using Eq. 9.9. Since this flow is more than pipe flow capacity of 1.26 m³/s, pipe flow rather than orifice flow controls. By use of SCS engineering computer programs (Appendix I) a similar flow was calculated.   ☐

☐ Example 9.2

Determine the capacity of the 762-mm (30-in.) culvert of Example 9.1 if the elevation of the outlet invert is 125.15 m (410.6 ft), and the tailwater elevation is 124.36 m (408.0 ft).

Solution. Assume pipe flow and calculate the discharge by Eq. 9.8 with $H = 129.54 - (125.15 + 0.46) = 3.93$ m (12.9 ft). Note that $0.6d = 0.46$.

$$q = 0.456 \left[\frac{2 \times 9.8 \times 3.93}{1 + 0.5 + (0.112 \times 18.29)} \right]^{1/2}$$

$$= 2.12 \text{ m}^3/\text{s (75.0 cfs)}$$

$$\text{Neutral slope } s_n = \frac{0.112 \times 2.12^2}{2 \times 9.8 \times 0.456^2} = 0.12$$

$$\text{Actual slope } s = (127.92 - 125.15)/18.29 = 0.15$$

Since the culvert is at greater than neutral slope, pipe flow will not exist. Entrance conditions will prevail, and the problem is solved by application of the orifice flow formula, Eq. 9.9. From Example 9.1, $q = 1.34$ m³/s (47.3 cfs).

An alternate solution may be obtained from the curve in Fig. 9.15 with head $H = 129.54 - 127.92 = 1.62$ m. For $H/D = 1.62/0.762 = 2.13$, read $q/D^{5/2} = 2.65$ (SI units). Solving for q, which is $2.65 \times 0.762^{5/2} = 1.34$ m^3/s, the discharge is thus the same as computed from Eq. 9.9. □

□ Example 9.3

Determine the capacity of a 1.52-m (5-ft)-diameter concrete culvert 30.5 m (100 ft) long. The culvert entrance is square-edged. Elevation of the inlet invert is 157.76 m (517.6 ft), and the elevation of the outlet invert is 155.60 m (510.5 ft). Headwater elevation is 158.80 m (521.0 ft), and tailwater elevation is 152.40 m (500.0 ft).

Solution. Assume that the conduit controls and entrance conditions are not limiting. Neglect for the moment the loss of static head at the culvert entrance resulting from acceleration of the flow entering the culvert. Under these assumptions, the depth of flow in the culvert would be 1.04 m (3.4 ft). Calculating the flow by the Manning formula, $a = 1.33$ m^2 (14.3 ft^2), $n = 0.015$, $R = 0.445$ m (1.46 ft), and $s = 0.071$, for which $q = 13.76$ m^3/s (486 cfs). Checking in Fig. 9.15, $H/D = 0.68$ and $q = 1.98$ m^3/s (70 cfs). Since only this flow rate can enter the culvert, a flow of 13.76 m^3/s could not possibly occur; thus entrance conditions prevail and the capacity is 1.98 m^3/s. □

□ Example 9.4

Determine the capacity of a culvert as in Example 9.3, but having the outlet invert at an elevation of 157.75 m (517.55 ft).

Solution. As before, we note that the maximum possible flow through the entrance inlet is 1.98 m^3/s; however, in this case the culvert is on a flat slope, and conduit flow conditions may limit the flow. Assume a flow depth in the conduit of 0.76 m (2.5 ft). Then $a = 0.91$ m^2 (9.8 ft^2), $n = 0.015$, $R = 0.38$ m (1.25 ft), and $s = 0.0005$. Substituting in the Manning formula, $v = 0.78$ m/s (2.56 fps) and $q = 0.71$ m^3/s (24.9 cfs). Assume the approach velocity is negligible; then the loss of static head at the culvert entrance resulting from acceleration is $v^2/2g = 0.78^2/19.6 = 0.03$ m (0.102 ft). Depth of water at the entrance is 1.04 m (3.4 ft), and a loss of 0.03 m (0.102 ft) would give 1.01 m (3.3 ft), which does not correspond with our assumption of 0.76 m (2.5 ft). Thus, the first assumption of flow depth was in error. Now assume a flow depth in the culvert of 1.00 m (3.28 ft), for which $a = 1.26$ m^2 (13.5 ft^2), $n = 0.015$, $R = 0.44$ m (1.44 ft), and $s = 0.0005$. Substituting in the Manning formula, $v = 0.86$ m/s (2.83 fps) and $v^2/2g = 0.04$ m (0.12 ft). Subtracting the 0.04 m from the entrance depth of 1.04 m leaves a flow depth of 1.00 m, which agrees with the original assumption and is the correct depth of flow. Thus, flow is limited by the conduit, and the discharge capacity is

$$q = 1.26 \times 0.44^{2/3} \times 0.0005^{1/2}/0.015 = 1.08 \text{ m}^3/\text{s} \text{ (38 cfs)} \qquad \square$$

Drop Inlets. The discharge characteristics of a drop-inlet pipe spillway are given in Fig. 9.16. At low heads, the crest of the riser controls the flow, and discharge is proportional to $h^{3/2}$. Under this condition, the discharge should be calculated as outlined in Section 9.5. When this type of flow equals the capacity of the conduit or conduit inlet section, the flow becomes proportional to the square root of the total head loss through the structure or the head on the conduit inlet.

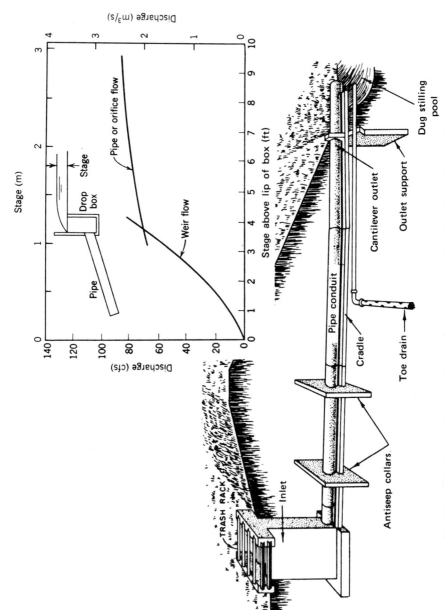

Fig. 9.16 Discharge characteristics and components of a drop-inlet pipe spillway.

Hood Inlets.   For farm pond mechanical spillways and similar small structures the hood inlet has largely replaced the drop inlet entrance. For slopes up to 30 percent the hood inlet, when provided with a suitable antivortex device, will cause the pipe to prime and flow full. Hood inlets shown in Figs. 9.17*a* and 9.17*b* were developed by Blaisdell and Donnelly (1958). Beasley et al. (1960) reported that model and field tests of a hood inlet with an end plate shown in 9.17*c* gave satisfactory performance although the entrance loss was somewhat higher than with the other two. The discharge characteristics of these three inlets are shown in Fig. 9.17*d* for a pipe length of 110D. For H/D less than 1, weir flow occurs, and up to H/D of about 1.4, the flow is rather erratic. At H/D of 1.4 the vortex is eliminated and pipe flow controls. The entrance coefficient for thin-wall pipe (ratios of wall thickness to pipe diameter below 0.04) is slightly higher than that for heavier pipe.

The design capacity is determined from the pipe flow equation as described for culverts. Entrance-loss coefficients are given in Appendix C. Although the approach conditions have little effect on spillway performance, the presence of the face of the dam will reduce somewhat the entrance-loss coefficient. High velocities near the hood inlet may erode the dam, but such a hole is small and becomes stabilized in a short time. For example, Blaisdell and Donnelly (1958) report that for the size of the bed material equal to 0.001D, the scour hole diameter is only 6D. Hood inlets are simple and easy to install and for small pipe diameters are much more economical than the reinforced concrete drop inlet. They are also available commercially.

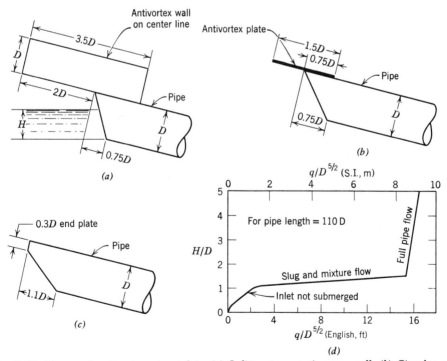

Fig. 9.17   Types of antivortex pipe inlets. (*a*) Splitter-type antivortex wall. (*b*) Circular or square antivortex top plate. (*c*) End antivortex plate. (*d*) Hood inlet discharge-head relationships for (*a*), (*b*), and (*c*). (*Source:* Blaisdell and Donnelly, 1958, and Beasley et al., 1960.)

The hood drop inlet shown in Fig. C.1 (in Appendix C) is a hood inlet on the pipe of a drop-inlet pipe spillway shown in Fig. 9.16. Because the hood inlet must have a head of 1.1D to prime, the pipe spillway on large structures will require a specific and significant rise in the reservoir level before the spillway will flow at design capacity. The hood drop inlet will thus reduce the height and cost of the dam compared with that for a hood inlet.

Outlet Protection.   For small culverts or drop-inlet pipe spillways, a canti-lever-type outlet is usually satisfactory. The straight apron outlet may be used in some instances. Large drop-inlet pipe spillways may be provided with the SAF stilling basin discussed in Section 9.7.

IRRIGATION AND DRAINAGE STRUCTURES

Many types of permanent structures are needed to control irrigation water. Most of these are placed in canals or farm distribution channels. Figure 9.18a shows an adjustable gate for controlling the flow into a farm ditch from a larger canal. A

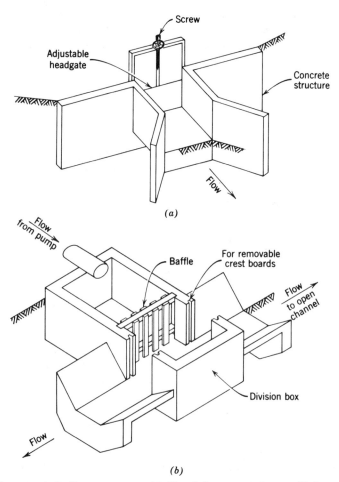

Fig. 9.18   Concrete irrigation structures. (a) Canal flow control gate. (b) Pump outlet and division box. (Redrawn from SCS, 1979.)

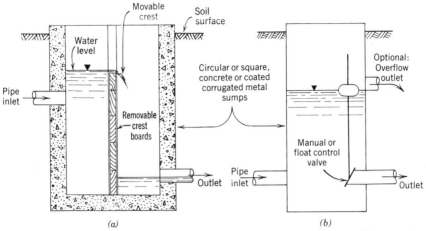

Fig. 9.19 Drainage structures for water level control. (*a*) Sump with removable crestboards (also called stop logs). (*b*) Sump with manual or float valve control for pipe drains.

concrete division box for distributing the flow from a well into several small channels is shown in Fig. 9.18*b*.

Control structures for regulating the water level in drainage systems are shown in Fig. 9.19. Crest boards shown in Fig. 9.19*a* are manually placed in vertical slots. They may also be installed in an open ditch as part of a weir structure (Fig. 13.7). In pipe systems, sumps as shown in Fig. 9.19*b* are suitable for controlled drainage (see Chapter 14). They may be installed near the outlet or in each drain line depending on the land slope. Float control may be replaced with water level sensors located either in the structure or in a measuring well placed midway between pipe drains in the field. Such automatic controls can maintain the water level between minimum and maximum levels, making these structures suitable for subirrigation (see Chapter 18). With both pipe drain structures shown in Fig. 9.19, sealed pipe is required near the sump to reduce seepage flow past it.

REFERENCES

Beasley, R. P., L. D. Meyer, and E. T. Smerdon (1960). "Canopy Inlet for Closed Conduits." *Agr. Eng.* **41**, 226–228.

Blaisdell, F. W. (1948). "Development and Hydraulic Design, Saint Anthony Falls Stilling Basin." *Trans. ASCE* **113**, 483–520.

Blaisdell, F. W., and C. A. Donnelly (1951). *Hydraulic Design of the Box Inlet Drop Spillway.* USDA SCS-TP-106. SCS, Washington, DC.

———(1958). *Hydraulics of Closed Conduit Spillways, Pt X: The Hood Inlet.* University of Minnesota, St. Anthony Falls Hydraulic Lab. Tech. Paper No. 8, Ser. B.

Blaisdell, F. W., and A. N. Huff (1948). *Report of Tests Made on Three Types of Flume Entrances.* USDA SCS-TP-70. SCS, Washington, DC.

Brater, E. F., and H. W. King (1976). *Handbook of Hydraulics*, 6th ed. McGraw-Hill, New York.

Mavis, F. T. (1943). *The Hydraulics of Culverts.* Pennsylvania Eng. Expt. Sta. Bull. 56. Pennsylvania State University, College Station, PA.

Morris, B. T., and D. C. Johnson (1942). "Hydraulic Design of Drop Structures for Gully Control." *Proc. ASCE* **68**, 17–48.

Murphy, G. (1950). *Similitude in Engineering*. Ronald Press, New York.

Smith, D. D. (1952). "A 20-Year Appraisal of Engineering Practices in Soil and Water Conservation. *Agr. Eng.* **33**, 553, 556.

U.S. Bureau of Reclamation (USBR) (1987). *Design of Small Dams*, 3rd ed. GPO, Washington, DC.

U.S. Soil Conservation Service (SCS) (1979). *Engineering Field Manual for Conservation Practices*, (litho.). Washington, DC.

Wooley, J. C., M. W. Clark, and R. P. Beasley (1941). *The Missouri Soil Saving Dam*. Missouri Agr. Expt. Sta. Bull. 434. University of Missouri, Columbia, MO.

PROBLEMS

9.1 What is the maximum capacity of a straight-drop spillway having a crest length of 3 m (10 ft) and a depth of flow of 0.9 m (3 ft)?

9.2 Determine the design dimensions for the drop spillway in Problem 9.1, using the Morris and Johnson outlet, if the drop in elevation is 1.5 m (5 ft). The waterway is 4.6 m (15 ft) wide.

9.3 Determine the crest length for a straight-inlet drop spillway to carry 7.5 m^3/s (265 cfs) if the depth of flow is not to exceed 0.9 m (3 ft). What should be the dimensions of a square-box inlet for the same conditions?

9.4 Determine the design dimensions for a 1.5 × 1.5-m (5 × 5-ft) box-inlet drop spillway to carry 7.0 m^3/s (250 cfs) if the end sill is 3 m (10 ft) in length.

9.5 What is the capacity of a formless flume 1.8 m (6 ft) wide when the flow depth is 0.6 m?

9.6 Determine the discharge of a 30-m (100-ft) 0.9 × 0.9-m (3 × 3-ft) concrete box culvert having a square entrance, $n = 0.013$, $s = 0.003$, and elevations of 10.97 m (36.0 ft) at the center of the conduit at the outlet, 12.89 m (42.3 ft) for the headwater, and 12.34 m (40.5 ft) for the tailwater.

9.7 Tabulate and plot the head discharge up to a 1.5-m (5-ft) depth above the crest for a 0.9 × 0.9-m (3 × 3-ft) drop inlet attached to a 457-mm (18-in.)-diameter concrete pipe ($n = 0.015$) 46 m (150 ft) in length. Assume pipe flow controls in the conduit; tailwater height is not higher than the center of the pipe at the outlet; radius of curvature of the pipe entrance is 0.09 m (0.3 ft); and the difference in elevation between crest and center of pipe at outlet is 5.5 m (18 ft).

9.8 Compute the critical depth of flow in a rectangular chute 1.8 m (6 ft) in width if the flow is 2.8 m^3/s (100 cfs). If the roughness coefficient is 0.015, what is the slope of the chute?

9.9 Determine the critical depth for a triangular-shaped channel having side slopes of 1:1 if the flow is 2.8 m^3/s (100 cfs).

9.10 If the discharge as controlled by pipe flow through a drop-inlet pipe spillway is 1.30 m^3/s (46 cfs) with a depth of flow of 0.6 m (2 ft) over the crest of the inlet box, what is the minimum size of a square-box inlet to keep the pipe flowing full? Under this condition, weir flow equals pipe flow. Assume the weir coefficient for the box is 3.2, and one side of the box serves as an antivortex wall.

9.11 Determine the capacity of a 610-mm (24-in.)-diameter culvert 30 m (100 ft) in length. $K_e = 0.25$, $K_c = 0.049$ (SI units), $s = 0.032$, headwater is 2.13 m (7 ft) above inlet invert, and tailwater is 0.37 m (1.2 ft) above outlet invert. What is the neutral slope assuming full pipe flow?

9.12 A 457-mm (18-in.)-diameter culvert 15 m (50 ft) in length is to be placed under a farm road to carry an estimated peak runoff of 0.28 m³/s (10 cfs). The invert elevation of the pipe inlet is 15.24 m (50 ft) and the outlet invert is 14.94 m (49 ft). Maximum tailwater elevation is 15.09 m (49.5 ft). The entrance is square with a concrete headwall flush with the end of the pipe. Corrugated metal pipe with $n = 0.025$ is to be installed. Assume that ponding of water above the culvert has no effect on runoff. To what elevation would the road have to be built so that the peak flow would not overtop the road?

9.13 Determine the discharge of a circular-plate hood-inlet pipe spillway through a dam if the conduit is a 305-mm (12-in.)-diameter corrugated metal pipe, 30 m (100 ft) long on a slope of 16 percent. The depth of water above the inlet invert is 0.91 m (3 ft).

9.14 Determine the minimum dimensions of a square-drop inlet and pipe spillway to cause the pipe just to flow full with a 0.3-m (1-ft) depth of flow above the spillway crest. The pipe is 610 mm (24 in.) in diameter and 30 m (100 ft) in length, and the total head causing pipe flow is 4.88 m (16 ft). Assume $K_e = 0.15$, $K_c = 0.066$ (SI units), and the weir coefficient is 3.2.

CHAPTER 10

Earth Embankments and Farm Ponds

In all land-use programs the availability of water for crops, livestock, and many other purposes is of primary importance. Farm ponds and reservoirs provide a logical source of such water, for they may be designed and adjusted to fit the individual farm. Conservation and protection of land also depend on the control of excess waters, such as flood control measures (see Chapter 11). Earth embankments in the form of dikes, levees, and detention dams are important protective structures.

The design of earth embankments that are effective and safe requires thorough integration of the principles of soil physics and soil mechanics with sound engineering design and construction principles.

EARTH EMBANKMENTS

Regardless of the structure size, the basic design principles to be described apply equally to all conservation structures, such as dikes, levees, farm ponds, and reservoirs, having a total height above ground level not exceeding 15 m. Although structures higher than 15 m may be found in upstream watershed projects, such structures have specialized design requirements not covered in this text.

10.1 Types of Earth Embankments

The discussions in this chapter are limited to the rolled-fill type of earth embankment in which the soil material has been spread in uniform layers and then compacted at optimum water content until maximum density is achieved. The selection and design of embankments for water control is predicated by (1) the foundation properties, that is, stability, depth to impervious strata, relative permeability, and drainage conditions, and (2) the nature and availability of the construction materials.

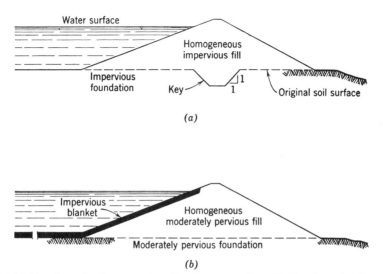

(a)

(b)

Fig. 10.1   (a) Simple embankment using "key" construction. (b) Simple embankment with an impervious "blanket" seal.

There are three major types of earth fill. The *simple embankment* type is constructed of relatively homogeneous soil material and either is keyed into an impervious foundation stratum, as shown in Fig. 10.1a, or is constructed with an upstream blanket of impervious material, as shown in Fig. 10.1b. This type is limited to low fills and to sites having sufficient volumes of satisfactory fill materials available. The *core* or *zoned type* of design places within the dam a central section of highly impermeable or puddled soil materials extending from above the water line to an impermeable stratum in the foundation. In some instances an upstream blanket is used in conjunction with this design. These designs, shown in Fig. 10.2, reduce the percentage of high-grade fill materials needed for construction. The *diaphragm* type uses a thin wall of plastic, butyl, concrete, steel, or wood to form a barrier against seepage through the fill. A "full-diaphragm" cutoff extends from above the water line down to and sealed into an impervious foundation stratum as shown in Fig. 10.3a. The "partial diaphragm" (Fig. 10.3b) does not extend through this full range and is sometimes referred to as a cutoff wall. The internal or buried diaphragm, particularly when constructed of rigid materials, has the disadvantage of being unavailable for inspection or repair if broken or cracked as a result of settlement in the foundation or the fill. The use of flexible film diaphragms of plastic and butyl rubber has partially overcome this problem.

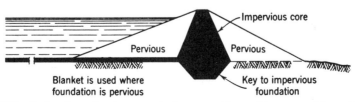

Fig. 10.2   Embankment using a central core and key of impermeable material.

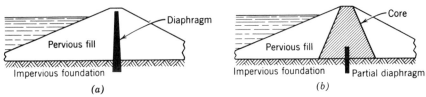

Fig. 10.3 (*a*) Full diaphragm. (*b*) Partial-diaphragm construction for seepage control.

10.2 General Requirements for Earth Embankments

To ensure an effective reservoir for water storage, six basic requirements must be met: (1) The topographic conditions at the dam site must allow economical construction; cost is a function of fill length and height, which determine volume content. (2) Soil materials must be available to provide a stable, impervious fill. (3) All storage embankments must have adequate mechanical and flood spillway facilities to maintain a uniform water depth during normal conditions and to safely manage flood runoff. (4) Large storage embankments should be equipped with a bottom drain pipe to facilitate maintenance and fish management. (5) Adequate safety equipment must be provided around drop-inlet structures and other hazardous portions of the dam. (6) All design specifications must be adhered to during construction, and a sound program of maintenance must be followed to protect against damage by wave action, erosion, burrowing animals, livestock, farm equipment, and careless recreational use. All of these criteria are necessary to ensure safety of the structure and to prevent damage to downstream property.

10.3 Foundation and Earth Fill Requirements

Earth dams and embankments may be built on a wide range of foundation conditions provided these conditions are properly considered in design.

On small dams these investigations may be limited to soil pits or auger borings. On larger structures the subsurface exploration should be more thorough. Wash borings, test pits, and other standard procedures should be employed to determine the underlying soil and geologic conditions.

In evaluating the physical and engineering properties of the foundation soils and the embankment materials, a number of standard soil mechanics tests, such as particle size distribution, liquid limit, plasticity index, shear strength, compressibility, and permeability, can be used. A Unified Soil Classification System has been developed to provide a logical grouping of soils based on these physical and mechanical properties. A full discussion of exploration procedures, soil tests, classification, and interpretation of the data in terms of their use appears in Terzaghi and Peck (1967), USBR (1987), Jansen (1988), Creager et al. (1945), Spangler and Handy (1981), and similar references. Foundation materials have been classified by Creager et al. (1945) in the following way:

(1) *Ledge rocks.* Under earth-filled dams, ledge rocks present a potential permeability hazard and frequently need grouting.

(2) *Fine uniform sands.* If below "critical density" (void ratio at which a soil can undergo deformation without change of volume) fine uniform sands must be consolidated to prevent flow when saturated under load.

(3) *Coarse sands and gravel.*   From the stability standpoint they will consolidate under load. An upstream blanket may be required to prevent seepage losses.

(4) *Plastic clays.*   They require careful analysis to ensure that shear stress imposed by the weight of the dam is less than the shear strength of the foundation material; flattened side slopes may be required to reduce shear stress. Knowledge of porous strata, preglacial gorges, geologic faults, and other hazardous conditions will be of value in design of the structure.

The design of a dam or embankment should be based on the most economical use of the available materials immediately adjacent to the site. For example, if satisfactory core materials are unavailable and must be hauled some distance, the hauling cost should be compared with the cost of a thin-section diaphragm of concrete, steel, or plastic.

Cross-section design depends on both the foundation conditions and the fill material available. Where depth to an impervious foundation is not too great and where supplies of quality core materials can be found, designs shown in Figs. 10.2 and 10.3b can be used. The combination core and blanket design shown in Fig. 10.2 is adapted to sites having extremely deep pervious foundations. Other designs may be developed to use diaphragms alone and, in combination with cores and other construction features, to meet specific conditions.

For optimum compaction, experiments and experience have shown that preparations of gravel, sand, silt, and clay that will compact to maximum density are, generally, not over 20 percent gravel, 20 to 50 percent sand, not over 30 percent silt, and 15 to 25 percent clay. Soils having high shrinkage and swelling characteristics and ungraded soils, when their use cannot be avoided, should be placed in the downstream interior of the embankment. Here they are subject to less changes in water content and, because of overburden weight, have less volume change than if placed elsewhere in the embankment. Soils having higher percentages of graded sandy materials resist changes in water content, temperature, and internal stresses. Organic soils should be entirely eliminated from the fill.

In levee construction, the selection of material is usually limited to that found adjacent to the structure. As these materials change in their characteristics along the route of the levee, it will be necessary to adjust the cross-sectional design of the embankment.

10.4   Seepage Through the Embankment

Although characteristics of the soil materials in both the foundation and the embankment may have a marked effect on the seepage losses through and beneath the embankment, many design factors may also have an influence [Cedergren (1989); Volpe and Kelly (1985)]. An understanding and knowledge of the position of the seepage line permit adjustments to be made in the design that will allow improved seepage control.

The seepage line is the upper line of seepage (Fig. 10.4). It is the "line above which there is no hydrostatic pressure and below which there is hydrostatic pressure." Some refer to it as the "phreatic line." It has been found that the seepage line is affected by (1) the permeability of the fill materials and of the foundation, (2) the position and flow of ground water at the site, (3) the type and

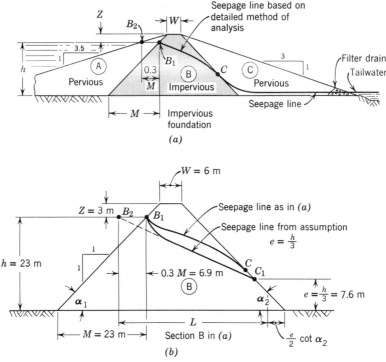

Fig. 10.4 Seepage line determination for a composite dam (*a*) by detailed method of analysis and (*b*) by approximation method. (Redrawn and revised from Creager et al., 1945.)

design of any core wall or cutoff within the embankment, and (4) the use of drainage devices to collect seepage in the downstream portion of the structure.

It is important to know at what point the seepage line intersects the downstream slope so that adequate toe drainage can be developed. If the intersection is well up the face of the dam and the rate of seepage is sufficient to move the soil, serious sloughing will result.

Where a dam is homogeneous and is located on an impervious foundation, the seepage line will cut the downstream face above the base of the dam. Its location is dependent on the cross section of the dam and is not affected by the permeability of the dam unless the permeability changes. The basic shape of the seepage line is that of a parabola varying at the ends as the intake and outflow conditions vary.

If certain assumptions are made, an approximate seepage line may be located as illustrated for the composite dam shown in Fig. 10.4*a*. The upstream and downstream portions of this dam are relatively pervious, whereas the center portion is impervious. When wide differences in the permeability of these two portions exist, it is usually sufficient to determine the position of the seepage line through the most impervious portion of the dam. The upstream shell will have little effect on the position of the line and the downstream portion will act as a drain, thus placing the line slightly above the tailwater.

In this rough solution, it is assumed that where the downstream side slopes are flatter than 1:1,

$$e = h/3 \tag{10.1}$$

where e = distance from the impervious base of the dam up to the intersection of the seepage line and the downstream face of the dam,

h = distance from the impervious base to the level of the water in the reservoir.

If Darcy's law (Chapter 3) is applied to a unit length of dam and a mean discharge area of $(h + e)/2$ is assumed, the formula for expressing the discharge through the dam is

$$q = \frac{K(h - e)}{L} \frac{(h + e)}{2} = \frac{K}{2} \frac{(h^2 - e^2)}{L} \tag{10.2}$$

where q = discharge rate per unit length of dam (L^3/T),

K = hydraulic conductivity of the material in the dam (L/T),

L = mean length of flow (L).

By trial and error or by differentiation, a value of e for which q is maximum may be computed. As indicated above, this is approximately $h/3$ for slopes flatter than $1:1$. Therefore, when $h/3$ is substituted for e in Eq. 10.2,

$$q = \frac{4Kh^2}{9L} \tag{10.3}$$

Estimates of e and the average flow area are usually more accurate than the hydraulic conductivity K, which is the most difficult to evaluate.

□ *Example 10.1*
_____

Construct the approximate seepage line and estimate the seepage rate through the dam shown in Fig. 10.4.

Solution.   Considering only the impervious section B, $e = 23/3 = 7.6$ m (25 ft) from Eq. 10.1. The intersection of the seepage line and the downstream face, point C_1, can then be plotted in Fig. 10.4b using $e = 7.6$ m (25 ft). From point B_2 on the water surface upstream from the face of the dam the seepage line is drawn to point C_1. Point B_2 on the water surface is equal to $0.3M$, where M is the horizontal projection of the wetted upstream slope. As shown in Fig. 10.4b, B_2 to $B_1 = 0.3 \times 23 = 6.9$ m (22.5 ft) $= 0.3M$. The upstream or ingress end of the seepage line may be sketched in as shown in Fig. 10.4b. The mean length of the seepage line from upstream to downstream is

$$L = 0.3M + Z \cot \alpha_1 + W + (Z + h - e/2)\cot \alpha_2$$

using symbols shown in Fig. 10.4.

By substituting the values computed above in the equation,

$$L = (0.3 \times 23) + (3 \times 1) + 6 + (3 + 23 - 7.6/2)1 = 38.1 \text{ m (125 ft)}$$

By substitution in Eq. 10.3, the discharge through the dam per unit length, assuming $K = 0.02$ m/day, is

$$q = \frac{4 \times 0.02 \times 23 \times 23}{9 \times 38.1} = 0.123 \text{ m}^3/\text{day per lineal meter of length}$$

(1.33 cfd per lineal ft of length)   □

The determination of the probable position of the seepage line in levees follows the same principles as those outlined above. Maximum flood stage is substituted for the water surface elevation. Determination of the seepage line in levees is important where sustained peaks are expected and where construction materials are particularly permeable. Levees controlling short-duration flood peaks rarely become sufficiently saturated to create a hazardous condition.

Frequently farm ponds are needed at sites where the storage area will not hold water. In such instances it is necessary to line the entire impounding area with an impervious clay blanket, as illustrated in Fig. 10.1b.

Blankets should not be applied to the upstream face of a dam where frequent, total, or rapid drawdown is anticipated because of the danger of slumping, resulting from the internal hydrostatic pressure in the saturated portion of the dam, and of cracking during drying. Uncontrolled stock watering or wading also causes disturbance of the blanket.

For earth embankments of the size covered in this text, a blanket thickness equal to 10 percent of the depth of the water above the blanket, but with a minimum of 1 m, is usually satisfactory if constructed from the same impervious materials as in the dam. Where blankets are used on small structures, they usually extend out from the toe of the dam 8 to 10 times the depth of the water.

Where the soil materials on the pond bottom approach the proportions of 70 percent sharp well-graded sand, 20 to 25 percent clay, and sufficient silt to provide good gradation of particle size, the seal can be accomplished by ripping the soil to a depth of about 0.3 m and then recompacting at optimum water content with a sheepsfoot roller until maximum density is attained. When proper soils are not available at the site, materials must be selected from adjacent borrow pits, mixed, spread, and compacted.

Support of the head of stored water is essential; therefore, consideration must be given to the necessary mantle thickness needed to prevent the water pressure from forcing the soil out through underlying rock crevices. Under most conditions, a minimum of 0.6 m of soil should exist between the compacted blanket and the rock formations. Soils with the particle size gradation outlined above have sufficient structural resistance to counteract the pressure of the water, thus preventing slippage and flow of the soil mantle into underlying rock crevices. Soils having a high clay content do not have this resistance. Therefore, to provide the needed strength, high-clay-content soil mantles must have greater thickness.

Where satisfactory construction materials are unavailable, seals may sometimes be accomplished by incorporating swelling clays, such as bentonite, with the soil. This material swells to fill the interstices between the soil particles, thereby decreasing the permeability. Bentonites alone will not resist the pressure of the stored water; therefore, they should be used only with soils containing a minimum of 10 to 15 percent sand, which will provide the strength. Rates of application must be adjusted to individual soil conditions. Similarly, polyphosphates and other chemical additives that tend to deflocculate the soil, thus reducing its porosity, must be carefully selected in relation to the soil type and its chemical characteristics.

Where special site or construction conditions dictate maximum precautions against seepage through the reservoir sides and bottom or where water costs dictate maximum water storage efficiency, consideration may be given to use of total lining techniques. Plastic and butyl films sealed into one-piece reservoir liners may be justified.

10.5   Foundation Treatment

Exposed rock foundations should be examined for joints, fractures, and permeable strata that might contribute to seepage and eventual structural damage through piping. The usual treatment is pressure grouting to a depth in the bedrock equal to the design head of water above the rock surface by techniques described in USBR (1987).

Where fine-grained impermeable material constitutes the foundation, minimum treatment should include removal of vegetative cover, topsoil having a high organic matter content, and soil that has been loosened through frost action or is of recent deposition. If no cutoff trench is necessary because of the impermeability of the foundation material, other steps should be taken to bond the impervious zone of the embankment to the foundation. This can be done with a *shallow cutoff trench* cut 0.6 to 1 m into the foundation (Fig. 10.2).

Where the foundation consists of porous material, seepage should be reduced by construction of a *cutoff trench* extending to bedrock or to an impervious soil stratum. In general, a minimum bottom width of 3 to 6 m should be maintained.

The minimum depth of a cutoff below original ground surface will depend on the position of the impermeable stratum. Sideslopes of the trench should not be less than 1:1.

Where satisfactory soil materials for cores are unavailable or where there is an excessive distance through the pervious foundation material to impeding strata, cutoff or core walls of steel sheet piles or reinforced concrete may be installed. The effectiveness of these devices is based on the care taken in installation and in keying the cutoff into the soil materials. Cement-bound curtain cutoffs and chemical grouting have been used with success where there is adequate economic justification for their use.

10.6   Drainage

In relatively impervious and homogeneous dams the seepage line frequently appears high on the downstream slope, resulting in a saturated unstable soil condition. The construction of a system of toe filters or drains will lower the seepage line until it intersects with the drain. The capacity of toe filters or drains should be at least twice the computed maximum seepage discharge through the dam. Figure 10.5 illustrates several types of effective drainage systems.

Where a saturated pervious foundation underlies an impervious top layer, consideration must be given to the possibility of piping resulting from uplift pressure that exceeds the pressure of the impervious mantle and the reservoir head. Where the overlying mantle is too thick to permit use of drainage ditches,

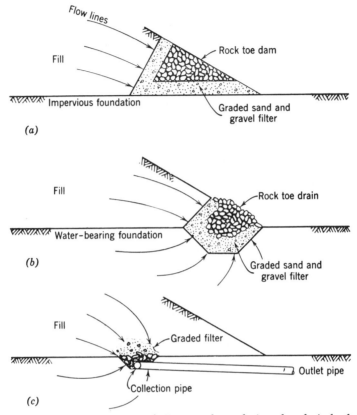

Fig. 10.5 (*a*) Simple rock-fill toe drain. (*b*) Deep rock toe designed to drain both the fill and the water-bearing foundation. (*c*) Pipe-type drain with collection pipe.

pressure-relief wells may be used to intercept the seepage and reduce the uplift pressure.

Drainage blankets and filters are designed to be (1) pervious so that seepage removal can be accomplished, (2) operational as a filter so as to prevent movement of soil particles from the foundation or embankment into the seepage discharge, and (3) of sufficient weight to overbalance the high upward seepage forces that might lead to downstream blowouts or related piping failures. (See Appendix H for filter design criteria.)

Adequate drainage is equally important in levee construction, for a saturated impounding structure is always potentially dangerous. A pipe drain or drainage ditch located near the land-side toe at a depth that will affect the lower end of the saturation line is a standard design. These drains are carried away from the levee and outletted through controlled relief wells.

10.7 Side Slopes and Berms

The sideslope of an earth dam or levee is dependent on the height of the structure, the shearing resistance of the foundation soil, and the duration of inundation.

On structures less than 15 m high with average materials, the sideslopes should be no steeper than 3:1 on the upstream face and 2:1 on the downstream face.

Coarse, uncompactable soils may need sideslopes of 3 : 1 or 4 : 1 to ensure stability. Where more precise designs are required and on dams exceeding 15 m in height, a complete soil analysis should be made to determine the horizontal shear strength of the materials.

In levee construction on major waterways having flood peaks of long duration that cause maximum saturation of the levee, sideslope ratios may range as high as 7 : 1. These flat slopes are necessary because levee materials are frequently unstable and receive no compaction.

Most conservation levee construction is limited to protecting farm lands on the upper tributaries. Since they are subjected to short-duration flood crests, minimum sideslopes of 2 : 1 may be provided if they are properly protected at critical points. Levees that protect land from back flow after pump drainage, as in certain muck areas, tidal marshes, and swamp areas, should be designed and constructed by the same criteria established for dams.

10.8 Top Width

Top widths of dams vary with the height and purpose. The minimum top width for dams up to 3.5 m in height should be 2.4 m. If the top serves as a roadway, this minimum should be increased to 3.6 m to provide a 0.6-m shoulder to prevent raveling. Road runoff should be controlled to prevent erosion of the sideslopes.

Top widths for dams exceeding 3.5 m in height may be designed by the empirical formula

$$W = 0.4H + 1 \tag{10.4}$$

where W = top width in m,
H = maximum height of embankment in m.

In levee construction top or crown width may vary from a minimum of 1 to 6 m or more. On levees subject to short-duration flood peaks, 1-m crowns are adequate. On levees subject to sustained peaks, the width usually equals twice the square root of the height of the structure.

10.9 Freeboard

The distance between the maximum designed high water or flood peak level in the reservoir and the top of the settled dam or embankment constitutes the net freeboard. Reference is sometimes made to gross freeboard, or "surcharge," which is the distance between the crest of the mechanical spillway and the top of the dam.

The net freeboard should be sufficient to prevent waves or spray from overtopping the embankment or from reaching that portion of the fill that may have been weakened by frost action. The depth of soil loosened by frequent alternate freezing and thawing is seldom over 0.15 m).

Wave height for moderate-size reservoir areas can be determined by

$$h = 0.014(D_f)^{1/2} \tag{10.5}$$

where h = height of wave from trough to crest under maximum wind velocity in m,

D_f = fetch or exposure in m.

All freeboards should be based on the water level at the maximum flood heights for which the dam and spillways are designed. For farm ponds and other detention reservoirs with small mechanical spillways, the design flood height will include a depth of flow in the spillway of not more than 0.3 m. The flood storage depth, which is the difference in elevation between the mechanical and the emergency spillway crests, can be accurately determined by flood routing procedures given in Chapter 11. For small storage reservoirs with a 200- to 250-mm-diameter mechanical spillway, this depth can be estimated from Fig. 10.7. The following example, as illustrated in Fig. 10.6, shows the procedure for estimating freeboard.

☐ Example 10.2

Determine the net and gross freeboard for a farm pond with a 0.6-ha (1.5-ac) water surface and an exposure length of 180 m (600 ft). Assume the frost depth is 0.15 m (0.5 ft), the 25-year design runoff peak is 4.00 m³/s (141 cfs), and the flow depth in the flood spillway is 0.3 m (1.0 ft).

Solution. From Eq. 10.5, the wave height

$$h = 0.014 \ (180)^{1/2} = 0.19 \text{ m } (0.6 \text{ ft})$$

From Fig. 10.7 read the flood storage depth of 0.6 m (2.0 ft).

$$\text{Net freeboard} = 0.15 + 0.19 = 0.34 \text{ m } (1.1 \text{ ft})$$
$$\text{Gross freeboard} = 0.34 + 0.3 + 0.6 = 1.24 \text{ m } (4.1 \text{ ft}) \qquad ☐$$

Additional freeboard should be added as a safety factor where lives and high-value property would be endangered by a dam failure. On large dams the net freeboard is 50 percent more than the wave height and is increased for high winds. Levee freeboard is designed on the same basis as dams.

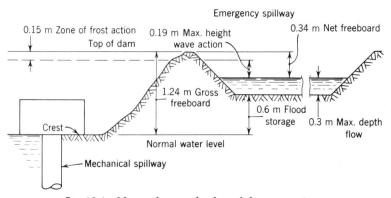

Fig. 10.6 Net and gross freeboard for reservoirs.

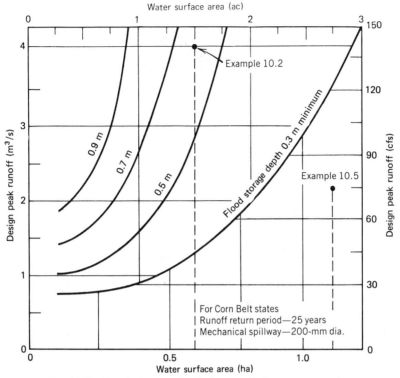

Fig. 10.7 Flood storage depth for small storage reservoirs.

10.10 Compaction and Settlement

The volume relationships of soil may be expressed by the formula

$$V = V_s + V_e \tag{10.6}$$

where V = total in-place volume (L³),
V_s = volume of solid particles (L³),
V_e = volume of voids, either air or water (L³).

The void ratio is expressed by

$$e = V_e/V_s \tag{10.7}$$

The soil density is defined as the oven dry mass per unit volume of soil in place.

There is an optimum water content at which maximum density occurs for a given amount of energy applied during the compaction process. A standard test developed for disturbed soils is known as the Proctor density test. A typical Proctor density curve is shown in Fig. 10.8. In conducting a test the soil is placed in a container and compacted in layers with a weight dropped from a certain height for a definite number of times. The water content after each compaction is determined, and then water is added for the next test. After several of these tests a curve can be drawn as shown in Fig. 10.8. The density increases with water content up to a certain point, above which the density decreases. The maximum or

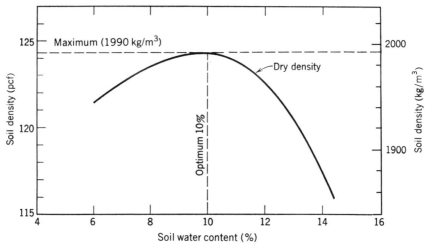

Fig. 10.8   A typical Proctor soil density curve.

Proctor density occurs at the optimum water content for compaction. This test is widely used in engineering construction.

A compacted earth fill is normally more dense than soil from the borrow area. If the volume of the borrow pit is desired, the above formulas apply as illustrated in the following example.

□ Example 10.3

In constructing an earth dam, 1200 m³ (1600 yd³) of soil is required in the settled fill. If the desired void ratio of the dam is 0.6, what is the volume required from the borrow pit provided its void ratio is 1.1?

Solution.   The volume of solids is the same in the fill as in the borrow pit, but the volume of voids will decrease in the fill. From Eq. 10.7, $V_e = 0.6V_s$ for the fill. Substituting in Eq. 10.6,

$$V = 0.6V_s + V_s = 1.6V_s = 1200 \text{ m}^3 \text{ (1600 yd}^3)$$
$$V_s = 750 \text{ m}^3 \text{ (1000 yd}^3)$$

For the borrow pit $V_e = 1.1V_s$. Substituting in Eq. 10.6,

$$V = 1.1V_s + V_s = 2.1V_s = 2.1 \times 750 = 1575 \text{ m}^3 \text{ (2100 yd}^3) \qquad \square$$

Rolled-fill embankments that have been placed in thin layers and compacted at optimum water content and that are on an unyielding foundation will not settle more than about 1 percent of the total fill height. On deep plastic foundations settlement may reach 6 percent. Fills for small farm ponds and reservoirs usually do not receive as thorough compaction as do the larger structures; therefore, the fill is usually constructed to a height 5 to 10 percent higher than the designed settled fill height for all points along the embankment.

In levee construction a minimum allowance for settlement of 20 to 25 percent is usually added because dragline or conveyor placement without intensive compaction does not give a high degree of consolidation.

10.11   Wave Protection

If the water side of dams or levees is exposed to considerable wave action or current, it should be protected either by a covering of nonerodible material called riprap or by energy dissipaters such as anchored floating logs. A dense sod of a suitable grass will usually give slope protection on levees and small ponds that are not subjected to excessive wave action.

Riprap may consist of hand-placed or dumped stone or various types of concrete block or slab construction. It should be carefully installed on a bedding or cushion layer of graded gravel or crushed stone 0.2 to 0.5 m deep, depending on quality. This layer prevents the clay and silt from being washed out by wave action, thus causing the riprap stone to settle into the embankment.

10.12   Stability of Earth Embankments

The factors previously discussed in this chapter have a direct bearing on the stability and safety of low-height embankments, but if these embankments are properly constructed and maintained, they should be permanently safe.

10.13   Construction

Construction according to plan and adjustment to soil conditions while the work is in progress are essential for rolled-fill dam success. All trees, stumps, and major roots should be removed from the site. Sod and topsoil should be removed and stockpiled for later spreading over the dam and exposed areas to establish good vegetation. If necessary, the fill should be wetted to the optimum water content for compaction. Proper moisture conditions make it possible to force some of the fill material into the original surface, thus eliminating any dividing plane between the fill and the foundation.

After the site has been cleared, the cutoff trench is excavated. The compacted fill should be constructed only of carefully selected materials placed in thin blankets at optimum water content. The thickness of layers for pervious soils should be limited to 0.25 m; the more plastic and cohesive soils should not exceed 0.15 m in thickness. On the dam, the fill material should be placed nearly horizontal, with a slope of 20:1 to 40:1 away from the center of the dam. If, on the approach of rain, the surface is left in a fairly smooth condition, the resultant surface drainage will keep the dam from becoming saturated.

The degree of compaction to be achieved is specified as a ratio, in percent, of the embankment density to a specified standard density for the soil. In earth dam construction, the degree of compaction should run 85 to 100 percent of the maximum Proctor density.

Soils with a high clay content should be carefully compacted, with the soil slightly drier than the lower plastic limit to prevent the formation of shear planes called *slickensides*. Pervious materials, such as sands and gravels, consolidate under the natural loading of the embankment; however, additional compaction does aid in increasing shear strength, and in limiting embankment settlement. Compaction of these noncohesive materials is best achieved when they are nearing saturation.

Selection of proper compaction equipment is important. The sheepsfoot roller is best suited for compacting fills. Its weight may be varied by adding water to the drum, and the compaction per unit area may be adjusted by varying the numbers and sizes of tamping feet.

Special precautions should be taken when compacting materials close to core walls, collars, pipes, conduits, and so on. All such mechanical structures should be constructed so that they are wider at the bottom than at the top; thus settlement of the soil will create a tighter contact between the two materials. Thin layers of soil, at water contents equal to the remainder of the fill, should be tamped into place next to all structures. This can best be done with hand-operated pneumatic or motor tampers. Heavy hauling equipment should be given varied routes over a fill to prevent overcompaction along the travel ways.

10.14 Protection and Maintenance

The entire reservoir area should be fenced to prevent damage to embankments, spillways, and banks. Where this is not practicable, the spillways and dam at least should be protected.

To minimize damage from sedimentation the entire watershed should be protected by adequate erosion control practices. Buffer or filter strips of dense close-growing vegetation should be maintained around the pond edge.

A properly designed and constructed earth dam, well sodded and protected, should require a minimum amount of maintenance; however, a regular inspection and maintenance program should be established. Particular attention should be given to surface erosion, the development of seepage areas on the downstream face or below the toe of the dam, the development of sand boils and other evidence of piping, evidence of wave action, and damage by animals or human beings. Early recognition and repair of such conditions will prevent development of dangerous conditions that will increase the cost of repair. Under no conditions should trees be permitted to grow on or near the embankment.

SPILLWAYS

Most farm ponds and reservoirs are protected from overtopping by vegetated flood or emergency spillways. This vegetation will not survive if base flow entering the pond and frequent minor flood-flows keep the spillways unduly wet. Therefore, to protect these spillways and to carry low flows, various types of mechanical spillways are provided.

10.15 Mechanical Spillways

Mechanical spillways (Chapter 9) have their entrance openings set at the elevation at which the water surface is to be maintained and they extend through the dam to a safe outlet point below the dam. The entrance elevation depends on the level of permanent storage desired.

Any durable material may be used for the mechanical spillway, provided it has a long life expectancy and is not subject to damage by settling, loads, or impacts. The portion passing through the dam should be placed on a firm, impermeable foundation, preferably on undisturbed soil, and should be located to the side of the original channel. When rigid conduits, such as concrete, cast iron, or clay tile, are placed on rock foundations, they must be cradled in concrete. Flexible conduits, such as plastic and corrugated metal pipe, may be cradled in compacted impermeable soil material.

Where conduits are laid on foundations or are passing through fills that are subject to more than nominal settlement, consideration should be given to laying the conduit with a camber to compensate for this settlement.

All structures passing through the dam should be covered by thin, well-compacted layers of high-grade material. Where the fill surrounding the pipe is of low quality, concrete or metal seepage collars, extending out a minimum of 0.6 m in all directions, should be used. The number of collars should be sufficient to increase the length of the creep distance at least 10 percent. The creep distance is the length along the pipe within the dam, measured from the inlet to the filter drain or to the point of exit on the backslope of the dam. The increase in creep distance for an antiseep collar is measured from the pipe out to the edge of the collar (Fig. 10.9a). For example, the increase in creep distance for a 1 × 1-m collar on a 300-mm-diameter pipe would be twice the distance out, $2(0.5 - 0.15) = 0.7$ m. For small-diameter water pipe through the dam two 0.6×0.6-m collars are adequate because of better compaction around the small pipe and the long creep distance. Where concrete or mortar joints are used, it is advisable to complete the backfilling and tamping before it has set, thus avoiding the danger of cracking the concrete. A minimum of 1.2 m of protective fill should be carefully placed over the structure before heavy earth-moving equipment is allowed to pass over it.

The mechanical spillway may, in some instances, serve as a drain for cleaning, repairing, and restocking the pond. One of several types of installations is shown

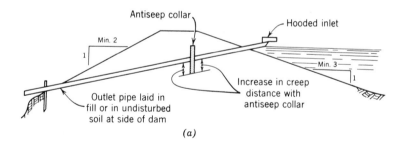

(a)

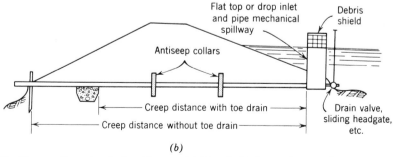

(b)

Fig. 10.9 Types of mechanical spillways. (a) Hooded inlet and pipe. (b) Vertical riser and drain pipe at the upper end of the outlet pipe.

Table 10.1 Mechanical Spillway Pipe Diameters for Farm Ponds[a]

Peak Runoff Rate for 25-year Return Period (m^3/s)	Water Surface Area at Normal Water Level (ha)				
	0.2	0.4	0.8	1.2	2.0
	Diameter (mm)				
≤0.5	200	150	150	—	—
0.5–1.0	200	200	150	150	—
1.0–1.5	250	250	200	200	150
1.5–2.0	250	250	250	250	200
2.0–2.5	—	300	300	300	250

[a]Suggested sizes of corrugated metal pipe primarily for ground water or seepage flow. For smooth wall pipes, use the next smaller diameter.

in Fig. 10.9*b*. All valve assemblies should be equipped to operate from above the surface of the water. All mechanical inlets should be protected with a screening device that prevents floating objects, turtles, and other foreign objects from entering and clogging the pipe. The device should be so designed that, after becoming loaded with debris, it will be self-cleaning. Outlets should be carried well beyond the toe of the dam and protected against scouring (see Chapter 9).

Mechanical spillways may be designed to carry a small percentage of the peak flow rate. For small watersheds and for flood spillways lined with concrete or other stable material, a pipe mechanical spillway may not be required. Where considerable flood storage volume is available, flood routing procedures should be followed to determine the pipe size. The mechanical spillway should always be large enough to carry the seepage or base flow so that the flood spillway will be dry and stable. For farm ponds with small watersheds, approximate pipe sizes are given in Table 10.1.

10.16 Flood or Emergency Spillways

All reservoirs must be equipped with an emergency spillway that will safely bypass floods that exceed the temporary storage capacity of the reservoir. Where a natural spillway, such as a depression in the rim of the impounding area, does not exist, it is necessary to construct a trapezoidal channel in undisturbed earth. As shown in Fig. 10.10, the flood spillway consists of an approach channel, a level control section, and an exit (grassed) section.

After careful estimation of the peak rates and amounts of runoff to be handled, application of the weir formula and vegetated waterway design procedures will determine the cross-sectional area required (see Chapters 4, 7, and 9). The following example illustrates the procedure for small structures.

☐ *Example 10.4*

The design capacity of a flood spillway for a dam is 3.4 m³/s (120 cfs). The exit section was designed as a grassed waterway with a bottom width of 12 m (39 ft). Assuming a maximum depth of flow of 0.3 m (1 ft), what width is required for the control section so that it does not restrict the flow, that is, increase the design level in the reservoir?

Solution.   From Chapter 9 the weir formula with $C = 3.2$ will give a satisfactory estimate for the width of the control section:

$$L = 3.4/(0.55 \times 3.2 \times 0.3^{3/2}) = 11.7 \text{ m (38 ft)}$$

The grassed approach channel should be flared out on each side at least 0.2 m per meter of length as measured along the center line of the channel upstream from the control section (Fig. 10.10).   □

Example 10.4 assumes that the flow passes through critical depth at the downstream end of the control section as shown in Fig. 10.10. For this condition to occur the slope in the exit section must be greater than the critical slope, otherwise the exit section could control the flow. The slope in the approach channel should not be less than 2 percent upstream to provide drainage and reduce inlet losses. The exit section should be adequate to carry the design flow, but at velocities not greater than those recommended for grassed waterways given in Chapter 7. Outlets should extend well beyond the dam or into an adjacent waterway and should be well protected against scouring.

The design capacity for flood spillways on small reservoirs and ponds is normally based on a return period of 25 years from the contributing watershed. For larger structures where high-value property may be damaged or where lives are endangered, a 100-year return period storm or greater may be considered. Flood

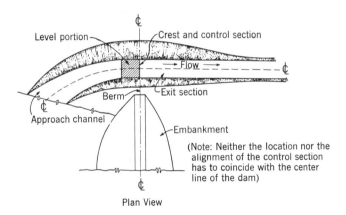

Plan View

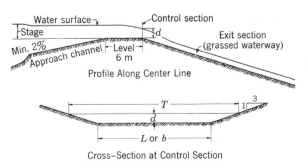

Profile Along Center Line

Cross-Section at Control Section

Fig. 10.10   Excavated flood spillway for reservoirs. (Redrawn from SCS, 1979.)

spillway design flow may be reduced if the capacity of the mechanical spillway is large compared with the flood flow.

FARM PONDS

10.17 Types of Ponds and Reservoirs

The three types of reservoirs in common use are (1) dugout ponds fed by ground water, (2) on-stream ponds fed by continuous or intermittent flow of surface runoff, streams, or springs, and (3) off-stream ponds.

Dugout Ponds. Dugout ponds are limited to areas having slopes of less than 4 percent and a prevailing reliable water table within 1 m of the ground surface. Design is based on the storage capacity required, depth to the water table, and stability of the sideslope materials.

On-Stream Ponds. The on-stream type of reservoir depends on the runoff of surface water for replenishment. The designed storage capacity must be based both on use requirements and on the probability of a reliable supply of runoff. Where heavy usage is expected, the design capacity of the pond must be adequate to supply several years' needs to ensure time for recharge in the event of a sequence of one or more years of low runoff. Spring- or stream-fed ponds consist of either an excavated basin below a spring or a reservoir formed by a dam across a stream valley or depression below a spring. The dam may be placed across a depression rather than a definite stream to catch diffuse surface water. A spring-fed pond should be designed to maintain the water surface below the spring outlet, which eliminates the hazard of diverting the spring flow caused by the increased head from the pond. When the spring flow is adequate to meet use requirements, surface waters should be diverted from the pond to reduce sedimentation and to reduce spillway requirements.

Off-Stream Storage Pond. The off-stream or bypass pond is constructed adjacent to a continuously flowing stream, and an intake, through either a pipe or an open channel, diverts water from the stream into the pond. Controls on the intake permit reduction in sedimentation, particularly if all flood water can be diverted from the pond. Water may be supplied by a pump. The pond should be protected against stream overflow damage.

10.18 Requirements for Ponds and Permanent Storage Reservoirs

In addition to the requirements listed in Section 10.2 these structures must be placed so as to provide an adequate and reliable supply of surface or ground water, free from pollutants. The site must also be convenient for use of the water.

10.19 Site Selection

Dams for livestock water, irrigation, or fire protection must be located where the impounded water may be used most effectively with a minimum of pumping and piping. Topographic features must be carefully studied to eliminate the need for

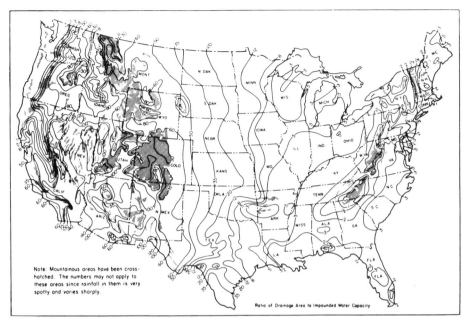

Fig. 10.11   Approximate drainage areas in acres required to yield 1 ac-ft of water for farm ponds. (Redrawn from SCS, 1979.) 1 ac = 0.40 ha; 1 ac-ft = 0.1233 ha-m.

excessively large structures. As previously discussed, suitable soil for the embankment is important, as is choice of a site requiring a minimum of fill. Channel slopes above the fill normally range from 3 to 6 percent to provide adequate depth and water surface area.

The site for water storage structures is also dependent on a contributing watershed capable of supplying the necessary runoff. The probable rates and volumes of runoff should be determined by the methods outlined in Chapter 4. Approximate volumes of runoff may be determined from Fig. 10.11. These are general values that should be checked with local hydrologic data.

10.20   Water-Storage Requirements

The storage capacity of a pond will depend on water needs, evaporation from the water surface, seepage into the soil or through the dam, storage allowed for sedimentation, and amount of carry-over from one year to the next. Two to four meters of depth are needed for fish to survive where the water surface freezes. Water needs for domestic uses, livestock, spraying, irrigation, and fire protection may be estimated from Table 10.2. Evaporation from a pond-water surface can be estimated by multiplying local pan evaporation values published by the Weather Bureau by a factor of about 0.7. Evaporation can be reduced by selecting a site having a small surface area and deep depth. Seepage losses depend on soil properties and construction techniques. For this reason they are difficult to predict. Studies have shown that ponds with large ratios of storage volume to watershed area provide a reliable supply even in dry years. Thus, within reasonable limits, maximum capacity for a given site is desirable. Minimum storage is usually com-

Table 10.2 Water Requirements for Farm Uses

	Average use	
Type of Use	*L/day*	*ha-m/year*
Household, all purposes, per person	190–380	0.01
Fire protection		0.03
Dry cow or steer, per 450 kg weight	75–115	0.003
Milk cow, 450 kg including milkhouse and barn sanitation	130–170	0.006
Horse or mule, 450 kg	45	0.0017
Turkeys per 100 head	60	0.002
Chickens per 100 head	35	0.001
Swine per 45 kg	10–20	0.0006
Sheep per 45 kg	8	0.0003
Orchard spraying, per year of tree age per application	4	
Irrigation (humid regions) per ha		0.3–0.45 m
Irrigation (arid regions) per ha		0.3–1.5 m

Source: Adapted from Midwest Plan Service (1987).

puted by estimating the total annual needs and allowing 40 to 60 percent of the total storage for seepage, evaporation, and other nonusable requirements.

10.21 Farm Pond Design

The following example illustrates the steps and procedure for designing a small farm pond (SCS, 1982).

☐ *Example 10.5*

Design a pond in northeastern Missouri to provide water for 140 steers, for irrigation of a 0.2-ha (0.5-ac) garden, and for a family of four. A suitable site for the dam has a drainage area of 27 ha (67 ac), and the estimated design peak runoff for a 25-year return period storm is 2.12 m³/s (75 cfs). Seepage and evaporation losses from the pond are estimated as 57 percent of the storage capacity.

Solution. From Table 10.2 the annual water requirements are

$$
\begin{array}{ll}
\text{140 steers } (140 \times 0.003) & 0.42 \text{ ha-m} \\
\text{Irrigation } (0.2 \times 0.45) & 0.09 \\
\text{Household use } (4 \times 0.01) & \underline{0.04} \\
\qquad\qquad\qquad \text{Total} & 0.55 \text{ ha-m (4.46 ac-ft)}
\end{array}
$$

The storage requirements allowing 57 percent losses are

$$0.55 \times 100/(100 - 57) = 1.28 \text{ ha-m (10.4 ac-ft)}$$

From Fig. 10.11, read 6 ac of drainage area for each acre-foot of storage. Required watershed size is $10.4 \times 6/2.47 = 25.3$ ha (62.4 ac), which is adequate.

From a contour map of the reservoir area and a field investigation of the soils, a dam site was selected as shown in Fig. 10.12. By measuring the area within each

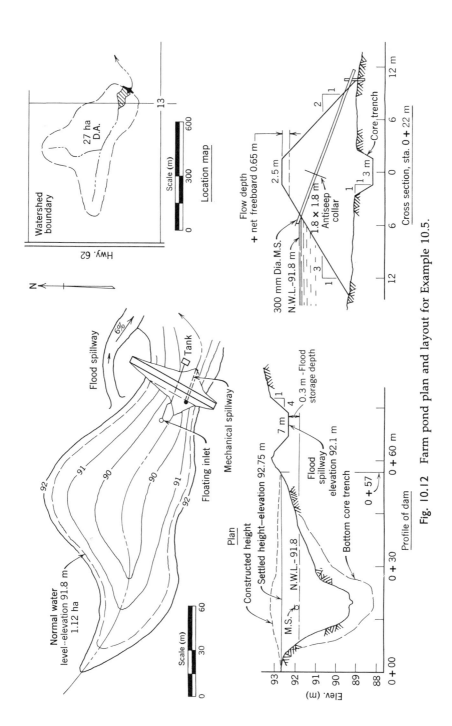

Fig. 10.12 Farm pond plan and layout for Example 10.5.

Table 10.3 Volume of Pond-Water Storage[a]

Contour Elevation (m)	Area Within Contour Line to Center Line of Dam (ha)	Average Area (ha)	Contour Interval (m)	Volume of Storage (ha-m)
89.5	0.05			
		0.15	0.5	0.08
90	0.25			
		0.45	1.0	0.45
91	0.64			
		0.94	1.0	0.94
92	1.24		Total	1.47 (11.9 ac-ft)

To elevation 91.8, storage is $1.47 - (0.94 \times 0.2/1) = 1.28$ ha-m and
water surface area $= 1.24 - 0.2 (1.24 - 0.64) = 1.12$ ha (2.8 ac).

[a]Computed for the pond in Fig. 10.12 and Example 10.5. The number of contour lines is reduced to simplify computations.

contour line, the water-level height was determined (Table 10.3). The area was measured to the center line of the dam. This procedure is sufficiently accurate because most of the soil in the dam is usually taken from the reservoir area. (An approximate volume may be obtained by multiplying the water-surface area by 0.4 times the maximum water depth.) Storage at this site could be increased by raising the water level or by moving the dam farther downstream. If by so doing, sufficient storage could not be obtained, another site would have to be selected.

By interpolation between contours at elevations of 91 and 92 m, 1.28 ha-m of storage is available at an elevation of 91.8 m, which is the normal water surface (Table 10.3). The following specifications are determined:

- Crest elevation of mechanical spillway: 91.8 m
- Flood storage depth from Fig. 10.7 for 1.12-ha surface area and 2.12-m³/s peak runoff rate (minimum):   0.3 m
- Flood spillway elevation: 92.1 m

   From Fig. 10.12 measure the water surface fetch, 200 m.
- Substituting in Eq. 10.5, $h = 0.014 (200)^{1/2} = 0.2$-m wave height.
- If a frost depth of 0.15 m is assumed, the net freeboard $= 0.2 + 0.15 =$ 0.35
- Flow depth in flood spillway (assumed):   0.3
- Elevation of top of dam (settled height):   92.75 m
- Allowance for settlement at sta. 0 + 22:

$$10\% \times (92.75 - 89.00) = 0.38 \text{ m (1.2 ft)}$$

- Top width of dam from Eq. 10.4:

$$W = 0.4 (92.75 - 89.00) + 1 = 2.5 \text{ m (8.2 ft)}$$

- Select side slopes of 3:1 upstream and 2:1 downstream.

Table 10.4 Volume of an Earth Dam[a]

Station along Center Line	Height of Dam (m)	Cross-Sectional Area[b] (m²)	Average Cross-Sectional Area (m²)	Length of Section (m)	Volume of Section (m³)
0 + 00 m	0	0			
0 + 08	0.9	4.28	2.14	8	17.12
0 + 15	3.6	41.40	22.84	7	159.88
0 + 22	3.8	45.60	43.50	7	304.50
0 + 27	1.7	11.48	28.54	5	142.70
0 + 37	1.0	5.00	8.24	10	82.40
0 + 57	0	0	2.50	20	50.00
				Total fill:	756.60

Core trench volume, average depth 1 m, average width 4 m, and length 40 m:

	160.00
Total fill and core trench:	916.60 m³
	(1199 yd³)

[a]Computed for dam in Fig. 10.12 and Example 10.5.
[b]Cross-sectional area = $(3h + 2h)(h/2) + 2.5h$, where h is the dam height.

- From Table 10.1, read 300-mm mechanical spillway diameter for 2.12 m³/s and 1.12-ha water surface.

- Antiseep collar for estimated 15-m pipe length and 10 percent increase in creep distance requires one 1.8 × 1.8-m collar or two 1 × 1-m collars.

- For livestock water pipe, select 32-mm (1¼-in.)-diameter steel pipe approximately 30 m (100 ft) in length.

- Volume of fill in dam and in core trench is computed in Table 10.4 from fill height measured from profile in Fig. 10.12.

- Flood spillway width at control section (Eq. 9.6):

$$L = 2.12/(0.55 \times 3.2 \times 0.3^{3/2}) = 7.3 \text{ m (24 ft)}$$

- Flood spillway below dam with 6 percent slope, maximum velocity of 1.5 m/s (5 fps), and 4:1 side slopes for trapezoidal vegetated waterway will require a bottom width of about 10.7 m (35 ft) (see Chapter 7). □

REFERENCES

Cedergren, H. R. (1989). *Seepage, Drainage, and Flow Nets*, 3rd ed. Wiley, New York.

Creager, W. P., J. D. Justin, and J. Hinds (1945). *Engineering for Dams*. Vols. I–III. Wiley, New York.

Jansen, R. B. (1988). *Advanced Dam Engineering for Design, Construction, and Rehabilitation*. Van Nostrand Reinhold, New York.

Midwest Plan Service (1987). *Private Water Systems*. MPS Iowa State University, Ames, IA.

Spangler, M. G., and R. L. Handy (1981). *Soil Engineering*, 4th ed. Harper & Row, New York.

Terzaghi, K., and R. B. Peck (1967). *Soil Mechanics in Engineering Practice*, 2nd ed. Wiley, New York.

U.S. Bureau of Reclamation (USBR) (1987). *Design of Small Dams*, 3rd ed. GPO, Washington, DC.

U.S. Soil Conservation Service (SCS) (1979). *Engineering Field Manual for Conservation Practices*, (litho.). Washington, DC.

——— (1982). *Ponds Planning, Design, Construction*. Agr. Handbook No. 590. SCS, Washington, DC.

Volpe, R. L., and W. E. Kelly (eds.) (1985). *Seepage and Leakage from Dams and Impoundments*. Proceedings of ASCE Symposium, Denver, CO. ASCE, New York.

PROBLEMS

10.1 Determine the distance to set the slope stakes from the center line of a dam 9.0 m (30 ft) high with sideslopes of 3 : 1 upstream and 2 : 1 downstream. The top width of the dam is 4.0 m (13 ft) and the ground slope perpendicular to the center line is level. Compute these distances if the ground slope is 15 percent.

10.2 If the critical frost depth is 0.15 m (0.5 ft), maximum exposure of the water surface is 400 m (1300 ft), depth of flow in the flood spillway is 0.3 m (1 ft), and flood storage depth is 1.2 m (4 ft), determine the net and the gross freeboard.

10.3 Determine the gross freeboard and top width for a farm pond. The elevation of the mechanical spillway crest is 24.0 m (79 ft), and the elevation of the base of the dam is 19.0 m (62 ft). Net freeboard is 0.46 m (1.5 ft) and elevation of the flood spillway crest is 25.0 m (82 ft). Flow depth in the flood spillway is 0.15 m (0.5 ft). Allowing 5 percent for settlement, what should be the elevation of the top of the dam before and after settlement?

10.4 Compute the volume of fill per unit length for an earth dam 9.0 m (30 ft) high, assuming a top width of 2.4 m (8 ft), side slopes of 3 : 1 upstream and 2 : 1 downstream, and a level base perpendicular to the center line of the dam.

10.5 A perforated drain pipe to control the seepage line in a dam has openings 14 mm (9/16 in.) in diameter. Using the criteria given in Appendix H and the following particle sizes obtained from distribution curves of available filter materials, will the filter meet these criteria? Show why.

Layer	D_{15} Size (mm)	D_{85} Size (mm)
Soil in dam	0.004	0.07
Graded sand	0.08	1.2
Graded gravel	3.0	36.0

Note: All size distribution curves are nearly parallel.

10.6 An earth dam is to be constructed having a volume after settlement of 1150 m³ (1500 yd³). The desired wet Proctor density of the dam is to be 1920 kg/m³ (120 pcf) with an optimum water content of 20 percent. If the void ratio of the dam is 0.75, how many cubic meters (cubic yards) of borrowed soil with a void ratio of 1.1 is required? What is the dry mass of the soil in the dam in Mg (tons)?

10.7 A farm pond is to provide 0.42 ha-m (3.4 ac-ft) of water for 100 milk cows; seepage and evaporation losses are 45 percent of the stored volume; drainage area is 12.0 ha (30 ac), settlement allowance for fill is 5 percent, and flood spillway flow depth is 0.3 m (1.0 ft). Determine the elevation of (1) normal water level, (2) flood spillway, and (3) the top of the dam after settlement and (4) before settlement at the highest point where the base of the dam is at elevation 24.0 m (79 ft). Assume flood peak flow is 1.5 m³/s (53 cfs) and water surface area is 0.45 ha (1.1 ac). Frost depth is 0.15 m. Data are shown below.

Contour Elevation		Length of Water Surface		Accumulated Storage	
m	*ft*	*m*	*ft*	*ha-m*	*ac-ft*
24.4	80	0	0	0	0
25.6	84	30	100	0.025	0.2
26.8	88	91	300	0.123	1.0
28.0	92	183	600	0.395	3.2
29.2	96	244	800	0.888	7.2

10.8 Immediately after construction of a dam, the embankment had a void ratio of 0.40. What is the settlement of the dam as a percentage of the height before consolidation if the void ratio after consolidation is 0.30?

10.9 Compute the seepage rate per unit length through a homogeneous earth dam resting on an impervious foundation if the top width is 4.0 m (13 ft), the height of the dam 13 m (43 ft), the upstream sideslope 3:1, the downstream sideslope 2:1, the hydraulic conductivity 0.00012 m/h (0.0004 fph), and the net freeboard 0.9 m (3 ft).

10.10 Assuming a storage loss of 50 percent by seepage and evaporation, determine the storage capacity of a farm pond to supply water for 100 steers, 50 milk cows, 50 hogs, 1000 chickens, and irrigation of a 0.8-ha (2-ac) garden.

CHAPTER 11

Headwater Flood Control

Headwater flood control includes all measures that will reduce flood flows in watersheds of small rivers and their tributaries. The maximum size of a watershed for a headwater area is arbitrarily selected as 2500 km². Floods may be classified as either small- or large-area floods. In this text a headwater flood is considered to be a small-area flood. All flood control activity downstream from headwater areas constitutes downstream engineering, and includes levees, large dams, etc.

Headwater flood control measures are considerably different than downstream measures. In headwaters the average area rainfall intensity is higher and the variation in those factors that affect runoff is much more extreme than in larger watersheds. The effects of crops, soils, tillage practices, and conservation measures are more important since the surface condition of the entire headwater area can often change completely from season to season and from year to year. Headwater floods are typically flash floods of short duration that occur rather frequently, often associated with severe thunderstorm activity. The effectiveness of headwater control measures decreases rapidly with distance downstream. Since downstream floods are more spectacular and damages more evident, the headwater flood is too often neglected.

Flood control is a relative term as it is not economical to provide protection for the largest flood that will occur. Since civilization began, floods have been one of the best recorded of all natural phenomena. After describing a series of historical floods, Hoyt and Langbein (1955) concluded the prologue to their book with the following concept of flood control that is too often overlooked:

> Somewhere in the United States, year 2000 plus or minus. Nature takes its inexorable toll. Thousand-year flood causes untold damage and staggering loss of life. Engineers and meteorologists believe that present storm and flood resulted from a combination of meteorologic and hydrologic conditions such as may occur only once in a millennium. Reservoirs, levees, and other control works which have proved effective for a century, and are still effective up to their design capacity, are unable to cope with enormous volumes of water

involved. This catastrophe brings home the lesson that protection from floods is only a relative matter, and that eventually nature demands its toll from those who occupy flood plains.

Although the federal government develops and supervises most flood control programs through such agencies as the Corps of Engineers and the U.S. Soil Conservation Service and Forest Service, there are exceptions where local groups have organized and financed flood control programs. Outstanding among these is the Miami Conservancy District in Ohio, which is described by Morgan (1951). It was completed in 1923 at a cost of about $31 000 000. In general, state governments have not financed or individually administered flood control programs.

11.1   Economic Aspects of Flood Control

Flood control is an economic problem in which the protection of life and property, as well as the public benefits, must be evaluated in balancing the annual savings from flood control against construction and maintenance costs. According to the U.S. Federal Emergency Management Agency (from a 1990 letter containing biennial report data), more than 23 000 000 people in the United States resided and/or worked on land subject to occasional overflow.

Damage.   Flood damage may be classified as (1) *direct* losses to property, crops, and land that can be determined in monetary values; (2) *indirect* losses, such as depreciated property, traffic delays, and loss of income; and (3) *intangible* losses not subject to monetary evaluation, including community insecurity, health hazards, and loss of life. Since the distinction between direct and indirect losses is one of degree, there has been a lack of uniformity in evaluating damages. Indirect losses are difficult to determine and are often estimated as a percentage of the direct losses. Although the damage to land frequently goes unnoticed, soil erosion from the uplands; sedimentation in reservoirs, stream channels, and flood plains; and pollution of water supplies greatly affect the economy of the entire watershed.

The flood losses from severe floods in the United States are shown in Fig. 11.1. These losses per unit watershed area are high for small watersheds and, in general, decrease with an increase in drainage area. Damage from small areas often goes unnoticed and does not receive the publicity that goes with the more spectacular large flood.

Annual flood losses for urban and agricultural land are shown in Table 11.1. Agricultural losses are much greater in headwater areas than downstream, whereas the reverse is true in urban areas. Without wise land-use control in floodplains, the estimated national flood damage is expected to increase from $3.8 to $6 billion by the year 2000. Goddard (1976) reported that 16 percent of urban lands were in floodplains (53 percent were developed) by 1973, but 7 percent were subject to the 100-year flood.

Flood water causes damage by inundation and by high velocities. Though in some instances sediment deposits may be beneficial to farmlands, more frequently deposits of fine soil particles and sand have a damaging effect. Bridges, buildings, roads, farmlands, and stream channels are often destroyed by flood waters having high velocities. Some of these damages may be classed as nonrecurrent, depending on the nature of replacements and repairs. For example, a bridge replaced well above the high-water level will probably not be in danger of subsequent damage;

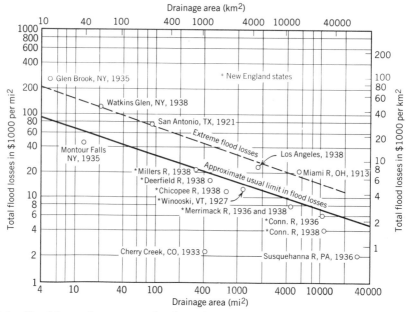

Fig. 11.1 Flood losses from severe floods in the United States. (Data from Barrows, 1948.)

however, most damage, such as that from inundation and damage to land, is recurrent in nature.

Flood damage cannot always be prevented, but may be reduced by flood forecasting and by use of proper flood control measures. On large rivers flood warnings can be issued several days in advance of the flood. In this connection the U.S. Corps of Engineers employs large-scale models of the Mississippi River for predicting downstream river stages. Longer-range forecasts can be made by using hydrologic data. About 15 percent of the annual flood loss could be prevented by accurate forecasting. Flood insurance is subsidized by the federal government under the National Flood Insurance Program. It is administered by the U.S. Department of Housing and Urban Development for approved areas that meet suitable elevation and flood-proofing criteria. By 1990 more than 17 000 communities had qualified for the program. Insurance will reduce only the risk of economic loss, not the damage itself.

Benefits. In general, benefits from flood control are based on reduction of losses. For example, when a reservoir reduces all floods by a given depth, the benefits are determined from the difference in damage at the two stages and by the

Table 11.1 U.S. Annual Flood Losses in Millions of Dollars

Land Use	Headwater	Downstream	Total
Urban and built-up	$ 330	$ 990	$1320
Agricultural	1130	682	1812
Other	340	378	718
Total	$1800	$2050	$3850

Source: Goddard (1976).

number of floods during a specified period. Aside from benefits to individuals and industries, there are public benefits, such as economic stability and better social conditions.

11.2 Downstream Flood Control

Flood control activities on all navigable streams and on many tributaries that are not navigable would be classified as downstream. The U.S. Corps of Engineers is largely responsible for downstream measures.

Control on Lower Reaches. Levees, channel improvement, and diversion floodways are the principal methods employed to reduce damage from floods. On the Mississippi River levees were the only means of control for many years. As levees were built farther upstream, valley storage was reduced and flood heights were increased, resulting in a continuous process of raising and straightening of levees. In more recent years it has been necessary to include such procedures as stabilization and protection of channel banks (gabions, riprap, and stable lining materials), straightening of the channel by cutoffs, construction of floodways, and dredging operations to maintain a navigable stream and to provide flood protection.

Control on Upper Reaches. On watersheds of about 13 000 km² or less, reservoirs may provide adequate flood protection. As in the lower reaches, levees for local protection, channel improvement, channel straightening, and stream bank protection are applicable flood control measures. Such methods may be more economical than a system of reservoirs, but in general, levees speed the flow and accentuate the problem in unimproved areas downstream.

Reservoirs for flood control may be classified as natural or artificial. Lakes, swamps, and other low areas on the land surface have a tendency to reduce flood heights by providing water storage. Even a lake that is full at the beginning of a storm will have a regulating effect on stream flow. The extent of this effect depends on the ratio of the surface area of the lake to the size of the watershed. A good example of a lake-regulated stream is the St. Lawrence River. The Great Lakes, which drain into the river, control the flow to such an extent that the maximum is not more than 20 percent greater than the minimum flow. This may be compared with the Missouri River which has a maximum stage 2900 percent greater than the minimum flow. If single-purpose flood control reservoirs are to be economically feasible, the protected area must be of high value. In Ohio the Miami River and Muskingum Valley Conservancy Districts and in New England the Connecticut River and Merrimack River projects are examples of flood reservoir projects.

Where water is stored for two or more uses, as illustrated in Fig. 11.2, reservoirs are classified as multiple-purpose structures. Requirements for flood control conflict to some extent with other objectives. For maximum flood prevention the reservoir must be empty at the beginning of the storm period and the stored water must be discharged as rapidly as the capacity of the stream below will allow. Water for power, irrigation, navigation, and water supply is withdrawn gradually at times when it can be used to best advantage. For the most part, irrigation water is needed only during the growing season. Water requirements for generating

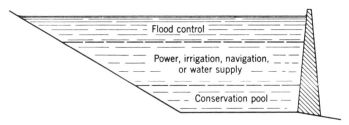

Fig. 11.2 Storage components in a multiple-purpose reservoir.

electrical energy vary with daily and seasonal loads. Water released during the summer for navigation also aids in alleviating stream pollution and may provide a more uniform water supply for cities downstream. Where a reservoir can be constructed with a capacity greater than required for other purposes, the additional capacity may be available for flood control. The Hoover Dam on the Colorado River, the Grand Coulee on the Columbia, Fort Peck Dam on the Missouri River, and Norris Dam in Tennessee are examples of multiple-purpose reservoirs, although not all of them are for flood control. The conservation pool is permanent storage for recreation, wildlife, and sediment accumulation.

11.3 Types of Floods

A flood may be defined as an overflow or inundation from a river or other body of water. The distinction between normal discharge and flood-flow is generally determined by the stage of the stream when bankful. As shown in Table 11.2, floods are a normal part of a stream's history and should not be looked on as an unexpected occurrence. Most floods occur on the flood plains adjacent to rivers and streams and result from such natural causes as excessive rainfall and melting snow. Occasionally, tidal waves or hurricanes cause flooding. With respect to loss of life, reservoir failures have produced some of the worst floods, but fortunately these seldom occur.

Large-Area Floods. Large-area floods occur from storms of low intensity having a duration of a few days to several weeks. Since melting snow may also contribute to or even cause large-area floods, seasons of maximum total precipitation do not necessarily coincide with the time of occurrence of large-area floods. In many parts of the country these floods come in the late winter or early spring. In many of the arid regions of the West these floods occur in late spring; in Florida they come in the autumn. As illustrated in Fig. 11.3 for Ohio, the greatest number

Table 11.2 Flow Characteristics of a Stream

Relative Flow	Frequency of Occurrence
10% of channel capacity	50% of the time
50% of channel capacity	5% of the time
Bankful	Equaled or exceeded about twice a year
Moderate flooding	About once in 10 years
Severe flooding	About once in 50 years

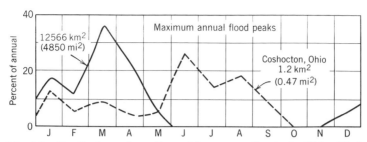

Fig. 11.3 Monthly distribution of small- and large-area floods in Ohio. (Redrawn from Harrold, 1961.)

of large-area annual floods occurred in March and April, but the small-area floods were most prevalent in June, July, and August.

Small-Area Floods. Small-area floods occur from storms of high intensity having a duration of 1 day or less. In the Midwest small-area floods occurred in the summer between the months of May and September. Ninety percent of the annual amount of rain, falling at rates greater than 25 mm/h, fell during this 5-month growing season. Since 85 percent of the annual soil loss occurs during this period, such storms cause great damage to agricultural land through soil erosion, which in turn results in sediment accumulation in rivers and reservoirs. These storms usually do not produce high runoff on large streams but often cause serious local damage.

11.4 Flood Plain Zoning

Any flood control program should be well coordinated among the many private and governmental interests. Regulation of land use in the flood plain by local governments can be more economical and effective than structural measures. Such action cannot be taken by private interests, but is a governmental responsibility.

Until recent years, the federal government gave practically no attention to nonengineering methods of reducing flood losses, such as by flood plain zoning. Occupancy of the flood plains in many areas has increased the potential damage. As an example, a study in Ohio showed that a recurrence of the rainfall that caused the 1913 flood would have caused more damage in 1960 than in 1913, despite the fact that millions of dollars had been spent for flood control.

11.5 Flood Control Legislation

Prior to 1936 there was no definite federal policy for the control of floods. With the exception of reclamation projects, federal undertakings were principally concerned with navigation. Laws enacted in several states created watershed districts primarily for watershed protection and flood control. A federal flood control policy was established in 1936 whereby the federal government authorized flood control activities on navigable streams and their tributaries. Federal investigations of watersheds and measures for runoff and waterflow retardation as well as erosion control were delegated to the USDA. Investigations and improvements on rivers and waterways for flood control and allied purposes were delegated to the U.S.

Corps of Engineers. The program under the USDA was carried out through the SCS for ranch- and farmlands and through the U.S. Forest Service for forest- and rangelands, both in cooperation with local, state, and other federal agencies. Public Law 566 (U.S. Congress, 1954), along with subsequent amendments, authorized the USDA to assist local organizations to prepare plans for engineering works and land treatment measures for the reduction of flood damage in headwater streams. This act was the first major authorization by the federal government for engineering works in headwater areas. Several states have enacted laws to authorize special-purpose districts to carry out works of improvement under this law.

Flood insurance legislation in 1968 was authorized for private owners on the condition that the participating community adopt and enforce government floodplain management standards. Subsidized insurance covers only direct losses to buildings and their contents. Motor vehicles, growing crops, land, livestock, and so on are not insurable under the government program.

11.6 Conservancy Districts

Conservancy districts are legal enterprises organized for the purpose of soil conservation and flood control. In some states flood control works constructed by the U.S. Corps of Engineers or the USDA under legislation, such as Public Law 566, are sponsored by and function under a conservancy district established for that purpose. The district may arrange for the necessary local support and cooperation among landowners and provide the necessary maintenance after the project is completed. The Miami Conservancy District in Ohio is a major privately financed project organized under the conservancy laws. Conservancy districts are known by a variety of names in different states, such as watershed districts, water-management districts, and other special-purpose titles. Conservancy districts are authorized in at least 24 of the 50 states.

METHODS OF HEADWATER FLOOD CONTROL

Two general classes of headwater flood control measures are (1) those that retard flow or reduce runoff by land treatment or reservoirs and (2) those that accelerate the flow by channel improvement, channel straightening, and levees.

Reducing Flood-Flows

Methods of reducing flood-flows include watershed treatment in which the storage of water is increased on the surface and in the soil profile; flood control reservoirs; and underground storage. Underground storage is accomplished by spreading the flow over a considerable area. This method is applicable only in special situations, particularly in arid regions.

Measures that retard the flow or reduce runoff are economically and physically more desirable because (1) all visible evidence or danger of the flood is removed; (2) the flow in the stream is more uniform, thus providing greater recharge of the ground water and a more adequate water supply; (3) an important step toward the conservation of natural resources is achieved; (4) higher crop production results, especially in areas where conservation of water is important; and (5) reduction of sedimentation in lower tributaries is accomplished.

11.7 Watershed Treatment

Watershed treatment includes all practices applied to the land that are effective in reducing flood runoff and controlling erosion. Proper land-use and conservation practices are necessary for adequate watershed control.

Land management may increase the amount of surface storage, rate of infiltration, and capacity of the soil to store water. Runoff retardation by land management is dependent largely on vegetative cover and favorable soil surface conditions. For example, in Fig. 11.4a the results of soil differences show about three times as much runoff occurring from heavy soils in Mississippi as from permeable loess soils in Iowa. The effect of vegetation on runoff for varying amounts of rainfall is shown in Fig. 11.4b.

The effect of land management on the time distribution of flood-flows is shown in Fig. 11.5. The land before treatment was severely eroded and about two thirds of the watershed was in scrawny woodland that had been burned and grazed. After the land was retired from use and reforested, the peak flow from a comparable storm about 10 years later was reduced to only 15 percent of that before treatment. The volume of flow represented by the area under the hydrograph was about the same for the two storms. Land management is thus effective in reducing flows in small watersheds from short-duration storms, but its effectiveness diminishes rapidly with increased drainage area and magnitude of the storm.

Conservation practices, such as contouring, strip cropping, and terracing, will reduce flood peaks as well as total runoff from small to medium storms during the growing season. For larger storms, the effect of contouring and strip cropping on runoff is usually much less than the reduction in soil losses discussed in Chapter 5. Level or absorptive-type terraces can have a considerable effect on flood peaks of rather large watersheds. For example, a 1953 flood from a 2287-km² watershed in northwestern Iowa could have been reduced 54 percent by installing level terraces on 37 percent of the watershed. If only 17 percent of the area was terraced, the flow could be reduced by 32 percent. Graded terraces generally will reduce runoff volumes less than 25 percent. Land-management and conservation practices have little, if any, effect on floods from winter storms and on catastrophic floods from large watersheds.

On flat watersheds subsurface drainage may be beneficial for flood control. In Ohio peak flows were reduced 7 percent or more and the number of floods was reduced 46 percent. Earlier studies by Woodward and Nagler (1929) on two large watersheds, the Des Moines and Iowa Rivers, concluded that flood peaks were unchanged because of extensive artificial drainage improvements from about 1903 to 1923.

11.8 Reservoirs

The two types of reservoir are the flood-regulated storage reservoir and the detention reservoir. The principal difference is that the detention reservoir operates automatically by discharging through one or more fixed openings in the dam, whereas the regulated reservoir discharges through adjustable gates. The chief advantage of the regulated reservoir is its flexibility of operation.

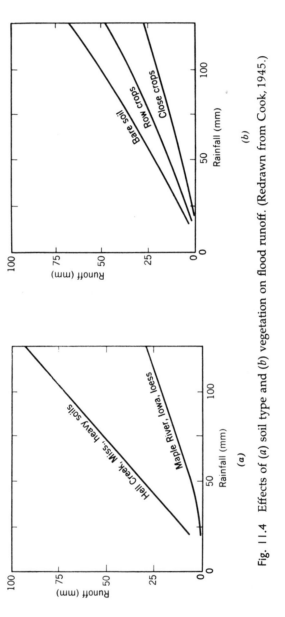

Fig. 11.4 Effects of (*a*) soil type and (*b*) vegetation on flood runoff. (Redrawn from Cook, 1945.)

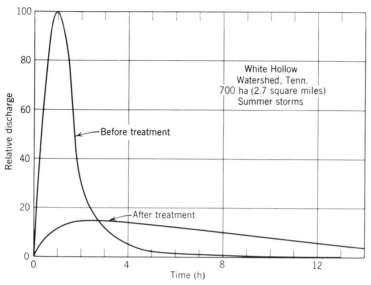

Fig. 11.5 Effect of land management on flood-flow. (Data from Tennessee Valley Authority.)

Detention reservoirs such as shown in Fig. 11.6 may have one or more discharge openings of fixed dimensions. These reservoirs have emergency spillways to handle runoff in excess of the design flood. Practically all headwater flood control reservoirs are of the detention type. Reservoirs on the watersheds of the Trinity River in Texas, Sandstone Creek in Oklahoma, the Little Sioux River in Iowa, as well as in the Miami Conservancy District, are examples of such structures. The principal advantages of the detention reservoir are its simplicity and automatic operation.

Reservoirs for flood control reduce flood peaks, but not flood volumes. This reduction in peak flow diminishes rapidly with distance downstream since the main stream receives an increasing percentage of its runoff from other tributaries. The principal disadvantages of reservoirs are that some land must be flooded to

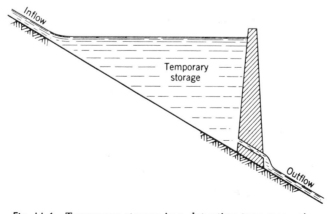

Fig. 11.6 Temporary storage in a detention-type reservoir.

protect other lower land and that the annual and initial costs are high. The annual cost may be reduced by controlling erosion and, thus, sedimentation, which prolongs the useful life of the reservoir. Initial cost of storage per unit volume, in general, decreases with increasing reservoir storage.

The effect of a series of small headwater reservoirs on flood runoff will be illustrated by the hypothetical watershed shown in Fig. 11.7. The 1500-km² watershed consists of ten 150-km² tributaries. From a storm that produced 25 mm of runoff in 4 h, the assumed hydrograph of flow at the mouth of each 25-km² watershed has a peak flow of 28 m³/s. By flood routing the runoff from watershed 1 to point B, the peak flow is reduced from 28 to 17 m³/s (curve 1 in Fig. 11.8). The area under the hydrograph or volume essentially remains unchanged. In Fig. 11.8 hydrographs 2 and 3 represent the flow from the 25-km² watersheds 2 and 3 each routed to point B, the outlet of the 150-km² area. The three hydrographs are then added together to obtain the composite hydrograph for the three areas as shown by the dashed line in Fig. 11.8. If the flow from the remaining 75 km² is added to the previous composite hydrograph, the resulting hydrograph at point B (top curve) is obtained.

Assume now that a reservoir with adequate capacity is constructed on each of the three 25-km² watersheds (1, 2, and 3), and that the outflow is controlled to 2.8 m³/s at each dam. The flow from the remaining one half of the area, or 75 km², is not controlled because of lack of good reservoir sites or other reasons. The composite hydrograph for the three controlled watersheds, when routed to point B, is shown in Fig. 11.9. The flow from the uncontrolled 75 km² is unchanged. When the uncontrolled area flow is added to the controlled flow, the resulting hydrograph with dams is obtained. The peak flow for this hydrograph is 56 m³/s compared with 96 m³/s without the dams, or a reduction of 42 percent.

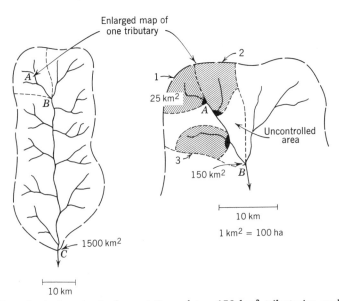

Fig. 11.7 Hypothetical watershed consisting of ten 150-km² tributaries and a series of reservoirs, each with a watershed of 25 km². (Revised and redrawn from Leopold and Maddock, 1954.)

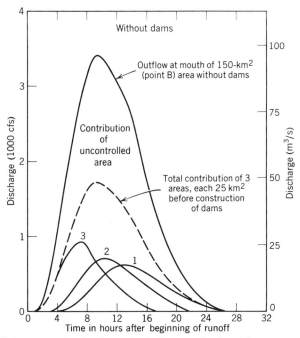

Fig. 11.8 Runoff hydrographs for hypothetical watersheds without reservoirs. (Redrawn from Leopold and Maddock, 1954.)

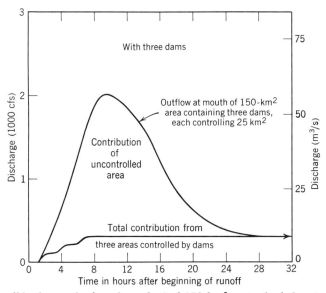

Fig. 11.9 Runoff hydrographs for a hypothetical 150-km² watershed showing the effect of reservoirs in reducing peak flows. (Redrawn from Leopold and Maddock, 1954.)

The percentage reductions in peak flood-flows resulting from reservoirs on these 25-km² watersheds for an 8-h rather than a 4-h storm producing 25 mm of runoff are 88 percent for one reservoir on a 25-km² watershed, 42 percent for three reservoirs on a 150-km² watershed, and 36 percent for 30 reservoirs on the 1500-km² area. Each dam would provide 88 percent reduction at its outlet and give good protection for a designated reach of the channel downstream from the dam. The effect of the reservoir rapidly diminishes downstream and thus further works are needed. These other downstream works have no benefit upstream. Both headwater and downstream works are needed for complete protection.

The effect of size and location of flood control reservoirs on peak flow reduction at the outlet of a large watershed is shown in Table 11.3. Although the large reservoir has the greatest effect downstream, it does not reduce flows within the watershed as do the small and medium reservoirs. The 12 small and medium dams give only an additional 5 percent (76 − 71) reduction downstream. Headwater reservoirs and downstream reservoirs serve different purposes and they are not generally interchangeable.

Increasing Channel Capacity

The purpose of increasing channel capacity is to decrease height and duration of floods and reduce flood damage. Increasing the capacity of a stream may be accomplished by (1) channel improvement, (2) channel straightening, and (3) levees.

11.9 Channel Improvement

Channel improvement here includes those measures that increase the channel capacity, namely, enlarging the cross-sectional area and increasing the velocity. In narrow flood plains, channel improvement and channel straightening are usually more economical than levees.

Increasing Cross Section. Increasing the cross section may be accomplished by deepening or widening the channel and by removing trees and sandbars from the watercourse. On small streams removal of trees and sandbars may increase the flow rate as much as one third to one half, but in large streams the increase is considered negligible.

Table 11.3 Flood Reduction by Reservoirs

Relative Size of Dam and Watershed			Reduction in Peak Flow of the Large Watershed (%)
Small	Medium	Large	
	Number of Dams		
7			40
	5		38
7	5		45
		1	71
7	5	1	76

Source: Black (1972).

Increasing Velocity.   Removing debris and vegetation has a greater effect on the roughness coefficient in small streams than in large streams. The hydraulic radius and resulting velocity can be increased by widening or deepening the channel. For the same increase in cross-sectional area, deepening the channel is more effective than widening. The depth may be increased by using levees as well as by dredging or cleaning out the channel. The two ways of increasing the channel slope are (1) deepening the channel or lowering the water level at the outlet and (2) straightening. Deepening the channel or lowering the water level at the outlet can be accomplished by increasing the cross-sectional area at the outlet and removing debris or sandbars. Bank slopes may be increased with permanent linings, especially on large channels.

11.10   Channel Straightening

The principal method of straightening streams is to provide cutoffs. A cutoff is a natural or artificial channel that shortens a meandering stream, as shown in Fig. 11.10. The purpose of a cutoff is to increase the velocity, to shorten the channel length, and to decrease the length of levees. The length of the stream channel may be shortened as much as one half the original length.

Cutoffs are desirable where (1) the stream capacity in the bend is less than the capacity in other parts of the channel, (2) the capacity of the entire channel is to be increased with levees, (3) construction of the cutoff is more economical than increasing the capacity around the bend, and (4) the cutoff does not detrimentally affect the flow characteristics of the stream but may cause channel scour upstream.

The effect of cutoffs is not always desirable. On alluvial streams cutoffs alone do not necessarily solve flood problems. They may cause serious streambank erosion above and below the cutoff and sediment deposits below. In some streams, cutoffs cause extreme lowering of the channel and its tributaries. In others, the increased gradient results in flow rates in excess of downstream channel capacity.

Cutoffs increase the velocity in the affected portion of the stream by increasing the hydraulic gradient. Because some water was formerly stored in the channel and in the flood plain along the bend BGE in Fig. 11.10, the cutoff also increases the stage downstream.

In a meandering stream, cutoffs may occur naturally. Since erosive forces are a function of the velocity and depth of the stream as well as of the angle between the banks and the direction of flow, these forces are continually changing. Where there are bends in the stream, the soil is eroded from the concave bank, carried downstream on the same side, and deposited at the end of the bend, forming a bar that deflects the current to the other side of the channel. As the cutting continues, the stream becomes more crooked and a natural cutoff results.

The effect of a single cutoff, either natural or artificial, for steady flow conditions in the channel is shown in Fig. 11.10. Before the cutoff was made, the stream flowed on a uniform slope around the bend BGE. After the stream was straightened, it flowed from B directly to E, which was about one fourth of its previous length. Because of the increased slope from B to E, the velocity was increased from v_1 to v_2 with a corresponding decrease in the depth of flow. The decreased depth caused the water surface to be lowered to point A upstream. In section CD flow takes place at a uniform depth with velocity v_2. At D the velocity began to decrease but was still greater at E than at A, and likewise the depth was less at E than at A.

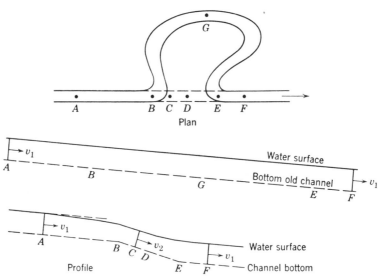

Fig. 11.10 Streamflow before and after a cutoff.

Below E there was a concave upward surface known as the backwater curve. At some point F, the lower end of the backwater curve, the velocity and depth were the same as at A. The cutoff was effective in lowering the stage through the cutoff section as well as above and below, to points A and F, respectively. The effect extended farther upstream than downstream. Since cutoffs are usually short, the points C and D may overlap so that the velocity in the cutoff section does not correspond to the channel slope.

The effect of cutoffs becomes more complicated for unsteady flow and for a series of cutoffs. During a storm period the flow is constantly increasing and points A, C, D, and F change with the stage of the stream. Unless the slope or cross section of a stream is changed between cutoffs, a series of cutoffs acts similarly to a single cutoff.

11.13 Levees

Levees are embankments along streams or on flood plains designed to confine the river flow to a definite width for the protection of surrounding land from overflow. Levees may be designed either to confine the river flow for a considerable distance or to provide local protection.

The effect of confining water between levees is (1) to increase the velocity through the leveed section, (2) to increase the water surface elevation during floods, (3) to increase the maximum discharge at all points downstream, (4) to increase the rate of travel of the floodwave, and (5) to decrease the surface slope upstream. Levees for protection of local areas have less effect on floodflow; however, the end result of any levee system is a reduction in valley storage and possible reduction of wetlands for wildlife.

The location, spacing, and height of levees must be adjusted to provide adequate capacity between the levees, to provide protection to the flood plain area, and to be economical in cost. The design and construction of levees are discussed in Chapter 10.

PREVENTIVE MAINTENANCE

Preventive measures for maintaining the capacity of the stream channel include those that affect erosion in the channel itself and those that reduce sediment from upper tributaries. Maintenance in the channel is required to prevent the collection of debris and to reduce sediment from eroding banks.

11.12 Bank Protection

The two classes of bank protection are (1) those that retard the flow along banks and cause deposition and (2) those that cover the banks and prevent erosion.

Retarding the flow along stream banks is desirable to control meandering, to protect the bank, thereby reducing deposition below, and to protect highways, railroads, and other structures near the channel. A common method of control is to build retards extending into the stream from the banks. Materials to construct retards include piles, trees, rocks, and steel framing. Retards, sometimes referred to as jetties, serve to decrease the velocity along the concave bank and, hence, increase deposition of sediment.

A method of locating retards is shown in Fig. 11.11. The first major retard at A is located by the intersection of the projected center line of flow with the concave bank. In locating the second major retard C a line HB is drawn parallel to the above projected center line and through the end of the retard A. The intersection of this line with the concave bank locates point B. AC is then made equal to twice AB. Additional retards are located by the intersection of a line connecting the end points of the two previous retards with the concave bank (see D). An auxiliary retard at K is located a distance AB upstream from A and is extended into the stream about one half the length of the other retards. The retards should extend into the stream at an angle of 45 degrees for a distance of about 30 percent of the channel width. On small streams the spacing of the retards may be made equal to the stream width, and the length, 0.25 times the spacing. On 30-degree (English units) curves or larger, continuous bank protection should be provided rather than retards (see Chapter 13).

Vegetative or mechanical control measures are methods of preventing stream-bank erosion. Plants suitable for vegetative control are grass, shrubs, and trees. Mechanical measures to cover the stream bank include such devices as wood and concrete mattresses, rock or stone, riprap, gabions (coarse rock in wire baskets), asphalt, geotextiles, and sacked or monolithic concrete.

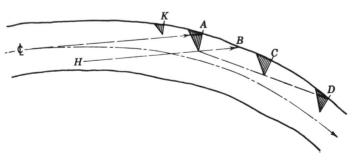

Fig. 11.11   Design and location of retards in a stream channel. (Redrawn from Saveson and Overholt, 1937.)

11.13 Reduction of Sediment and Debris

Sediment and debris in stream channels can be reduced by deposition in suitable settling basins or by land treatment.

Sediment from high-velocity streams in cultivated watersheds is deposited on flood plain areas and in the stream channels. Such sediment reduces the effectiveness of drainage ditches and the productivity of agricultural land. Although settling basins are often satisfactory, good land management accompanied by channel cleanout may be more practical.

Sedimentation and debris basins have three essential features: an inlet, a settling basin, and an outlet. Sediment-laden water from a stream may be diverted into a large settling basin where a portion of the sediment is deposited as a result of greatly reduced velocities. At the lower end of the basin the flow is then returned to the stream channel. Such settling basins are eventually filled with sediment, thus necessitating cleanout or the use of a new area. In western Iowa settling basins have been used to reduce sedimentation in channels across a wide flood plain.

The barrier system of removing debris and sediment from mountain streams was developed in Utah. Large debris is deposited as the flood spreads out at the mouth of the canyon and the finer material settles out in a settling basin. Additional features of the system consist of (1) a barrier or cross dike, (2) lateral dikes, and (3) temporary drift dams.

FLOOD ROUTING

Flood routing is the process of determining the stage height, storage volume, and outflow rate from a reservoir or a stream reach for a particular inflow hydrograph.

11.14 Principles of Flood Routing

Flood routing involves inflow into the reservoir or stream reach, outflow through the dam or stream reach, and storage in the reservoir or stream valley and channel. In routing a flood through a reservoir, the objective is to determine the height of the dam for a given size outlet structure. The problem in flood routing is to determine the relationships among inflow, outflow, and storage as a function of time. This problem can be solved by the following continuity equation for unsteady flow:

$$i\,dt = o\,dt + s\,dt \tag{11.1}$$

where i = inflow rate for a small increment of time (L^3/T),
o = outflow rate for a small increment of time (L^3/T),
s = storage rate for a small increment of time (L^3/T),
dt = time differential (T).

This equation must be satisfied at any and all times during the period from the beginning of inflow until outflow has stopped. Hence, it must also be satisfied for any given time interval between the above limits. Equation 11.1 may also be written

$$\Delta t \frac{i_1 + i_2}{2} = \Delta t \frac{o_1 + o_2}{2} + S_2 - S_1 \tag{11.2}$$

where $\quad S =$ volume of storage (L^3),

1 and 2 = subscripts denoting beginning and end of the time interval, respectively.

Evaporation and seepage losses are assumed negligible. In applying the equation to large reservoirs, the backwater effect assumed increases reservoir storage. This effect is negligible in small headwater reservoirs.

11.15 Methods of Flood Routing

Of the two general methods of flood routing, the first method involves the division of the inflow hydrograph into time intervals of short duration so that during each period inflow and outflow rates may be assumed to be constant. Storage- and outflow-stage relationships must be known. Either observed or developed inflow hydrographs are necessary in the solution. The second method involves analytical integration procedures in which a given flood hydrograph is replaced by an equivalent flood of uniform intensity. The available storage curve and outflow rates are represented by empirical formulas. The first flood routing method gives more accurate results, although analytical integration methods provide a more direct solution to the continuity equation (Eq. 11.1).

The many procedures for solving Eq. 11.1 include manual, graphical, and computer methods. Numerical and graphical procedures are basically trial-and-error solutions.

11.16 Elements of Flood Routing

In any flood routing procedure the factors that must be considered are (1) inflow hydrograph, (2) outflow or spillway discharge, (3) available storage, and (4) outflow hydrograph.

Inflow Hydrograph. Inflow hydrographs are the same as runoff hydrographs, discussed in Chapter 4.

Outflow. The outflow curve represents the depth-discharge relation of the reservoir spillway structure or the lower end of a stream reach. It may also include the flow through the emergency spillway. The rate of discharge from a reservoir is influenced by the hydraulic head and by the type and size of spillway. The most common types of structures for detention reservoirs are drop-inlet pipe spillways with box or orifice inlets (see Chapter 9). In the case of stream routing the outflow for the reach above is the same as the inflow to the next reach below.

Available Storage. The available storage curve represents the depth-capacity relation of a reservoir above the elevation of the mechanical spillway crest, or for a stream above the elevation of some arbitrarily selected stage. The volume of storage for various stages is determined from the topography of the storage basin or stream valley. Where considerable accuracy is desired, the volume in a reservoir can be computed from a contour map either by the average end area method or by the prismoidal formula given in Appendix G. On streams several valley cross sections are often determined to obtain storage.

Outflow Hydrograph. The outflow hydrograph shows the rate of outflow (spillway discharge) as a function of time. This hydrograph must be determined by flood routing.

11.17 Reservoir Flood Routing Procedure

Many graphical and numerical procedures have been developed for solving Eq. 11.2, the continuity equation. The procedure to be described is partly numerical and partly graphical. The procedure will apply to either reservoir or stream flood routing.

In most flood routing procedures the size of the mechanical spillway cannot be determined directly. A size is selected and the flood is routed through the reservoir. If the peak outflow rate or the maximum water stage in the reservoir does not meet site or cost limitations, another size spillway is selected and the flood routing procedure is repeated. With most drop-inlet pipe spillway structures, the discharge rate when the pipe first flows full will be only slightly less than the maximum outflow rate. With peak runoff, total runoff, drainage area, and available storage known, the rate of outflow and consequently the approximate size of the pipe may be determined from an equation developed by Culp (1948):

$$\frac{q_0}{q} = 1.25 - \left(\frac{1500V}{RA} + 0.06\right)^{1/2} \tag{11.3}$$

where q_0 = rate of outflow when the pipe first flows full in m³/s,
q = peak inflow in m³/s,
V = available storage in ha-m,
R = runoff in mm,
A = drainage area in ha.

The procedure for reservoir flood routing will be illustrated in Example 11.1.

☐ *Example 11.1*
_____

Design a combination flood control reservoir and farm pond for a site that has a drainage area of 48.58 ha (120 ac). The total runoff for a 50-year return period is 88.9 mm (3.5 in.) and the peak runoff rate is 5.38 m³/s (190 cfs). A depth of 2.44 m (8 ft) for the pond is available below an elevation of 29.26 m (96 ft). The storage capacity of the reservoir above 29.26 m (0 stage) is shown in Fig. 11.12. A box-inlet spillway and circular concrete outlet pipe are to be used in the outlet structure. The maximum allowable stage in the reservoir is 1.62 m (5.3 ft) at elevation 30.88 m (101.3 ft). By flood routing procedure, determine the size of the outlet structure, the actual water stage, the elevation of the flood spillway crest, and the maximum height of the dam, allowing a net freeboard of 0.61 m (2 ft) and a flow depth of 0.30 m (1 ft) in the flood spillway.

Solution. With a stage of 1.62 m (5.3 ft) above the crest of the mechanical spillway, the storage from Fig. 11.12 is 5.86 m³/s-h (207 cfs-h) or 2.109 ha-m (17.1 ac-ft). From Eq. 11.3 the rate of outflow when the pipe first flows full is

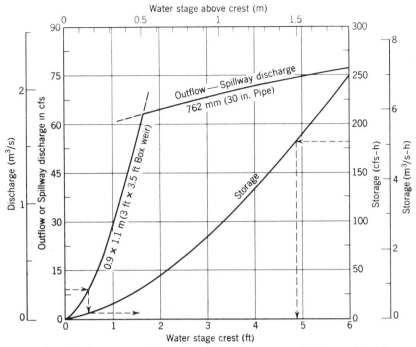

Fig. 11.12 Reservoir storage and outflow curves for Example 11.1.

$$q_0/5.38 = 1.25 - [(1500 \times 2.109)/(88.9 \times 48.58) + 0.06]^{1/2}$$
$$q_0 = 5.38 \times 0.36 = 1.94 \text{ m}^3/\text{s (68.4 cfs)}$$

Assume a 0.9×1.1-m (3×3.5 ft) box inlet (crest length 2.9 m) and a 762-mm (30-in.) outlet pipe. (Box-inlet area should be about twice the area of the pipe.) Using the weir formula with $C = 3.0$ and the pipe flow formula with $K_e = 1.0$, $n = 0.014$, and $L = 33.5$ m (110 ft), compute the spillway discharge curve shown in Fig. 11.12. q_0 is 1.81 m³/s (64 cfs) and is close enough to the estimated value to be satisfactory. Assume an inflow hydrograph as shown in Fig. 11.13. Such a

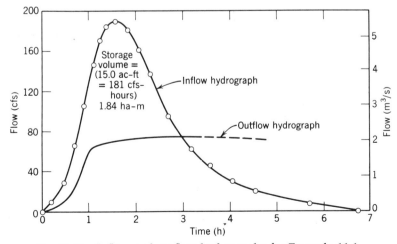

Fig. 11.13 Inflow and outflow hydrographs for Example 11.1.

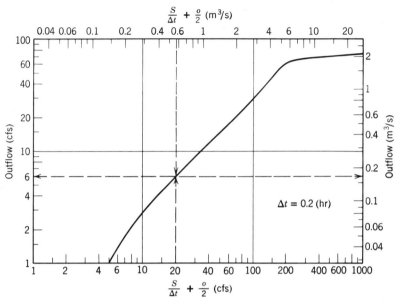

Fig. 11.14 Routing curve for Example 11.1

hydrograph may be developed as described in Chapter 4. The volume of runoff, which is the area under the hydrograph, is 4.32 ha-m ($88.9 \times 10^{-3} \times 48.58$) (35 ac-ft).

Compute the routing curve shown in Fig. 11.14 for 12-min ($\Delta t = 0.2$ h) time increments. The procedure is given in Table 11.4. The expression $S/\Delta t + o/2$ is to be evaluated to simplify the solution of Eq. 11.4 as will become clear later in the example. The time interval should be short enough to define the inflow with the desired accuracy (SCS, 1972). It may be estimated as 10 to 15 percent of the time of

Table 11.4 Computations for Routing Curve in Figure 11.14

Outflow or Spillway Discharge (m^3/s)	Storage, S (m^3/s-h)	$S/\Delta t^a$ (m^3/s)	$o/2$ (m^3/s)	$S/\Delta t + o/2$ (m^3/s)
(1)	(2)	(3)	(4)	(5)
0	0	0	0	0
0.17	0.10	0.48	0.09	0.57
0.28	0.17	0.85	0.14	0.99
0.57	0.34	1.70	0.28	1.98
0.85	0.48	2.41	0.42	2.83
1.13	0.62	3.11	0.57	3.68
1.42	0.76	3.82	0.71	4.53
1.70	0.93	4.67	0.85	5.52
1.81	1.13	5.66	0.91	6.57
1.93	2.27	11.33	0.96	12.29
2.04	3.88	19.40	1.02	20.42
2.15	6.17	30.87	1.08	31.95

[a] $\Delta t = 0.2$ h.

peak. Select sufficient number of outflow rates to adequately define the spillway discharge curve. From Fig. 11.12 read the storage (Column 2 in Table 11.4) for the corresponding outflow rate. For example, an outflow rate of 0.28 m³/s (10 cfs) in line 3 corresponds to a stage height of 0.15 m (0.5 ft), and the storage to this height is read as 0.17 m³/s-h (6 cfs-h). In column 3 and line 3 dividing the storage by the time interval gives 0.17/0.2 = 0.85 m³/s (30 cfs). By adding one half the outflow rate 0.14 m³/s (5 cfs) to this flow, the value, 0.99 m³/s (35 cfs) in Column 5, is obtained. The remaining values are computed and then plotted against the out-flow rate as shown in Fig. 11.14.

For plotting the routing curve select log-log paper or arithmetic scales that will give the desired accuracy throughout the range of work. This curve greatly simpli-fies the computations.

Determine the outflow hydrographs as illustrated in Table 11.5. Values in Column 2 are obtained from the inflow hydrograph in Fig. 11.13. The average inflow for each time interval is computed in Column 3. The computations are simplified by writing Eq. 11.2 in the form

$$\frac{S_2}{\Delta t} + \frac{o_2}{2} = \frac{i_1 + i_2}{2} + \left(\frac{S_1}{\Delta t} + \frac{o_1}{2}\right) - o_1 \tag{11.4}$$

The expression on the right side of Eq. 11.4 at 0.2 h in Table 11.5 is $0.14 + 0 - 0$ $= 0.14$, and the corresponding outflow rate for this expression in Column 4, as

Table 11.5   Computations for Flood Routing in Example 11.1

Time (h) $\Delta t = 0.2$	i	$(i_1 + i_2)/2$	$S/\Delta t + o/2$ (All units in m³/s)	o
(1)	(2)	(3)	(4)	(5)
0	0		0	0
0.2	0.28	0.14	0.14	0.03
0.4	0.62	0.45	0.56	0.17
0.6	1.36	0.99	1.38	0.40
0.8	2.55	1.95	2.93	0.88
1.0	3.74	3.14	5.19	1.61
1.2	4.76	4.25	7.83	1.87
1.4	5.21	4.98	10.94	1.90
1.6	5.38[a]	5.30	14.34	1.95
1.8	5.15	5.27	17.66	1.98
2.0	4.67	4.93	20.61	2.01
2.2	4.11	4.39	22.99	2.07
2.4	3.54	3.82	24.74	2.10
2.6	2.97	3.26	25.90	2.10
2.8	2.46	2.72	26.52	2.12
3.0	2.01	2.24	26.64	2.12[b]
3.2	1.78	1.90	26.42	2.12
3.4	1.50	1.64	25.94	2.10

[a]Maximum inflow.
[b]Maximum outflow is about equal to inflow rate. Time of maximum water level in reservoir and maximum storage.

read from Fig. 11.14, is 0.03 m³/s (1 cfs). The next value in Column 4 is 0.45 + 0.14 − 0.03 = 0.56, and o = 0.17 m³/s (6 cfs). The routing is continued until the outflow rate exceeds the inflow rate that occurs at about 3.0 h. If desired, the routing may be continued until the outflow is zero; however, in this case it is not necessary to do so. Maximum storage may be computed from Column 4 where

$$S = (26.64 - 2.12/2)0.2$$
$$= 5.12 \text{ m}^3/\text{s-h or } 1.84 \text{ ha-m (181 cfs-h or 15.0 ac-ft)}$$

After plotting the outflow hydrograph from Columns 1 and 5 from Table 11.5 in Fig. 11.13, storage volume may be checked by measuring the area between the inflow and the outflow hydrographs.

From Fig. 11.12 the maximum water level height corresponding to a storage of 5.12 m³/s-h (181 cfs-h) is 1.49 m (4.9 ft). The crest elevation of the flood spillway is 29.26 + 1.49 = 30.75 m (100.9 ft). Maximum settled height of the dam is 2.44 + 1.49 + 0.61 + 0.30 = 4.84 m (15.9 ft). The maximum water level height and the desired maximum outflow rate (2.12 m³/s) are close enough to the design requirements to give a satisfactory solution to the problem. If the flood is to be routed above the crest of the emergency spillway in the above example, the depth-distance curve must be changed to include both the outflow rate through the mechanical spillway and the emergency spillway.

11.18 Streamflow Routing Procedure

The basic principles of routing a flood in a stream channel are the same as for a reservoir. Reservoir storage is replaced by valley storage that increases rapidly with depth, especially when the water level exceeds the bankful stage. In streamflow routing the channel is divided into reaches of convenient length. Such factors as changes in cross sections, channel gradient, and location of tributaries affect the selection of reaches. Several cross sections are necessary within the reach to adequately determine the stage-storage relationship. The routing interval may be estimated as the time of travel through the reach at bankful stage. The outflow of a reach becomes the inflow of the reach below if routing is continued. Local inflow into the stream within the reach being routed, if large enough to be important, should be added to the inflow hydrograph or to the outflow hydrograph. The choice will depend on the nearness of the local flow to the upper or lower end of the reach. Detailed procedures are described in many references, such as SCS (1972) and Linsley et al. (1982).

REFERENCES

Barrows, H. K. (1948). *Floods: Their Hydrology and Control.* McGraw-Hill, New York.

Black, P. E. (1972). *Flood Peaks as Modified by Dam Size and Location.* Water Resources Bulletin 8 (No. 4).

Cook, H. L. (1945). "Flood Abatement by Headwater Measures." *Civil Eng.* **15**, 127–130.

Culp, M. M. (1948). "The Effect of Spillway Storage on the Design of Upstream Reservoirs." *Agr. Eng.* **29**, 344–346.

Goddard, J. E. (1976). "The Nation's Increasing Vulnerability to Flood Catastrophe." *J. Soil Water Cons.* **31**, 48–52.

Harrold, L. L. (1961). "Hydrologic Relationships on Watersheds in Ohio." *Soil Cons.* **26**, 208–210.

Hoyt, W. G., and W. B. Langbein (1955). *Floods*. Princeton Univ. Press, Princeton, NJ.

Leopold, L. B., and T. Maddock (1954). *The Flood Control Controversy*. Ronald Press, New York.

Linsley, R. K., M. A. Kohler, and J. L. H. Paulhus (1982). *Hydrology for Engineers*, 3rd ed. McGraw-Hill, New York.

Morgan, C. E. (1951). *The Miami Conservancy District*. McGraw-Hill, New York.

Saveson, I. L., and V. Overholt (1937). "Stream Bank Protection." *Agr. Eng.* **13**, 489–491.

U.S. Congress (1954). "Watershed Protection and Flood Prevention Act." Public Law 566, 83rd Congress, H.R. 6788, August.

U.S. Department of Housing and Urban Development (HUD) (1978). *Questions and Answers on National Flood Insurance Program*. Pamphlet HUD-471-FIA(2), May. Washington, DC.

U.S. Interagency Task Force on Floodplain Management (1989). *A Status Report on the Nation's Floodplain Management Activity*. Federal Emergency Management Agency, Washington, DC.

U.S. Soil Conservation Service (SCS) (1972). "Hydrology." In *National Engineering Handbook*, Sect. 4, Chap. 17 (litho.). Washington, DC.

Woodward, S. M., and F. A. Nagler (1929). "The Effect of Agricultural Drainage upon Flood Run Off." *Trans. ASCE* **93**, 821–839.

PROBLEMS

11.1 Before a cutoff was made on a meandering stream the length of the channel around the bend (*BGE* in Fig. 11.10) was 1280 m (4200 ft), and the stream gradient was 0.08 percent. After the cutoff was made, the distance *BE* was 640 m (2100 ft). If the velocity in the old channel was 0.6 m/s (2 fps), how much has the cutoff reduced the time of flow from *B* to *E*, assuming the same hydraulic radius and roughness coefficient for the old and the new channel?

11.2 Design a system of retards for a stream 15 m (50 ft) wide where the channel makes a 35-degree turn on an 8-degree curve (English units). Determine the length and spacing of retards by making a scale drawing of the stream.

11.3 By flood routing, determine the maximum water level for a flood control reservoir that has a watershed area of 48.58 ha (120 ac). The runoff depth for a 50-year return period storm is 89 mm (3.5 in.), and the peak runoff rate is 5.4 m³/s (190 cfs). The outlet structure is a box-inlet spillway, 0.9 m (3 ft) square (one side is the headwall) with a weir coefficient of 3.0 and a circular concrete outlet pipe, $L = 33.5$ m (110 ft), $K_e = 1.0$, n = 0.014, and $d = 610$ mm (24 in.) in diameter. The water elevation at zero storage (crest of the box inlet) is 26.8 m (88 ft), and the elevation of the pipe center at its outlet is 25.6 m (84 ft). The accumulated storage available at each 0.6-m (2-ft) stage above the crest is 0.06, 0.17, 0.41, 0.76, 1.32, 2.10, and 3.22 ha-m (0.5, 1.4, 3.3, 6.2, 10.7, 17.0, and 26.1 ac-ft).

11.4   Design the outlet structure (inlet and pipe size) for Problem 11.3, using the hydrograph described below and the runoff volume and peak of 80.8 mm (3.18 in.) and 6.12 m³/s (216 cfs), respectively. Maximum elevation of the flood spillway (maximum water level in the reservoir) should not exceed 30.2 m (99 ft).

Time (min)	q [m³/s (cfs)]		Time (min)	q [m³/s (cfs)]
0	0	(0)	87	5.86 (207)
10	0.31	(11)	97	5.21 (184)
23	0.91	(32)	110	4.42 (156)
32	2.15	(76)	129	3.06 (108)
42	3.43	(121)	151	2.01 (71)
52	4.70	(166)	171	1.47 (52)
58	5.52	(195)	193	0.99 (35)
64	5.95	(210)	217	0.68 (24)
74	6.12	(216)	270	0.25 (9)
			322	0 (0)

11.5   Route a flood through a 3000-m (10 000-ft) reach of a stream, whose bankful capacity is 23 m³/s (800 cfs). From a unit hydrograph the inflow hydrograph was obtained, and the outflow-storage data below were developed from weighted averages of four cross sections in the reach. Route the flood until the outflow is less than 2.8 m³/s (100 cfs). How much was the flood peak reduced within the reach?

Inflow Hydrograph			Outflow [m³s (cfs)]		Storage [m³/s-h (cfs-h)]	
Time (min)	q (avg) [m³/s(cfs)]					
0	0	(0)	0	(0)	0	(0)
0.75	21.24	(750)	1.42	(50)	1.98	(70)
1.50	63.72	(2250)	4.25	(150)	4.64	(164)
2.25	106.20	(3750)	8.50	(300)	7.02	(248)
3.00	148.68	(5250)	22.66	(800)	18.44	(651)
3.75	157.18	(5550)	42.48	(1500)	36.87	(1302)
4.50	131.69	(4650)	99.12	(3500)	93.46	(3300)
5.25	106.20	(3750)	141.60	(5000)	128.58	(4540)
6.00	80.71	(2850)	198.24	(7000)	159.16	(5620)
6.75	55.23	(1950)				
7.50	29.74	(1050)				
8.25	8.50	(300)				
9.00	0	(0)				

CHAPTER 12

Surface Drainage and Land Forming

Land forming is defined as the process of changing the natural topography so as to control the movement of water onto or from the land surface. It includes one or a combination of practices, such as land leveling for irrigation; land grading or shaping for irrigation, drainage, and water conservation; and shallow field ditches, which can be crossed with farm machinery. Land forming also includes grading work for erosion control, for example, contour benching or earthwork for parallel terracing. Land smoothing is generally referred to as the final operation of removing the minor differences in elevations that result from the operation of scrapers or other large earth-moving equipment. Land grading, shaping, and leveling are synonymous.

Land grading is essential to the development of surface irrigation systems. This practice has been adopted in more humid regions as a method for improving surface drainage on flat land (Coote and Zwerman, 1970). Grading land for both surface irrigation and drainage is entirely practical (Phillips, 1958) and compatible.

SURFACE FIELD DITCHES

The selection of surface drainage facilities for individual field areas depends largely on the topography, soil characteristics, crops, and availability of suitable outlets. Surface drainage systems must be suitable for mechanized operations on various types of topography, such as pothole areas, flat fields, and gently sloping land. In irrigated areas such drains are needed to collect wastewater from surface irrigation. Pothole areas are frequently found in glaciated regions where the topography is relatively flat and geologic erosion has not had time to develop natural outlets. Flat or level land having impermeable subsoils with shallow topsoil frequently requires surface drainage because subsurface drains are not

245

practicable or economical. Claypan, hardpan, and tight alluvial soils are examples. On these flat fields water may accumulate because of excess rainfall, flooding from uplands, or overflow from streams. Flat land is defined as land with slopes less than 2 percent, the major portion of which is less than 1 percent. The two primary methods of surface drainage are land grading and field ditches. Field ditches include (1) bedding, in which the plow deadfurrows serve as drains; (2) field ditches with wider spacings than deadfurrows; and (3) parallel open ditches that cannot be crossed with farm machinery.

The importance of surface drainage is indicated by Gain (1964) who estimated that over 40 500 000 ha in the eastern United States and about 3 200 000 ha in the eastern provinces of Canada would benefit from surface drainage. The extent of these drainage areas is shown in Fig. 12.1.

Excessive surface water can be drained by one or more of the following processes: natural or constructed channels, infiltration, or evaporation and transpiration. Evaporation is usually inadequate, and, if the soil is impervious, surface drainage is the only remaining method. Shallow surface drains cannot remove ground water and give the benefits incident to good subsurface drainage; however, surface drainage may be required even though subsurface drains are installed.

12.1   Random Field Drains

These drains are best suited to the drainage of scattered depressions or potholes (Fig. 12.2) where the depth of cut is not over 1 m. In cross section they are a flat "V" or parabolic in shape.

The design of field drains is similar to the design of grass waterways, as discussed in Chapter 7. Where farming operations cross the channel, the side-slopes should be flat, that is, 8 : 1 or greater for depths of 0.3 m or less and 10 : 1 or greater for depths over 0.6 m. Minimum sideslopes of 4 : 1 are desired if the field is farmed parallel to the ditch. The depth is determined primarily by the topography of the area, outlet conditions, and capacity of the channel. The grade in the channel should be such that the velocity does not cause erosion or sedimentation. Maximum allowable velocities for various soil conditions are given in Chapter 13. Minimum velocities vary with the depth of flow; however, these range from about 0.3 to 0.6 m/s for depths of flow less than 1 m. The maximum grade for sandy soil is about 0.2 percent and for clay soils, 0.5 percent. The minimum grade is 0.05 percent. The roughness coefficient in the Manning equation may be taken as 0.04 if more reliable coefficients are not available. The capacity of the drain is usually not considered for areas less than 2 ha provided the minimum design specifications are met; however, where the area is larger than 2 ha, the capacity should be based on a 10-year return period storm, making allowances for minimum infiltration and interception losses. Since most field crops are able to withstand inundation for only a short period without damage, it is desirable to remove surface water within 12 to 24 h.

The layout of a typical random field drain system is shown in Fig. 12.2. Normally, the channel should follow a route that provides minimum cut and least interference with farming operations. Where possible several potholes should be drained with a single ditch. The outlet for such a system may be a natural stream, constructed drainage ditch, or protected slope if no suitable ditch is available. Where the outlet is a broad, flat slope, the water is permitted to spread out on the land below. This type of outlet is practical only if the drainage area is small.

Fig. 12.1 Extent of lands needing surface drainage in eastern United States and Canada. (From Gain, 1964.)

Legend

Need for surface drainage insignificant—areas small, scattered or less than 5% of total area

Need significant—areas 5 to 20% of total area

Need extensive—areas 20 to 50% of total area

Need dominant—areas 50 to 100% of total area

500 km

12.2 Bedding

Bedding is a method of surface drainage consisting of narrow-width plow lands in which the deadfurrows run parallel to the prevailing land slope (Fig. 12.3). The area between two adjacent deadfurrows is known as a bed. Bedding is most practicable on flat slopes of less than 1.5 percent where the soils are slowly permeable and pipe drainage is not economical. Studies in southern Iowa showed that leveled land gave slightly better yields than bedded land.

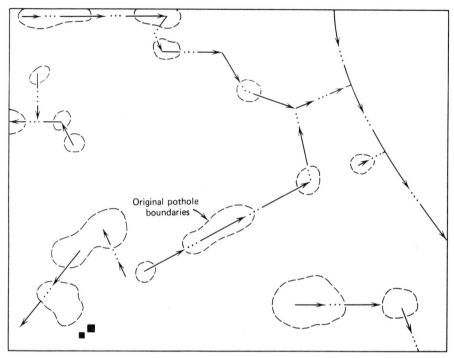

Fig. 12.2 A random field drain system in central Iowa.

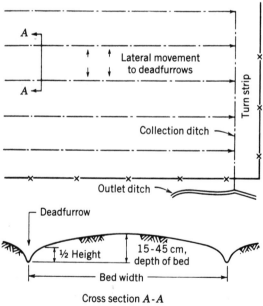

Cross section *A-A*

Fig. 12.3 Bedding system of surface drainage.

The design and layout of a bedding system involves the proper spacing of deadfurrows, depth of bed, and grade in the channel. The depth and width of bed depend on land slope, drainage characteristics of the soil, and cropping system. Bed widths recommended for the Corn Belt region of the United States vary from 7 to 11 m for very slow internal drainage, from 13 to 16 m for slow internal drainage, and from 18 to 28 m for fair internal drainage. The length of the beds may vary from 90 to 300 m. In the bedded area the direction of farming operations may be parallel or normal to the deadfurrows. Tillage practices parallel to the beds have a tendency to retard water movement to the deadfurrows. Plowing is always parallel to the deadfurrows.

12.3 Parallel Field Drain System

Parallel field drains are similar to bedding except that the channels are spaced farther apart and may have a greater capacity than the deadfurrows. This system is well adapted to flat, poorly drained soils with numerous small depressions that must be filled by land grading.

The design and layout are similar to those for bedding except that drains need not be equally spaced and the water may move in only one direction. The layout of such a field system is shown in Fig. 12.4. As in bedding, the turn strip is provided where ditches border a fence line. The size of the ditch may be varied, depending on grade, soil, and drainage area. The depth of the ditch should be a minimum of 0.2 m and have a minimum cross-sectional area of 0.5 m^2. For trapezoidal cross sections the bottom width should be 2.4 m (ASAE, 1986). The sideslopes should be 8:1 or flatter to facilitate crossing with farm machinery. As in bedding, plowing

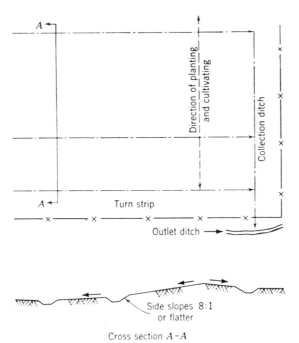

Fig. 12.4 Parallel field drain system of surface drainage.

operations must be parallel to the channels, but planting, cultivating, and harvesting are normally perpendicular to them. The rows should have a continuous slope to ditches. The maximum length for rows having a continuous slope in one direction is 180 m, allowing a maximum spacing of 360 m where the rows drain in both directions. In very flat land with little or no slope, some of the excavated soil may be used to provide the necessary grade; however, the length and grade of the rows should be limited to prevent damage by erosion. On highly erosive soils that are slowly permeable, the slope length should be reduced to 90 m or less.

The cross section for field drains may be V-shaped, trapezoidal, or parabolic. The W-drain shown in Fig. 12.5 is essentially two parallel single ditches with a narrow spacing. All of the spoil is placed between the channels, making the cross section similar to that of a road. The advantages of the W-drain are that it (1) allows better row drainage because spoil does not have to be spread, (2) may be used as a turn row, (3) may serve as a field road, (4) can be constructed and maintained with ordinary farm equipment, and (5) may be seeded to grass or row crops. The disadvantages of the W-drain are that (1) the spoil is not available for filling depressions, (2) a greater quantity of soil must be moved, and (3) a larger area is occupied by drains. The minimum width for W-drains varies from about 5 to 30 m, depending on the size. The W-drain is best adapted to relatively flat land where the rows drain from both directions.

12.4 Parallel Lateral Ditch System

The parallel lateral ditch system is similar to the field drain system except that the ditches are deeper. These drains, illustrated in Fig. 12.6, cannot be crossed with farm machinery. For clarity the minimum size for open ditches is 0.3 m deep and sideslopes steeper than 6:1. The purpose of lateral open ditches is to control the ground water table and to provide surface drainage. These ditches are applicable for draining peat and muck soils to obtain initial subsidence prior to subsurface

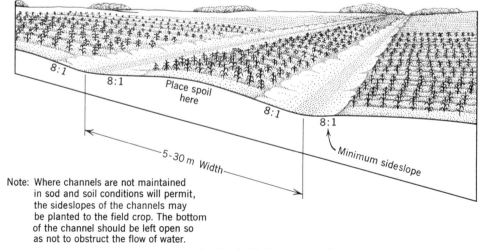

Note: Where channels are not maintained in sod and soil conditions will permit, the sideslopes of the channels may be planted to the field crop. The bottom of the channel should be left open so as not to obstruct the flow of water.

Fig. 12.5 W-drain or double field drain for surface drainage.

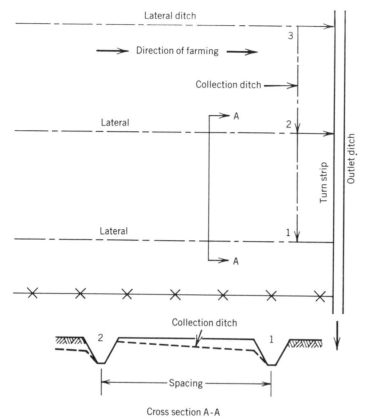

Cross section A-A

Fig. 12.6 Parallel lateral ditch system for water table control and surface drainage. (Adapted from Beauchamp, 1952.)

drainage. For the same depth to water table, ditches provide the same degree of subsurface drainage as pipe drains.

The design specifications for lateral ditches are given in Table 12.1. Since lateral ditches are considerably deeper than collection ditches, overfall protection must be provided at outlets 1 and 2 indicated in Fig. 12.6. This protection may be obtained

Table 12.1   Ditch Specifications for Water Table Control

	Sandy Soil	Other Mineral Soils	Organic Soils, Peat, and Muck
Maximum spacing	200 m	100 m	60 m
Minimum sideslopes	1:1	1½:1	Vertical to 1:1[a]
Minimum bottom width	1.2 m	0.3 m	0.3 m
Minimum depth	1.2 m	0.8 m	0.9 m

[a]Vertical for raw peat to 1:1 for decomposed peat and muck.
Source: Beauchamp (1952).

with a suitable permanent structure, by providing a gradual slope near the outlet, or by establishing a grassed channel.

Since these ditches are too deep to cross with farm machinery, farming operations must be parallel to the ditches. A collection ditch, row drain, or quarter drain should be provided for row drainage. As in other methods of drainage on flat land, the surface must be graded and smoothed and large depressions filled or drained by random field ditches.

For water table control during dry seasons, dams with removable crestboards are placed at various points in the open ditches to maintain the water surface at the required level. During wet seasons the crestboards are removed and the system provides surface drainage. In highly permeable soils, such as sand, peat, and muck, crop yields may be increased by controlled drainage. Deep permeable soils underlain with an impervious material provide the best conditions for successful water table control. The water level may be regulated by gravity, pumping, or a combination of gravity drainage and pumping. The depth at which the water table is to be maintained depends largely on the crop to be grown, soil, seasonal conditions including the quantity of water available, topography, and climatic conditions (see Chapters 13 and 14). In organic soils a high water table is desirable to provide water for plant growth, to control subsidence, and to reduce fire and wind erosion hazards. In these soils the water level should be maintained from 0.5 to 1.2 m below the surface, depending on the crop.

12.5 Cross-Slope Ditch System

The drainage of sloping land may be feasible with cross-slope ditches. Such channels usually function both for surface drainage and for erosion control. Where designed specifically for the control of erosion, these drains are called terraces. Diversion ditches (see Chapter 8) are sometimes used to divert runoff from low-lying areas, thus reducing the drainage problem.

The cross-slope ditch system is adapted primarily to soils with poor internal drainage where subsurface drainage is not practicable and for land with slopes of 4 percent or less having numerous shallow depressions. This land is generally too steep for bedding or field drains since farming up and down the slope results in excessive erosion.

12.6 Construction

The selection of equipment and procedure for construction of field ditches varies with the depth of cut and quantity and distribution of excavated soil. For depths of cut up to 0.75 m, blade graders, scrapers, and heavier terracing machines are suitable. Construction of field ditches by the above methods may require spreading the spoil from both sides of the drain to prevent ponding back of the spoil. For deep cuts over 0.75 m, bulldozers equipped with push or pull-back blades and carryall scrapers may be used to fill the pothole area or other depressions near the point of excavation.

The field layout of a random surface drain is shown in Fig. 12.7. Stakes are usually set every 15 m along the center line. Slope stakes are placed about 2 m farther out from the center line than the computed distance so as not to be removed by construction equipment.

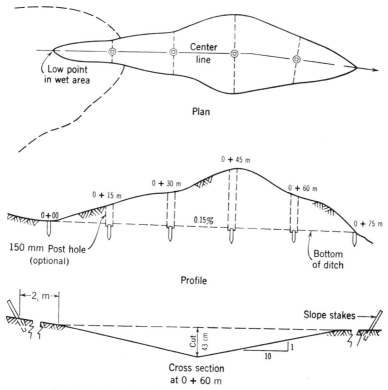

Fig. 12.7   Construction layout for a random field drain.

☐ Example 12.1

Determine the volume of cut for the random field drain shown in Fig. 12.7 (formulas given in Appendix G).

Solution.   The cuts below were determined by instrument survey and were computed for a channel grade of 0.15 percent and a sideslope of 10:1.

Sta. (m)	Cut (m)[a]	One Half Top Width (m)	Cross-Sectional Area (m²)	Average Cross-Sectional Area (m²)	Distance (m)	Volume of Cut (m³)
0 + 00	0	0	0			
				0.27	15	4.05
0 + 15	0.23	2.3	0.53			
				1.07	15	16.05
0 + 30	0.40	4.0	1.60			
				3.62	15	54.30
0 + 45	0.75	7.5	5.63			
				3.74	15	56.10
0 + 60	0.43	4.3	1.85			
				0.93	15	13.95
0 + 75	0	0	0			
					Total volume	144.45 m³

[a]If the spoil is not placed in the pothole area, all cuts should be increased by 0.15 m.   ☐

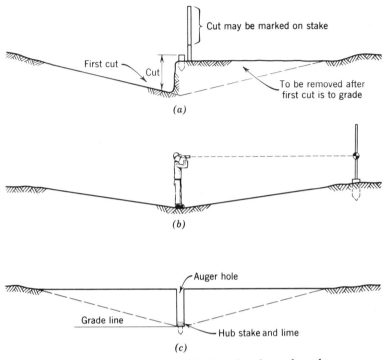

Fig. 12.8 Methods of establishing the channel grade.

Several methods of establishing the grade in the channel are shown in Fig. 12.8. The laser system described in Chapter 15 is generally used by contractors. Final grade should be checked with a surveying instrument. The method shown in Fig. 12.8*a* is convenient for shallow drains constructed with blade equipment. The hub stake in Fig. 12.8*b* may be left in place for later reference. This procedure is also suitable for terrace construction. The procedure in Fig. 12.8*c* is convenient for the operator, but considerable labor is required for layout.

12.7 Maintenance

Field surface drains can usually be maintained by normal tillage operations. Such maintenance is particularly important on flat land since a very small obstruction in the channel may cause flooding of a sizable area. Tillage implements should be lifted when crossing ditches to avoid blocking the channel. If this procedure is not followed, the channel should be kept open by dragging or shaping a smooth channel in the bottom of the drain. When the soil is wet, equipment should not cross deadfurrows, field drains, or grass waterways. Livestock may also damage such channels during rainy seasons. Pasturing at other times, however, is desirable. Plowing parallel to shallow surface drains, leaving the deadfurrow in the channel is usually adequate for maintenance. Minor depressions between drains should be filled by land grading.

LAND GRADING

Although land leveling is the term generally associated with surface irrigation, land grading is synonymous but somewhat more descriptive. For most conditions, a sloping plane surface rather than a level surface is desired.

12.8 Factors Influencing Design

Slopes, cuts, and fills are influenced by the soil, topography, climate, crops to be grown, and method of irrigation or drainage. The major problem with land grading is the effect of removing the topsoil and its influence on plant growth. Reduced growth may occur on the fill areas, although the exposure of subsoil in the cuts is usually a more serious problem. Stockpiling the topsoil and placing the spoil over the cut areas is a practical solution, where the cost can be justified.

Establishment of a uniform design slope is more important for surface irrigation than for drainage. In Fig. 12.9 the graded surface for drainage is variable with less cut and fill than for the uniform grade for irrigation. Having a variable slope for drainage is not usually objectionable, provided flow velocities are not erosive. The topography thus places a severe limitation on the length and degree of slope as well as the location of slope changes. Row lengths and slopes for irrigation are discussed in Chapter 19. The required accuracy of leveling will depend largely on its effect on crop production. For crops sensitive to excesses or shortages of water, greater precision is required. For flood irrigation the land slope in both directions may be restrictive, whereas for furrow irrigation the length-of-run and furrow grade are the most critical. In semihumid to humid climates land grading can be made compatible for both drainage and irrigation. For irrigation purposes, the largest flow occurs at the upper end of the slope; for drainage, rainfall enters along the entire slope length with the highest runoff at the lower end. These factors should be carefully considered in design.

Several methods of computing cuts and fills for land grading are described. Field data are normally obtained from a survey with ground elevations taken to the nearest 0.01 m on a 30-m square grid for horizontal control. Elevations are taken at other critical points, such as highs and lows between grid stakes, and the

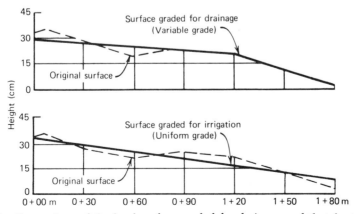

Fig. 12.9 Comparison of the land surface graded for drainage and that for irrigation.

water surface in the supply ditch or in the drainage outlet. After obtaining the desired balance between cuts and fills, the cut volume and fill volume are computed.

12.9 Plane Method

This method assumes that the area is to be graded to a true plane. The average elevation of the field is determined, and this elevation is assigned to the centroid of the area. The centroid is located by taking moments about two perpendicular reference lines, as shown in Fig. 12.10 for an irregularly shaped field. This procedure has been simplified by locating the grid system so that each grid point is at the center of the grid square and represents an area equal to 30 × 30 m. In Fig. 12.10 the centroid is located at $X_c = 3.75$, $Y_c = 2.84$, and the average elevation is 8.38 m. Any plane passing through the centroid will produce equal volumes of cut and fill. The general equation for a plane surface is

$$E = a + S_x X + S_y Y \tag{12.1}$$

where
E = elevation at any point (L),
a = elevation at the origin (L),
S_x and S_y = slope in the x and y directions, respectively (L/L),
X and Y = distance from the origin (L).

The slope of any line in the X or Y direction is determined by the statistical least-squares procedure presented by Chugg (1947). The least-squares plane by definition is that which gives the smallest sum of all the squared differences in elevation between the grid points and the plane. It is called the plane of best fit.

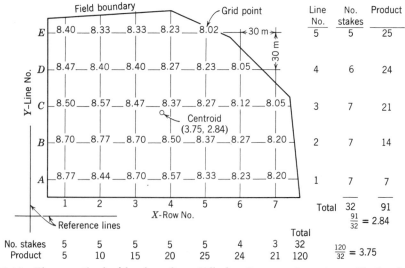

Fig. 12.10 Plane method of land grading. (All elevations are in meters with the decimal points at the grid intersections.)

This method does not necessarily provide the best slope for irrigation. These slopes can be computed from two simultaneous equations,

$$(\Sigma X^2 - nX_c^2)S_x + [\Sigma (XY) - nX_cY_c]S_y = \Sigma (XE) - nX_cE_c \tag{12.2}$$

$$(\Sigma Y^2 - nY_c^2)S_y + [\Sigma (XY) - nX_cY_c]S_x = \Sigma (YE) - nY_cE_c \tag{12.3}$$

where n = total number of grid points,
  X_c = X distance to the centroid,
  Y_c = Y distance to the centroid,
  E_c = elevation of the centroid (average elevation of all points).

For rectangular fields the terms involving XY become zero, and

$$S_x = \frac{\Sigma (XE) - nX_cE_c}{\Sigma X^2 - nX_c^2} \tag{12.4}$$

The slope S_y can be obtained from Eq. 12.4 by substituting Y for X. Although Eq. 12.4 is valid only for rectangular fields, a satisfactory solution can often be obtained by taking one or more arbitrarily selected rectangular areas within the field and extending the slopes of the plane to the remaining areas. The following example will illustrate computational procedure for an irregularly shaped field.

□ *Example 12.2*
___

Determine the equation for the plane of best fit for the field shown in Fig. 12.10. All elevations are in meters.

Solution.   As shown in Fig. 12.10, $n = 32$, $X_c = 3.75$, and $Y_c = 2.84$.

$E_c = (8.77 + 8.70 + 8.50 + \cdots + 8.20 + 8.20 + 8.05)/32 = 8.38$ m
$\Sigma X^2 = 1^2 + 1^2 + 1^2 + 1^2 + 1^2 + 2^2 + \cdots + 6^2 + 7^2 + 7^2 + 7^2 = 566$
$nX_c^2 = 32 \times 3.75^2 = 450$
$\Sigma (XY) = (1 \times 1) + (1 \times 2) + (1 \times 3) + \cdots + (7 \times 1) + (7 \times 2) + (7 \times 3)$
   $= 327$
$nX_cY_c = 32 \times 3.75 \times 2.84 = 340.8$
$\Sigma (XE) = (1 \times 8.77) + (1 \times 8.70) + (1 \times 8.50) + \cdots + (7 \times 8.20) +$
   $(7 \times 8.20) + (7 \times 8.05)$
   $= 996.69$
$nX_cE_c = 32 \times 3.75 \times 8.38 = 1005.60$

Substituting in Eq. 12.2,

$$(566 - 450)S_x + (327 - 340.8)S_y = 996.69 - 1005.60$$

From similar calculations and substituting in Eq. 12.3,

$$(327 - 340.8)S_x + (319 - 258.1)S_y = 759.14 - 761.57$$

These two simultaneous equations result in

$$S_x = -0.084/30 = -0.28 \text{ percent}$$
$$S_y = 0.059/30 = -0.20 \text{ percent}$$

Since the plane of best fit must pass through the centroid, substituting the above values in Eq. 12.1 gives the elevation of the origin as 8.86 m. Thus, the plane of best fit is

$$E = 8.86 - 0.0028X - 0.002Y$$

Elevations at each grid point are then computed to the nearest 0.01 m. After subtracting the settlement allowance, to be discussed later, cuts and fills are computed to the nearest 0.01 m (1 cm) for each grid point. Computer programs were developed by Smerdon et al. (1966). Software is available commercially (Appendix I) (Spectra-Physics, 1981) to solve land grading problems.   □

12.10 Profile Method

Ground profiles are plotted that are adapted to making changes in design slopes. A grade is established that will provide a desirable ratio of cuts to fills, as well as reduce haul distances to reasonable limits. The grid elevations for the field in Fig. 12.10 are plotted in Fig. 12.11 to illustrate the profile method. The profiles are drawn along the X direction with elevations along each line A, B, and so on. The design slopes and grade lines are selected by trial and error so the desired cut–fill ratio can be obtained (see Sect. 12.12). In Fig. 12.11 lines C and D by coincidence have the same slope as the plane of best fit in Example 12.2. Cuts and fills can be adjusted by either changing the design grade elevations or the slope of the plane as was done for lines A, B, and E. If needed, profiles can be plotted at right angles to the original lines. In this way proposed grade lines can be adjusted to meet cross-slope and downgrade criteria.

12.11 Other Methods

The plan-inspection and contour-adjustment methods may be preferred in some situations. They were described in the previous edition of this text and other references, such as SCS (1961).

12.12 Earthwork Volumes

The average end area or the prismoidal formulas given in Appendix G are suitable for making earthwork calculations, but these are time consuming. A more common procedure, called the four-point method by the SCS (1961), is sufficiently accurate for land grading. Volume of cuts for each grid square is

$$V_c = \frac{L^2(\Sigma C)^2}{4(\Sigma C + \Sigma F)} \tag{12.5}$$

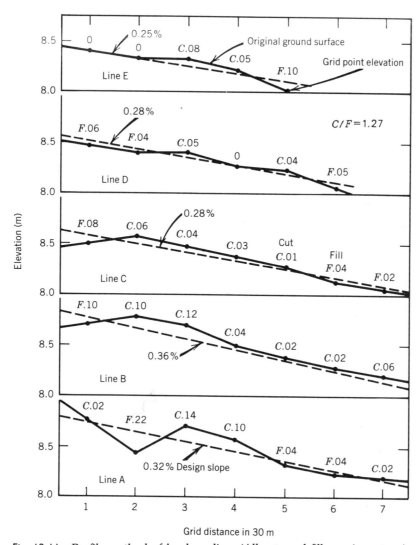

Fig. 12.11 Profile method of land grading. (All cuts and fills are in meters.)

where V_c = volume of cut (L³),
       L = grid spacing (L),
       C = cut on the grid corners (L),
       F = fill on the grid corners (L).

For computing V_f the volume of fills, $(\Sigma C)^2$ in the numerator of Eq. 12.5 is replaced by $(\Sigma F)^2$. If the volume for only a portion of a grid square is desired, the full grid volume is reduced in proportion to the reduced area. For example, if V_c is 34 m³ on a 30 × 30-m grid, the volume for a 15 × 30-m area would be 17 m³. Another more approximate method is to assume that the cut or fill at a grid point represents the average for 15 m from the point in four directions. This method is suitable only for preliminary estimates.

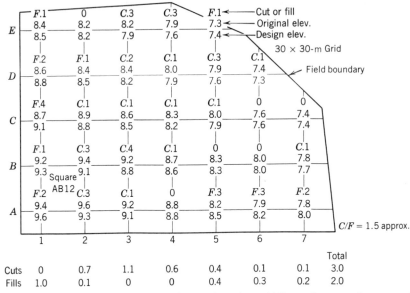

Fig. 12.12 Grid point cuts or fills for land grading. (All numbers are in meters.)

Experience has shown that, in leveling, the cut–fill ratio should be greater than 1. Compaction from equipment in the cut area, which reduces the volume, and also compaction in the fill area, which increases the fill volume needed, are believed to be the principal reasons for this effect. Marr (1957) stated that on level ground between stakes the operator has an optical illusion of a dip in the middle, and therefore in filling, crowning often occurs. The ratio of cut to fill volume is usually 1.3 to 1.6, but may range from 1.1 to 2.0. With the plane method of computing cuts and fills, a settlement correction for the whole field is more convenient to apply. The settlement allowance or the amount of lowering of the elevation may range from 0.003 to 0.01 m for compact soils and from 0.015 to 0.05 m for loose soils. A small change in elevation will cause a considerable change in the cut–fill ratio. If extra quantities of soil are needed outside the area to be leveled, such as for a roadway or depression, the plane surface can be lowered by the amount of earthwork required. In computing costs normally the volume of cut is the basis for computation. The following example illustrates earthwork computations.

□ *Example 12.3*

Determine the volumes of cuts and fills and the cut–fill ratio for the field in Fig. 12.12 using the four-point method. All elevations are in meters. Grid square $A1$, $B1$, $A2$, $B2$ is designated ($AB12$), and so on.

Solution. For all the full size 30×30-m grid squares (within $A1$, $E1$, $E5$, $D6$, $C7$, and $A7$), substituting the cuts and fills in Eq. 12.5 from the lower left ($AB12$, $AB23$, and so on) to upper left ($ED12$) grid squares in Fig. 12.12 gives

$$\Sigma V_c = \frac{30 \times 30}{4}\left[\frac{(0.3 + 0.3)^2}{(0.3 + 0.3 + 0.1 + 0.2)} + (0.3 + 0.4 + 0.1 + 0.3) + \cdots + 0\right]$$
$$= 1863 \text{ m}^3$$
$$\Sigma V_f = \frac{30 \times 30}{4}\left[\frac{(0.1 + 0.2)^2}{(0.1 + 0.2 + 0.3 + 0.3)} + 0 + \cdots + (0.1 + 0 + 0.1 + 0.2)\right]$$
$$= 603 \text{ m}^3$$

For the partial grid squares, the cut or fill at the field boundary is assumed the same as the nearest grid cut or fill. Summation of these volumes starting at $A1$ using the lower left one-fourth grid square (fill is 0.2 m at all four corners and cut is zero) and ending at the rectangle between $A1$ and $B1$ gives

$$\Sigma V_c = \frac{30 \times 30}{4}\left[0 + \tfrac{1}{2}\frac{(0.3 + 0.3)^2}{(0.3 + 0.3 + 0.2 + 0.2)} + \cdots + 0\right] = 527 \text{ m}^3$$
$$\Sigma V_f = \frac{30 \times 30}{4}\left[\tfrac{1}{4}\frac{(0.2 \times 4)^2}{(0 + 0.2 \times 4)} + \tfrac{1}{2}\frac{(0.2 + 0.2)^2}{(0.2 + 0.2 + 0.3 + 0.3)} + \cdots\right.$$
$$\left. + \tfrac{1}{2}(0.1 + 0.1 + 0.2 + 0.2)\right] = 899 \text{ m}^3$$

Total $V_c = 1863 + 527 = 2390 \text{ m}^3$
Total $V_f = 603 + 899 = 1502 \text{ m}^3$
C/F ratio $= 2390/1502 = 1.6$

The cut–fill ratio of 1.6 is more accurate than the estimated 1.5 in Fig. 12.12 because it was based on the four-point method. In this example the fill volume of the partial grids was greater than the cut volume. To have ignored the partial grids would have resulted in too high a cut–fill ratio, 3.1 (1863/603). □

12.13   Laser Surveying and Accuracy

Elevations of the land surface may be obtained by standard grid surveys with a surveying instrument or with the scraper receiving unit and laser transmitter shown in Fig. 12.13. The receiving unit mounted on the scraper references the scraper blade height from the laser plane beam. With the scraper blade just touching the field surface and using the survey mode of the laser system, the operator drives the equipment back and forth across the field along lines for which the elevations are desired. The elevations are hand-recorded at the desired points or they are automatically recorded for direct use by the computer. Some computer software systems will determine the plane of best fit, a cut–fill map, earthwork volumes, and three-dimensional and profile views of a field. With the laser system, grid points do not have to be staked or preserved as with the operator-controlled system shown in Fig. 12.14; however, the location of laser survey lines should be known by the operator. The accuracy of leveling with laser-controlled equipment, as reported by Dedrick (1979), was within 0.015 m compared with 0.033 m for conventionally leveled basins without laser-controlled equipment. The 0.015-m deviation from the design value is considered reasonably accurate. Accuracy is more critical for irrigation than for drainage, but no reverse grade is permitted for either.

Fig. 12.13 Laser transmitter and scraper with laser receiver for grade control. (Courtesy Spectra-Physics, Construction and Agricultural Division.)

On a field with a design slope in both the X and Y directions, the laser transmitter must be set on the maximum slope (also called compound slope) and at the appropriate horizontal angle (A) between the direction of maximum slope and the direction of the major slope (Fig. 12.15). The tangent of this angle is the ratio of the minor (S_x) to the major (S_y) slope. With knowledge of the angle (A) the maximum slope is ($S_x \sin A + S_y \cos A) = (S_y/\cos A)$, which is the slope to set the laser rotating beam transmitter (Schwab, 1988). For example, if the minor slope is 0.003 and the major slope is 0.004, the maximum slope is 0.005 and the angle is 36.87 degrees. Some lasers have slope settings in both the X and Y directions, which permit slopes to be set directly on the instrument and do not require these calculations.

12.14 Construction and Maintenance

Prior to making the survey, heavy vegetative growth and residue should be removed. The surface of the soil should be fairly compact and as smooth as possible. Heavy carrier-type scrapers or pans especially designed for accurate

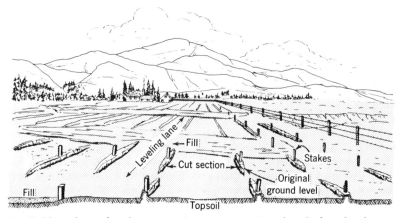

Fig. 12.14 Field surface after heavy equipment operation, but before land smoothing. (Stakes not required for laser plane grade control.)

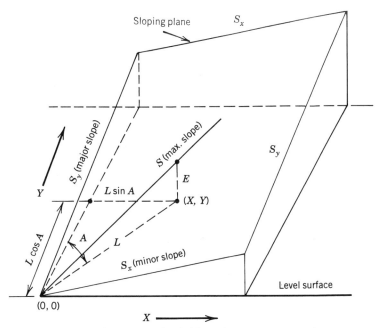

Fig. 12.15   Geometry of a field to obtain maximum slope.

depth control at shallow cuts and with laser grade control systems are most satisfactory. These are powered with crawler or rubber-tired units. Grades should be maintained as accurately as possible. For operator-controlled equipment, cuts and fills are made between grid stakes as shown in Fig. 12.14. After the grade is checked, the ridges or islands are removed and the land smoothed. In land smoothing, leveling equipment should go over the area several times. For example, levelers should be operated in both directions across the field and then diagonally both ways. Smoothing operations may be required for several years since fill material has a tendency to settle.

REFERENCES

American Society of Agricultural Engineers (ASAE) (1986). *Design and Construction of Surface Drainage Systems on Farms in Humid Areas*. EP302.3. ASAE, St. Joseph, MI.

Anderson, C. L., A. D. Halderman, H. A. Paul, and E. Rapp (1980). "Land Shaping Requirements." In *Design and Operation of Farm Irrigation Systems*, M. E. Jensen (ed.). ASAE Monograph 3, Chap. 8, pp. 281–314. ASAE, St. Joseph, MI.

Beauchamp, K. H. (1952). "Surface Drainage of Tight Soils in the Midwest." *Agr. Eng.* **33**, 208–212.

Chugg, G. E. (1947). "Calculations for Land Gradation." *Agr. Eng.* **28**, 461–463.

Coote, D. R., and P. J. Zwerman (1970). *Surface Drainage of Flat Lands in the Eastern U.S.* Cornell University Ext. Bull. 1224. Cornell University, Ithaca, NY.

Dedrick, A. R. (1979). *Land Leveling-Precision Attained with Laser Controlled Drag Scrapers*. ASAE Paper No. 79-2565. ASAE, St. Joseph, MI.

Gain, E. W. (1964). "Nature and Scope of Surface Drainage in Eastern United States and Canada." *ASAE Trans.* **7**, 167–169.

Marr, J. C. (1957). *Grading Land for Surface Irrigation.* California Agr. Expt. Sta. Circ. 438 (rev.). University of California, Davis, CA.

Pavelis, G. A. (ed.) (1987). *Farm Drainage in the United States: History, Status, and Prospects.* USDA Misc. Publ. 1455. GPO, Washington, DC.

Phillips, R. L. (1958). "Land Leveling for Drainage and Irrigation." *Agr. Eng.* **39**, 463–465, 470.

Schwab, G. O. (1988). "Land Grading." In *CRC Handbook of Engineering in Agriculture*, R. H. Brown (ed.). Vol. II, pp. 87–93. CRC Press, Boca Raton, FL.

Smerdon, E. T., K. R. Tefertiller, R. E. Kilmer, and R. V. Billingsley (1966). "Electronic Computers for Least-Cost Land-Forming Calculations." *ASAE Trans.* **9**, 190–193.

Spectra-Physics Inc. (1981). *Land Leveling Design System Software for Use with Apple II Computers.* Construction and Agricultural Division, Dayton, OH.

U.S. Soil Conservation Service (SCS) (1961). "Land Leveling." In *Irrigation, National Engineering Handbook*, Sect. 15, Chap. 12 (litho.). Washington, DC.

PROBLEMS

12.1 Compute the volume of soil to be excavated in cubic meters for a random field drain with $8:1$ sideslopes if the cuts at consecutive 15-m (50-ft) stations are 0.15 (0.5), 0.30 (1.0), 0.73 (2.4), 0.55 (1.8), 0.24 (0.8), and 0 m (0 ft). The first station is in a depression where a cut of 0.15 m is needed for adequate drainage.

12.2 Determine the equation for the plane of best fit by the least-squares procedure for the field in Fig. 12.10, using only the 25 grid point elevations on the first five lines in the X and in the Y directions. Compute the cut or fill for points A2 and C2. How does your solution compare with that in Example 12.2?

12.3 Determine the cut and fill volumes for the data shown in Fig. 12.11, by assuming the cut or fill at each grid point represents the average for the 30-m (100-ft) grid square. What is the cut–fill ratio?

12.4 Determine the cut and fill volumes for grid squares bounded by points A1, B1, A2, and B2 and by points D5, E5, D6, and E6 shown in Fig. 12.11 using the four-point method. Assume point E6 has a fill of 0.2 m and the grid square is 85 percent of a full grid.

12.5 Determine the cuts and fills by the plane method for a field assigned by the instructor, assuming a settlement allowance of 0.012 m (0.04 ft). Compute the cut and fill volumes.

12.6 Determine the slope (maximum) setting for the laser transmitter if the field slope in the X direction is 0.003 that in the Y direction is 0.007. Determine the angle between the direction of the major slope and the maximum slope to set the transmitter so as to obtain the correct plane of the laser beam for land grading operations.

CHAPTER 13

Open Channels

In this chapter open channels refer to open ditches for drainage and to canals for carrying irrigation water. Broad, shallow open ditches, which can be crossed with field machinery, are called surface drains and are discussed in Chapter 12. Open ditches provide outlets for pipe and surface drains and remove surface water directly. Open ditch systems generally drain large areas and often involve several property owners. The design of lined or earth canals that convey water from storage reservoirs or wells is basically the same as for open ditches; however, for irrigation, flows can be regulated and canals may be lined with stabilizing materials, giving greater flexibility in selecting a more efficient cross section.

The U.S. Bureau of the Census (1962) showed that more than 235 000 km of open ditches, mostly in organized drainage districts, have been constructed in 38 states. This length would be much greater if all privately owned drainage ditches were included. The length of irrigation canals would be quite large considering that about 20 000 000 ha is irrigated in the West.

An earth-lined open channel properly designed should provide (1) velocity of flow such that neither serious scouring nor sedimentation will result, (2) sufficient capacity to carry the design flow, (3) hydraulic grade at the proper depth for good water management, (4) sideslopes that are stable, and (5) minimum initial cost and maintenance. For carrying irrigation water additional requirements, such as low seepage loss, must be met. The engineer should have a knowledge of the capabilities and limitations of the various types of construction equipment and should consider these factors in the design of the system.

13.1 Channel Discharge Capacity

The discharge capacity of an open channel is computed from the Manning formula, which was discussed in Chapter 7. The depth of the constructed channel should provide a freeboard allowance ($D - d$ in Fig. 13.1) of 20 percent of the total depth, D. The freeboard provides a safety factor for overtopping and allows a reasonable depth for sediment accumulation. On large earth irrigation canals in the West the freeboard commonly varies from 30 to 35 percent, and for lined

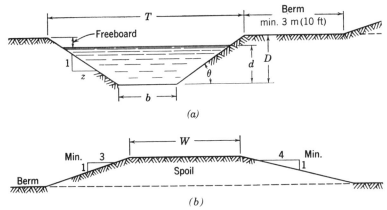

Fig. 13.1   Elements of an open channel (a) cross section and (b) spoil bank.

canals it ranges from 15 to 20 percent. The freeboard for lined canals refers only to the top of the lining. The total freeboard is the same for lined and unlined channels. The freeboard should be increased on the outside edge of curves.

Wherever practical, the channel should be designed for a high hydraulic efficiency. In earth channels the stability of the soil places limitations on channel grade and sideslopes. The topography and desired water levels may limit the design grade and the velocity of flow.

The roughness coefficient for open channels varies with the height, density, and type of vegetation; physical roughness of the bottom and sides of the channel; variation in size and shape of the cross section; alignment; and hydraulic radius. Primarily because of differences in vegetation, the roughness coefficient varies from season to season. In general, conditions that increase turbulence increase the roughness coefficient. Values of n can be obtained from Appendix B.

Experience and knowledge of local conditions are helpful in the selection of the roughness coefficient, but it is one of the most difficult problems in channel design. In general, a value of 0.035 is satisfactory for medium-sized earth ditches with bottom widths of 1.5 to 3 m, and 0.04 is suitable for smaller ditches with bottom widths of 1.2 m.

The design procedure for open channels is illustrated in the following example.

□ *Example 13.1*

Design an open channel to carry 4.4 m³/s (156 cfs) on a slope of 0.09 percent, assuming ditch sideslopes of 2:1 and a bottom width of 1.2 m (4 ft).

Solution.   From Table B.1, select a roughness coefficient of 0.04 for a small drainage ditch. Make a trial solution, using a flow depth of 1.5 m (5 ft) with the symbols in Fig. 13.1.

$$R = \frac{bd + zd^2}{b + 2d(z^2 + 1)^{1/2}} = \frac{1.8 + 4.5}{1.2 + 3(5)^{1/2}} = 0.80 \text{ m (2.6 ft)}$$

From Fig. B.1, read $v = 0.65$ m/s (2.1 fps) or substitute in the Manning formula, $v = R^{2/3}s^{1/2}/n$, and calculate:

$$q = av = 6.3 \times 0.65 = 4.10 \text{ (m}^3\text{/s) (0.3 m}^3\text{/s too low)}$$

Next try $d = 1.56$ m (5.1 ft), from which $R = 0.82$ m (2.7 ft), and determine $v = 0.66$ m/s (2.16 fps):

$$q = 6.74 \times 0.66 = 4.45 \text{ m}^3/\text{s (157 cfs)}$$

which is close to the desired capacity. If the freeboard is 20 percent, the total depth is $(1.56/100 - 20)100 = 1.95$ m (6.4 ft). The SCS engineering computer program (Appendix I) gave a depth of 1.55 m.   □

13.2 Cross Section

The design dimensions of an open channel cross section are shown in Fig. 13.1. Earth channels and lined canals are normally designed with trapezoidal cross sections. The size of the cross section will vary with the velocity and quantity of water to be removed.

Sideslopes. Channel sideslopes are determined principally by soil texture and stability. The most critical condition for sloughing occurs after a rapid drop in the flow level that leaves the banks saturated, creating a seepage drag force. For the same sideslopes, the deeper the ditch the more likely it is to slough. Sideslopes should be designed to suit soil conditions and not the limitations of construction equipment. Suggested sideslopes are shown in Table 13.1. Whenever possible, these slopes should be verified by experience and local practices. Very narrow ditches should have slightly flatter sideslopes than wide ditches because of greater reduction in capacity in the narrow ditches if sloughing occurs.

Bottom Width. After the channel grade, depth, and sideslopes are selected, the bottom width can be computed for a given discharge. The bottom width for the most hydraulic-efficient cross section and minimum volume of excavation is determined by the formula

$$b = 2d \tan \theta/2 = 2d/[z + (z^2 + 1)^{1/2}] \tag{13.1}$$

where $b =$ bottom width (L),
   $d =$ design depth (L),
   $\theta =$ sideslope angle (degrees),
   $z =$ sideslope ratio (L/1) (horizontal/vertical).

Table 13.1   Maximum Sideslopes for Open Channels

	Sideslopes — Horizontal : Vertical	
Soil	*Shallow Channels up to 1.2 m*	*Deep Channels, 1.2 m and Deeper*
Peat and muck	Vertical	¼ : 1
Stiff (heavy) clay	½ : 1	1 : 1
Clay or silt loam	1 : 1	1½ : 1
Sandy loam	1½ : 1	2 : 1
Loose sandy	2 : 1	3 : 1

Source: By permission, from *Land Drainage and Flood Protection*, by B. A. Etcheverry, copyright, 1931, McGraw-Hill Book Company.

For any sideslope it can be shown mathematically that, for a bottom width computed from Eq. 13.1, the hydraulic radius is equal to one half the depth. The minimum bottom width should be 1.2 m except in small laterals. It is not always possible to design for the most efficient cross section because of construction equipment limitations, allowable velocities, and increased maintenance.

Spoil Banks.   The excavated soil may be placed on one or both sides of the channel, depending on the type of equipment, use of the channel, and size of the cross section. If the spoil bank is to serve as a levee, the spoil is normally placed on only one side. For drainage ditches, the spoil bank should be spread until it blends into the adjoining field, thus permitting cultivation near the edge of the ditch. Where levees or access roads are desired, the spoil should be spread to conform to the cross section shown in Fig. 13.1. The berm width, or the distance from the edge of the ditch to the edge of the spoil, provides a degree of protection against sloughing and may be used as a place for the ditching machine to operate. The steeper the sideslopes, the greater should be the berm width. For ditches with sideslopes of 1 : 1 the berm width should be twice the depth, and for sideslopes of 2 : 1 the berm width should be equal to the depth. The minimum berm width should be 3 m. The depth and top width of the spoil bank vary with the size of the ditch. Unless the spoil bank is to be used as a levee, breaks or gaps that allow surface water to drain to the channel should be provided, or tube inlets may be installed through the spoil bank (see Chapter 9).

For irrigation canals spoil banks are often compacted and placed so as to raise the water level above the original ground surface. In this case the spoil sideslope should be the same as for the channel. If the soil is sufficiently stable or if the canal is lined, the berm is sometimes omitted. In deep cuts berms are provided for stability as well as for roadways. The land sideslope of canal spoil banks may be made steeper than that for drainage ditches, unless flat slopes are required for maintenance or control of seepage or if the land is to be farmed.

13.3   Velocities

In earth channels optimum velocities of flow are based on (1) selection of limiting velocities or on (2) computed values of the critical tractive force. Tractive force is defined as the hydraulic shearing force per unit area on the periphery of the channel.

No definite optimum velocity can be prescribed, but minimum and maximum limits can be approximated. Velocities should be low enough to prevent scour but high enough to prevent sedimentation. Usually an average velocity of 0.5 to 1.0 m/s for shallow channels is sufficient to prevent sedimentation. In channels that flow intermittently, vegetation may retard the flow to such an extent that adequate velocities at low discharges are difficult, if not impossible, to maintain. For this reason maximum or even higher velocities should be obtained, since scouring for short periods at high flows may aid in maintaining the required cross section.

Typical velocity distribution in an open channel is shown in Fig. 13.2. The maximum velocity occurs near the center of the stream and slightly below the surface. The average velocity in this stream, having a rather straight and relatively smooth channel, was 1 m/s with a maximum of 1.3. Around curves the maximum velocity and the velocity contours move toward the outer bank. Since turbulent flow in open channels occurs at Reynolds numbers above 500, velocities are almost always high enough for turbulent flow.

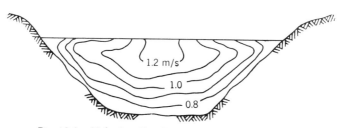

Fig. 13.2   Velocity distribution in an open channel.

Limiting Velocity Method.   From a survey of irrigation canals in the western United States, Fortier and Scobey (1926) recommended the limiting velocities shown in Table 13.2. Their data have been widely accepted. These velocities may be exceeded for some soils where the streamflow contains sediment because deposition may produce a well-graded channel bed resistant to erosion. Where a powerful abrasive is carried in the water, these velocities should be reduced by 0.15 m/s. For depths over 0.9 m Fortier and Scobey (1926) permitted velocities 0.15 m/s higher than shown. Where the channel is winding or curved, the limiting velocity should be reduced about 25 percent.

Tractive Force Method.   In 1950 the USBR began studies based on the tractive force theory for improving the design of earth-lined irrigation canals. As shown in

Table 13.2   Fortier and Scobey's Limiting Velocities with Corresponding Tractive Force Values (Straight Channels After Aging)

Material	Roughness Coefficient, n	Clear Water		Water-Transporting Colloidal Silts	
		Velocity (m/s)	Tractive Force (Pa)	Velocity (m/s)	Tractive Force (Pa)
Fine sand, colloidal	0.020	0.46	1.3	0.76	3.6
Sandy loam, noncolloidal	0.020	0.53	1.8	0.76	3.6
Silt loam, noncolloidal	0.020	0.61	2.3	0.92	5.3
Alluvial silts, noncolloidal	0.020	0.61	2.3	1.07	7.2
Ordinary firm loam	0.020	0.76	3.6	1.07	7.2
Volcanic ash	0.020	0.76	3.6	1.07	7.2
Stiff clay, very colloidal	0.025	1.14	12.4	1.53	22.0
Alluvial silts, colloidal	0.025	1.14	12.4	1.53	22.0
Shales and hardpans	0.025	1.83	32.1	1.83	32.1
Fine gravel	0.020	0.76	3.6	1.53	15.3
Graded loam to cobbles when noncolloidal	0.030	1.14	18.2	1.53	31.6
Graded silts to cobbles when colloidal	0.030	1.22	20.6	1.68	38.3
Coarse gravel, noncolloidal	0.025	1.22	14.4	1.83	32.1
Cobbles and shingles	0.035	1.53	43.6	1.68	52.7

Source: Adapted from Lane (1955).

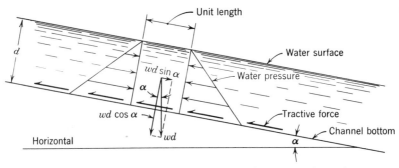

Fig. 13.3 Tractive force on the bottom of an open channel.

Fig. 13.3, the tractive force (shear) is equal to and in the opposite direction to the force the bed exerts on the flowing water. In a uniform channel of constant slope and constant flow, the water is moving in a state of steady, uniform flow without acceleration because the force tending to prevent motion is equal to the force causing motion. For small channel slopes, sin α is equal to tan α or the slope, and the tractive force,

$$T = wdsK \qquad (13.2)$$

where T = tractive force (F/L^2),
   w = unit weight of water (9800 N/m^3) (F/L^3),
   d = depth of flow (L),
   s = slope (hydraulic gradient) (L/L),
   K = ratio of the tractive force for noncohesive material necessary to start motion on the sloping side of a channel to that required to start motion for the same material on a level surface.

For an infinitely wide channel of uniform depth K is 1. The ratio is less than 1 on the sloping sides of a channel because the force of gravity is also acting on the particle, tending to roll or slide it down the slope. For noncohesive material the ratio K is further defined by Lane (1955) in terms of the angle of repose of the material with the horizontal and the sideslope of the channel. For cohesive and fine single-grain materials, the cohesive forces are so great in proportion to the gravity forces causing particles to roll down the slope that the gravity component may be neglected, in which case K equals 1. A tractive force equation has been derived by Smerdon and Beasley (1961) for nonuniform flow, such as would occur in a terrace channel or in an irrigation canal where water is removed along its length. They also found that the critical tractive force for cohesive soils was correlated with the plasticity index, dispersion ratio, particle size, and clay content. Laflen and Beasley (1960) concluded that the critical tractive force increased linearly as the void ratio decreased, but the values varied with different soil types.

Critical tractive force values were computed by Lane (1955) for the corresponding velocity of flow and roughness coefficient indicated in Table 13.2. In making the computations a depth of 0.9 m, bottom width of 3.1 m, and sideslope ratio of 1.5:1 were assumed. Thus, these tractive force values were not independently derived, but design may be based on these values if more reliable data are not available. The application of tractive force theory to design provides a theoretical rather than a purely empirical basis for design. Design values for noncohesive soils, have been well established. The USBR (1962) has developed equations for cohesive soils that involve soil tests more elaborate than those indicated above. The design procedure using critical tractive force is illustrated in the following example.

□ Example 13.2

Design a stable channel with a sideslope ratio of $1:1$ to carry a flow of 4.42 m³/s (156 cfs) on a slope of 0.25 percent. The soil is stiff clay, very colloidal, $n = 0.025$, and the water is transporting colloidal silts.

Solution.   From Table 13.2 the critical tractive force at the bottom of the channel when the particles are in a state of impending motion is 22.0 Pa. Since the soil is cohesive, tractive force on the sideslopes does not limit the design. If the soil were noncohesive, the tractive force on the sideslope would control in the analysis. Substituting in Eq. 13.2, the maximum depth with $K = 1$ is

$$d = \frac{22.0}{9800 \times 0.0025} = 0.9 \text{ m (2.95 ft)}$$

Assuming b = 2.44 m (8 ft), compute a = 3.01 m² (32.3 ft²) and p = 4.99 m (16.36 ft). Substituting in the Manning equation, $v = 1.43$ m/s (4.7 fps) and q = 4.30 m³/s (152 cfs), which is close enough to the required capacity of 4.42 m³/s (156 cfs).

An alternate solution can be obtained by using the maximum permissible velocity of 1.53 m/s (5.0 fps) in Table 13.2 rather than the critical tractive force. In this case, the 1.53-m/s (5.0-fps) velocity would result in a greater depth of flow and a smaller bottom width than shown above. For practical purposes either criterion is satisfactory. Selection of design criteria should be based on the best available data. For more detailed design procedure see Chow (1959, pp. 164–179).

□

13.4 Water Surface Profiles

For simple design problems the depth of flow and the velocity can be assumed constant. For this uniform flow condition the energy grade line, hydraulic grade line, and the channel bottom are all parallel. However, if an obstruction, such as a culvert, is placed in the channel, a concave water surface called a "backwater curve" develops upstream of the structure. A change from a gentle to a flat grade could produce the same effect. If the channel slope changed from a flat to a steep grade, a convex water surface would develop. These examples illustrate two common types of gradually varied flow conditions. Such a water surface profile for uniform-shaped channels can be computed from the energy equation

$$\frac{v_1^2}{2g} + d_1 + S_0x = \frac{v_2^2}{2g} + d_2 + Sx \qquad (13.3)$$

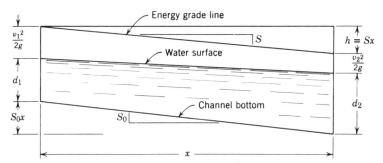

Fig. 13.4 Gradually varied flow in an open channel.

where v is the velocity of flow and all other symbols are defined in Fig. 13.4. Such an equation can be easily solved with computers.

Where the water surface profile is to be computed, the general procedure is to (1) determine the depth of flow at the control section; (2) assume a new depth at a short distance (reach) from the control section, (3) evaluate the slope of the energy grade line S and the velocity head, (4) substitute these values in Eq. 13.3, (5) compute the length of the reach, and (6) continue the trial and error procedure for each reach repeating Steps (1), (2), (3), and (4). The process is continued until the desired profile is obtained. Variations in depth should not result in velocity changes of more than 20 percent. Where the depth of flow is greater than the critical depth (see Chapter 9), profile computations should proceed upstream. Where the depth is less than critical, they should proceed downstream. The following example illustrates the procedure for a backwater curve.

☐ *Example 13.3*

Determine the water surface profile in a drainage ditch in which a culvert restricts the normal flow. The channel has a 1.2-m bottom width, sideslopes of 1:1, n of 0.04, a normal depth of flow of 0.9 m, and a channel slope of 0.2 percent. The culvert at Station 1000 causes the backwater to rise 0.2 m at the culvert. Elevation of the bottom of the ditch at the culvert is 16.15 m.

Solution. From the Manning equation, the velocity is 0.71 m/s at a depth of 0.9 m and the discharge is 1.34 m/s. Computation of the length of successive reaches between selected depths of flow is illustrated in Table 13.3.   ☐

Thirteen types of water surface profiles have been classified according to the nature of the channel slope and the relative depths of flow. Chow (1959) has developed a graphical-integration method for determining the shape of the flow profile that has broad application. Because the subject is beyond the scope of this text, the reader should refer to Chow (1959), Brater and King (1976), French (1985), and many other excellent hydraulics books for further details.

13.5 Alignment

Where changes in direction are necessary, gradual curves should be provided to prevent excessive bank erosion. The radius of curvature depends on the velocity of flow and the stability of the sideslopes. If gradual curves will not eliminate erosion

Table 13.3 Solution for Example 13.3

d (m)	Station (m)	Elevation Bottom (m)	Elevation Surface (m)	v (m/s)	$d + \dfrac{v^2}{2g}$ (m)	G	S (m/m)	$S_0 - S$ (m/m)	x (m)
(1)	(2)	(3)	(4)	(5)	(6)	(7)	(8)	(9)	(10)
1.10	1000	16.15	17.25	0.53	1.1143				
						1.116	0.0010	0.0010	48
1.05	952	16.25	17.30	0.57	1.0664				
						1.171	0.0012	0.0008	60
1.00	892	16.37	17.37	0.61	1.0189				
						1.224	0.0014	0.0006	67
0.96	825	16.50	17.46	0.65	0.9813				
						1.270	0.0017	0.0003	82
0.93	743	16.66	17.59	0.68	0.9533				
						1.311	0.0019	0.0001	228
0.90	515	17.12	18.02	0.71	0.9256				

Col. 1: Depths arbitrarily assumed depending on accuracy desired.
Col. 2: Line 1 given, other lines determined from Col. 10.
Col. 3: Computed for a 0.2 percent slope from the previous station.
Col. 4: Col. 1 plus Col. 3.
Col. 5: Flow rate (1.34 m³/s) divided by cross-sectional area.
Col. 6: Specific energy head for the given depth (Col. 1) and velocity head.
Col. 7: Ratio of bottom width to average depth $G = \dfrac{2b}{d_1 + d_2}$.
Col. 8: Relationship derived from Brater and King (1976); $S = \left(\dfrac{Qn}{b^{8/3}} \dfrac{(G + 2(z^2 + 1)^{0.5})^{2/3} \, G^{8/3}}{(z + G)^{5/3}} \right)^2$.
Col. 10: Solve for x in Eq. 13.3, which is the change in Col. 6 divided by Col. 9.

in the channel, it may occasionally be necessary to decrease the velocity by increasing the width or flattening the sideslopes or to provide bank protection. It is convenient to express the radius of a curve by the degree of curve, which is defined as the angle subtended at the center of a circle by a 100-m or 100-ft chord. The relationship between the degree of curvature and the radius of curvature shown in Fig. 13.5 is expressed by the equation

$$R = \frac{50}{\sin D/2} \tag{13.4}$$

where $R =$ the radius of curvature in m or ft,
$\qquad D =$ the degree of curvature for SI or English units (Fig. 13.5).

The SI radius of curvature can be obtained by making the chord length 100 m and by multiplying the English degree of curvature by 3.28 as shown in Table 13.4. Equation 13.4 is applicable for either system of units using the appropriate degree of curvature. As illustrated in Appendix G, layout of such a circular curve is simplified by using a chord length of 100 units. As shown in Fig. 13.5, the radius of the curve increases as the degree of curvature decreases. For small channels, the degree of curvature may be 60 degrees (SI units) or greater. It should be reduced as sideslopes and the velocity of flow are increased.

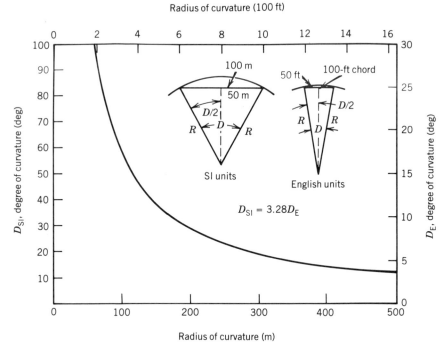

Fig. 13.5   Degree of curvature in SI or English units.

13.6   Junctions

The junction of one channel with another should be such that serious bank erosion, scour holes, or sedimentation does not occur. Where the general direction of the main is perpendicular to that of the lateral, the lateral may be curved near the junction. The angle at which the two channels join is less important for small channels having low velocities than for larger ditches having higher velocities.

The bottom of the main and the lateral should join at the same elevation. If the lateral is shallower than the main, the overfall at the junction may be eliminated

Table 13.4   Curvature Recommendations for Open Channels

Channel Top Width (m)	Slope (%)	Approximate Maximum Curvature (deg)[a]	
		D_E	D_{SI}
4–6	<0.06	19	62
4–6	0.06–0.12	14	46
5–11	<0.06	11	36
5–11	0.06–0.12	10	33
11	<0.06	10	33
11	0.06–0.12	7	23

[a]Degree of curvature, $D_{SI} = 3.28 \, D_E$ (English units).
Source: SCS (1973).

by increasing the grade of the lateral in the first 60 to 90 m or by increasing the slope on the entire lateral. Some engineers recommend excavating the first 15 to 90 m of the lateral at a level grade to provide a recessed area for sediment storage until the channel stabilizes. Since maximum velocities occur at the higher stages, the increased slope near the outlet may not be serious because the water in the main will back up into the lateral. The most serious condition exists when the main is at low flow and the lateral has a large discharge.

DRAINAGE DITCHES

For a given watershed area the required capacity of an open ditch is considerably different than the design capacity of a grassed waterway (see Chapter 7). Open ditches generally have flatter bed slopes, lower velocities, steeper sideslopes, and a greater depth of channel flow than do grassed waterways. Although grassed waterways are designed to carry peak runoff, open ditches are designed to remove water much more slowly but still rapidly enough to prevent serious damage to crops on adjacent land.

Open ditches are arbitrarily designated as mains, submains, laterals, and field ditches. The drainage plan should incorporate, as needed, levees, pump installations, field and open ditch systems, and pipe drains into a coordinated system.

13.7 Design Capacity

The rate of conveyance of water by open ditches is influenced by (1) rainfall rate, (2) size of the drainage area, (3) runoff characteristics including slope, soil, and vegetation, (4) potential productivity of the soil, (5) crops, (6) degree of protection warranted, and (7) frequency and height of flood waters from rivers and creeks. Although the degree of protection is very important in design, it is one of the most difficult factors to evaluate because costs must be compared with anticipated flood damage. More frequent flooding is permissible for agricultural land than for homes and building sites. Likewise, timberland requires less intensive drainage than does cultivated land. High-value croplands, such as those for vegetables, may require special consideration and designs for removing high flood-flows.

The runoff for open ditch design may be expressed as a drainage coefficient or rate of flow per acre, hectare, or square mile. The drainage coefficient is defined as the depth of water that is to be removed in a 24-h period from the entire drainage area.

A method of determining the required rate of water removal is by the empirical formula (ASAE, 1988)

$$q = 0.013CM^{0.833} \tag{13.5}$$

where q = runoff in m³/s,
C = a constant,
M = watershed area in km².

The constant C varies with the degree of drainage desired and the location (see Fig. 13.6). The SCS (1973) presented a general procedure for computing the constant C from rainfall excess. Runoff volume is determined for a 2- to 5-year return period

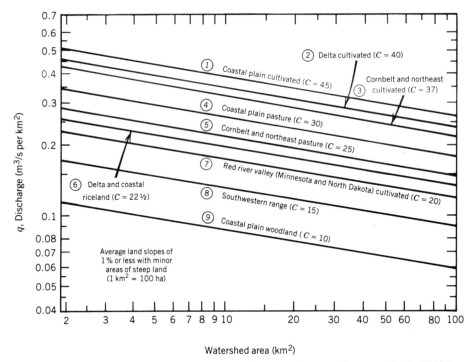

Fig. 13.6 Drainage design discharge curves for humid areas. (*Source:* ASAE, 1988.)

storm for a 48-h period. Rainfall excess is taken as one half this value for the 24-h depth.

Quite frequently watersheds with upper tributaries having rather steep slopes and high runoff-producing characteristics drain into an alluvial valley. Open ditches are then often necessary to carry the water across the flood plain from the hill area to the natural outlet. Relying on experience and judgment the engineer should modify Eq. 13.5 to meet such local conditions.

The design discharge below the junction of two ditches depends on the size of their drainage areas. An empirical procedure called the 20–40 rule can be applied to arrive at the design flow rate. When the watershed of one of the ditches is from 40 to 50 percent of the total watershed, the design flow below the junction should be the sum of the design discharges of the two ditches using their appropriate watershed areas. When the watershed of one lateral is less than 20 percent of the total watershed, the design flow rate is for the total watershed treated as a unit. For example, if two ditches at their junction have drainage areas of 1000 and 200 ha, respectively, the design flow should be based on a watershed area of 1200 ha below the junction. When the watershed of a lateral is in the range of 20 to 40 percent of the total watershed, the proper design flow rate should be somewhere between the previous two design rates as determined by interpolation (ASAE, 1988).

For arid regions, drainage and seepage discharge from irrigated lands may be estimated as a percentage of the irrigation water, expressed as (ASAE, 1988)

$$D_c = I(P + S)/(100T) \tag{13.6}$$

where D_c = drainage coefficient in mm/day,

P = deep percolation from irrigation and leaching based on the maximum area to be irrigated at the same time in percent of the irrigation application,

S = field canal seepage loss in percent,

I = irrigation depth of application in mm,

T = time between irrigations in days.

The drainage coefficient for collection ditches usually ranges from 3 to 6 mm/day. It may be reduced by water losses to trees, phreatophytes, and so on. In areas subject to seasonal high rainfall, design discharges may be based on 24-h rainfall, such as occurs during the monsoons in India. Designing for flood water control may be more critical than for discharges from irrigation. In the West the USBR (1978) recommends that ditches be designed for streamflow only and no allowance be made for irrigation waste. Streamflow may be estimated from rainfall for a 5-year return period. In all cases, design discharges should be based on local recommendations.

13.8 Grades

The engineer frequently has little choice in the selection of grades for open ditches since the grade is determined largely by the outlet elevation, elevation and distance to the lowest point to be drained, and depth of ditches. Where open ditches drain flat land, the grade should be as steep as possible, provided maximum permissible velocities are not exceeded.

The depth at all points along the channel should be sufficient to adequately drain the area. Where subsurface drains outlet into the ditch, a minimum depth of 1 to 2 m is required. In peat and muck soils the ditch should be made deeper to allow for subsidence. Because of reduced velocities, sediment accumulates more readily and vegetation grows more abundantly in shallow than in deep ditches. In some instances an allowance is made for accumulation of sediment, depending on channel velocities and soil conditions.

13.9 Controlled Drainage Structures

Controlled drainage with open ditches requires structures in the channel to maintain the water at the required level for optimum plant growth.

Several types of control structures are the burlap bag dam, timber or sheet piling cutoff, reinforced concrete structure, and a combination culvert and control gate. The elevation views of some of these structures are illustrated in Fig. 13.7. Since these structures are designed for organic soils, the cutoff wall is larger than required for less permeable soils. All control dams should be provided with an erosion-controlling apron constructed below the dam as for a drop spillway. Burlap bag dams with a facing of timber (Fig. 13.7a) are suitable for low heads and small drainage areas. These bags are filled with a weak concrete mixture and tamped together before wetting. Timber or sheet piling (Fig. 13.7b) may be placed to the side and under the ditch. This type of dam provides a greater barrier against seepage and is most suitable in deep organic soils or other highly permeable materials. Reinforced concrete dams (Fig. 13.7c) are better suited for handling large quantities of drainage flow. Where seepage is a problem, piling may be required

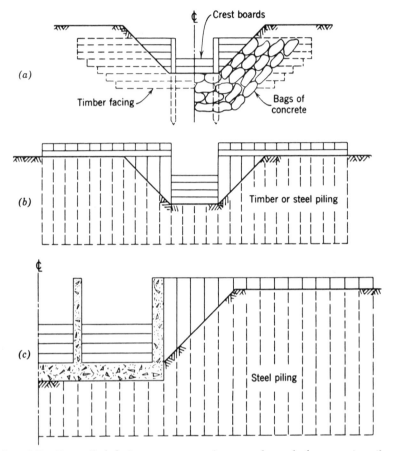

Fig. 13.7 Controlled drainage structures in open channels for organic soils.

both under and to the sides of the structure. Combination culvert and control gate structures (see Chapter 9) with crestboards at the upper end of the conduit may provide an economical and practical dam. Crestboards are placed in the openings when the water level is to be increased or removed when drainage is required.

13.10 Location and Layout

Preliminary Survey. The preliminary survey is an initial investigation of the proposed project. It should include an estimate of the potential productivity of the soil and the physical features of the watershed, as well as rainfall and runoff data, carried out in sufficient detail to provide a tentative design and to make possible a valid estimate of the economic feasibility of the plan.

Since the location of open ditches requires experience and good judgment combined with a careful study of local conditions, only a few general rules can be given: (1) follow the general direction of natural drainageways, particularly with mains and submains; (2) provide straight channels with gradual curves, especially for large ditches; (3) locate drains along property lines and utility rights of way, if practicable; (4) make use of natural or existing ditches as much as possible; (5) use the available grade to best advantage, particularly on flat land; and (6) avoid unstable soils and other natural conditions that increase construction and maintenance costs.

Location Survey.   After the ditch system has been designed and approved, and construction has been authorized, the engineer proceeds with the field layout. The general location and alignment of the channel have normally been determined by the preliminary survey. If the system is small, the preliminary and location surveys may be made at the same time. In staking the field layout, minor changes in location or design are sometimes desirable.

13.11   Legal Aspects

Mutual Drainage Enterprises.   Many states have laws that provide for the organization of mutual drainage enterprises, also called drainage associations. To establish such an enterprise the landowners involved must be fully in accord with the plan of operation and with the apportionment of the cost. After the agreement has been drawn up and signed, it must be properly recorded in the drainage record of the county or other political subdivision. The local court may be asked to name the district officials, sometimes called commissioners, who are responsible for the functioning of the district, or they may be named in the agreement. The principal advantage of the mutual district is that less time is required to establish an organization and the costs are held to a minimum. Because it may be difficult for several landowners to come to an agreement, particularly on the division of the costs, such districts are difficult to organize where the number of landowners is large or where considerable area is involved. Although a mutual enterprise cannot assess taxes, it may petition to become an organized district in some states to overcome this disadvantage. A large number of mutual enterprises are in existence and much drainage has been accomplished in this manner.

Organized Drainage Enterprises.   Such an enterprise, often called a county ditch or a drainage district, is a local unit of government established under state laws for the purpose of constructing and maintaining satisfactory outlets for the removal of excess surface and subsurface water. It is different from a mutual enterprise in that minority landowners can be compelled to go along with the project. For further details on drainage enterprises consult the laws or publications of the state involved. The method of organization, powers of district officials, and methods of assessing benefits, damages, and financing vary greatly from state to state.

IRRIGATION CANALS

The location and layout of laterals and farm ditches for distributing irrigation water are normally a part of a surface irrigation system, which is described in Chapter 19. The discussion to follow applies mainly to larger canals beyond the farm boundaries but may apply to farm ditches as well.

13.12 Design Flow

The design flow for irrigation canals should be adequate to supply the maximum rate of water use by plants. Peak use rates according to Houk (1956) are often estimated as 10 to 15 percent higher than maximum monthly rates. This maximum rate depends on the area irrigated, the climatic conditions, and the type and stage of growth of different crops within the area. Allowances should be made for additional flow from storm rainfall, for conveyance losses, and for unavoidable delivery and application losses. Canal seepage losses will be highest during the early part of the growing season. For large projects with diversified crops, maximum demand usually occurs during the middle of the growing season or slightly later (see Chapter 18).

Houk (1956) reported that seepage losses on large irrigation projects, where only small sections of the canals are lined, may vary from 15 to 45 percent of the water diverted. He recommended that the design flow of laterals per unit area should be 10 to 15 percent greater than for the main canals, and for sublaterals and farm ditches, 25 to 50 percent greater.

13.13 Location and Layout

Canals are located so that the water may flow by gravity from the canal to the point of use; however, under some conditions the topography or soil formation may make another location more economically feasible. Most problems in location arise because of varying surface and subsurface conditions. The water surface in the canal should be kept above the natural ground surface at the point of delivery and should permit the measurement of flow where required. Gradual curves are more important for canals than for drainage ditches since capacity flow must be carried for a longer period. Layout surveys are similar to those for drainage ditches (see previous section of this chapter and Appendix G).

13.14 Canal Linings

Irrigation canals are lined for several purposes: (1) to reduce seepage losses, thus increasing conveyance efficiency and decreasing drainage problems; (2) to ensure against uninterrupted operation resulting from breaks; (3) to provide a more efficient cross section by increasing sideslopes, by reducing the roughness coefficient, by eliminating weed and moss growth, and by increasing channel slope without danger of erosion; and (4) to reduce maintenance. In areas where water is in short supply one of the most important benefits from lining canals is the saving of water, which then becomes available for other beneficial uses. As discussed previously, losses in earth canals are sometimes a high percentage of the water diverted. One of the reasons for emphasizing linings is that canal losses are usually somewhat easier to control than losses resulting from poor application and water distribution on a field.

Canal linings may be constructed with a large number of materials, such as concrete, rock masonry, brick, colloid clay–soil mixtures, soil cement, asphalt, rubber, and plastic. Materials that are most satisfactory meet all the purposes described above. In laterals and farm ditches, concrete and asphalt are the two most common materials. The selection of a lining material will depend largely on cost and availability of materials, soil conditions, cross section and length of the canal, and comparative annual costs. Average annual cost, including maintenance and value of the water saved, is the best basis for making a decision.

Concrete more nearly meets all of the requirements for a lining than any other material. Its principal disadvantages are high initial cost and possible damage by soil chemicals and freezing and thawing. Concrete linings vary in thickness from 25 to 150 mm and may or may not be reinforced. Concrete may be placed by pouring in alternate panels, by shooting it on pneumatically (called shotcrete or Gunite), by plastering, by fitting precast slabs, or by continuous pouring with slip-form equipment. Because of its favorable cost, slip-form placement is being widely accepted for surface channels. Similar equipment has also been developed for placing underground distribution pipe. Precast slabs with tongue and groove joints filled with asphalt provide a slightly flexible lining that will adjust to minor movements of the subgrade. Stone and brick linings can be placed in a similar manner; however, the labor cost would be higher.

Asphalt linings may be applied as an asphaltic concrete that includes sand and gravel, hot-sprayed asphalt membranes, and prefabricated liners. Membrane-type linings may or may not be covered with soil or other material. Although asphalt is considerably lower in initial cost, it has a shorter life and is more subject to physical damage than concrete. Vegetation also grows through asphalt membranes, unless the soil is first sterilized. Lauritzen (1961) estimates that the life of exposed asphalt liners is probably less than 10 years. See USBR (1963) and Houk (1956) for further details on linings.

13.15 Conveyance and Control Structures

Most of the conservation structures discussed in Chapter 9 have application to irrigation. On-farm structures are discussed in Chapter 19. Inverted siphons, tunnels, flumes, and flow-regulating structures are often necessary for proper water control. In crossing natural depressions or canyons, flumes or inverted siphons may be constructed. Inlet and outlet transition sections between the structure and the canal must be carefully designed. In any lined channel where the construction cost is high, considerable effort should be made to design structures and facilities that have a high hydraulic efficiency. Model studies may be required in some cases.

CONSTRUCTION

Prior to construction open ditches or irrigation canals must be designed and the location marked in the field with center line and slope stakes. Where hub stakes must be preserved, a baseline offset from the channel at a suitable distance should be established. Slope stakes must be properly set to establish the desired sideslope ratio. Where the land is sloping or irregular normal to the center line, slope stakes must be located in the field by a trial-and-error procedure, such as illustrated in Appendix G.

13.16   Types of Equipment

As shown in Table 13.5, ditch construction equipment may be classified according to function and type of earth-moving action involved. With regard to function, construction equipment may either cut and carry the soil or cut, spread, and push the spoil. With regard to type of earth-moving action, it may be either continuous or intermittent. Several types of machines are illustrated in Fig. 13.8.

Continuous-action machines generally have a higher output than intermittent-motion equipment. Examples of continuous-action machines are the wheel and template excavators. The wheel excavator is similar to the wheel-type trenching machine (Fig. 15.2), except that the resulting sideslopes are less than vertical. The template excavator digs a somewhat larger channel than the wheel-type machine, but both types excavate channels whose cross sections cannot be varied except by modifying the equipment. Both machines construct the channel as they move along the ditch line. Plow-type ditchers are similar in construction and operation to an ordinary lister, but the ditcher is a much larger machine. Blade and elevating graders, though not precisely continuous in action, are suitable only in soils where sufficient traction can be obtained in the bottom of the ditch. Continuous-action machines for spreading the spoil include the bulldozer and the blade grader.

Intermittent-action machines are generally used for open channel construction (Table 13.5). For cleanout work, draglines are sometimes equipped with trapezoid-shaped buckets, with wide tracks for straddling the channel, and with a special gear to provide continuous movement of the machine. The hoe is known by a variety of names, such as back hoe, trench hoe, and drag hoe. In wet areas these machines may be mounted on floats. As land excavators, they may be equipped with rubber tires or with crawler-type tread. Since scrapers, shovels, and bulldozers require channel bottoms that are firm, they may not be suitable for drainage work; however, tractors with pull-back blades are satisfactory. The bulldozer is perhaps the most widely used machine for spreading soil banks, but scrapers may be satisfactory for small channels. Such machines as the dragline, clamshell, and large back hoe, however, may deposit the spoil so that very little, if any, spreading is required.

Table 13.5   Classification of Earth-Moving Equipment for Open Channel Construction

	Action	
Function	Continuous	Intermittent
Excavation	Wheel excavator (trench type)	Dragline (scraper-bucket excavator)
	Plow-type ditcher[a]	Clamshell
	Template excavator	Hoe (also template)
	Blade grader[a]	Shovel
	Elevating grader[a]	Scraper
	Hydraulic dredge	Bulldozer
	Rotary ditcher[a]	Pull-back blade
Spoil spreading	Blade grader[a,b]	Bulldozer[b]
	Tillage machines[a]	Scraper
	Terracing machines[a]	Pull-back blade[b]

[a]Continuous except for turning at the ends.
[b]Either continuous or intermittent, depending on the method of operation.

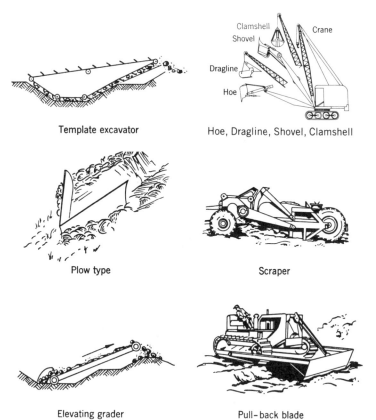

Fig. 13.8 Equipment for excavating open channels.

13.17 Factors Influencing the Selection of Equipment

The shape and design dimensions of the channel are influenced by the type of equipment available. For maximum efficiency, machinery should be selected to fit the requirements of the individual job. In many instances the engineer is called on to evaluate bids of contractors. Such bids may be rejected if the contractor does not have suitable equipment to do the work.

The three types of operations in earthwork construction are digging, hauling, and placing. A machine that can perform all three operations is desirable since such a machine usually provides the most economical construction. For example, a dragline or a large hoe can dig the soil, move it from the channel to the spoil bank, and place the spoil in the desired position.

The selection of equipment depends on such factors as soil water conditions, type of soil, degree of accuracy required, shape and dimensions of the channel and the spoil bank, moving requirements, volume of work, and financial considerations.

13.18 Explosives

Blasting with dynamite may be satisfactory for constructing new ditches or cleaning out old ones in such areas as swamps and wet natural channels that are not readily accessible with earth-moving machinery. It is generally more economical and practical to use machinery than to use explosives. At best, blasting is hazardous work and should be attempted only by experienced and qualified persons.

MAINTENANCE

Maintenance is a continuing problem and it may be required soon after construction is completed. Maintenance may be divided into two phases: preventive maintenance before failure, and corrective maintenance after partial or complete failure.

13.19 Causes for Deterioration of Open Channels

The deterioration of open channels usually results from one or more of three conditions: poor design, improper construction, and lack of adequate maintenance.

The major causes for deterioration of open channels are (1) sedimentation in the channel, (2) excessive growth of vegetation, (3) channel and bank erosion, (4) high sediment load in the water, (5) poor location and alignment, (6) improper depth or width, (7) inadequate culvert and bridge capacity, (8) failure to provide the necessary legal arrangements to fix responsibility and to collect maintenance expenses for channels involving more than one landowner, and (9) general lack of interest in maintenance by the public, landowners, and district officials.

13.20 Preventive Maintenance

Although the best possible design may have been developed and the channel may have been constructed according to plan, maintenance measures are always necessary for proper functioning. The control of excessive tree, brush, grass, and weed growth along the channels is an important preventive maintenance measure. The effect of weed growth on drainage ditch capacity is shown from a survey by Ramser (1947) in Missouri, Arkansas, Mississippi, and Illinois. The study showed that ditch capacities were reduced as much as 75 percent of the original capacity. The primary methods of control consist of spraying, mowing, grubbing and clearing, grazing, underwater cutting, and burning. A number of chemical sprays have largely replaced more costly hand-grubbing and clearing methods. The rate of application, method of application, and effectiveness of chemical sprays are adequately covered in many other publications. Burning vegetation along channels must be accomplished when the material is dry; flame burners are suitable when the vegetation is green. Burning and spraying of ditches may be required several times during the growing season. Since grazing may be detrimental, particularly in channels with steep sideslopes and in membrane-lined irrigation canals, this practice should be avoided.

Bank erosion resulting in channel sedimentation may be reduced by spreading and seeding the spoil bank. If the spoil banks are sufficiently flat, they may be farmed with the adjoining field. In organic soils where wind erosion causes serious filling of the ditches, permanent vegetation (grass) on the spoil banks is satisfactory. Proper land use combined with good conservation practices in the upper tributaries will greatly reduce maintenance of drainage ditches.

13.21 Corrective Maintenance

After the original installation, changes in cross section, grade, or alignment of the channel may be necessary for proper functioning of the system. An improper outlet or inadequate grade may cause sedimentation in the channel. Sideslopes that cave and fill the ditch need to be reshaped to a more stable slope. Sediment bars or sharp curves causing meandering may necessitate straightening the channel or flattening the curves. Widening of the channel and enlarging of culverts and other obstructions may also be necessary. Where serious scouring occurs, the grade may be reduced and drop spillways constructed if more economical measures are not adequate.

Even though all possible steps have been taken to reduce sedimentation, cleanout work is often eventually necessary. In general, equipment used for constructing the original ditch is suitable.

13.22 Maintenance Costs

Because the amount of maintenance varies so widely with different conditions, it is difficult to estimate these costs. Preventive maintenance measures are generally cheaper than corrective maintenance measures, and timely maintenance is more economical than delayed maintenance. Because of the small volume of soil to be moved, cleanout work per unit volume may be three to five times the original excavation cost. Where maintenance has been badly neglected, it may cost more to reclaim an old channel than to construct a new one. Where several landowners are involved, responsibility for maintenance must be clearly established and funds provided or assessed against the land to ensure that the work will be done. An allowance for maintenance for the first 2 years after construction is generally desirable.

REFERENCES

American Society of Agricultural Engineers (ASAE) (1988). *Agricultural Drainage Outlets—Open Channels*. EP407. ASAE, St. Joseph, MI.

Brater, E. F., and H. W. King (1976). *Handbook of Hydraulics*, 6th ed. McGraw-Hill, New York.

Chow, V. T. (1959). *Open-Channel Hydraulics*. McGraw-Hill, New York.

Fortier, S., and F. C. Scobey (1926). "Permissible Canal Velocities." *Trans. ASCE* **89**, 940–984.

French, K. A. (1985). *Open Channel Hydraulics*. McGraw-Hill, New York.

Houk, L. E. (1956). *Irrigation Engineering*, Vol. II. Wiley, New York.

Laflen, J. M., and R. P. Beasley (1960). *Effects of Compaction on Critical Tractive Forces in Cohesive Soils*. Missouri Agr. Expt. Sta. Res. Bull. 749. University of Missouri, Columbia, MO.

Lane, E. W. (1955). "Design of Stable Channels." *Trans. ASCE* **120**, 1234–1260.

Lauritzen, C. W. (1961). *Lining Irrigation Laterals and Farm Ditches*. USDA Agr. Inform. Bull. 242. GPO, Washington, DC.

Ramser, C. E. (1947). *Vegetation in Drainage Ditches Causes Flooding*. SCS-TP-62. SCS, Washington, DC.

Smerdon, E. T., and R. P. Beasley (1961). "Critical Tractive Forces in Cohesive Soils." *Agr. Eng.* **42**, 26–29.

U.S. Bureau of the Census (1962). *Irrigation of Agricultural Lands*, Vol. III. 1960 U.S. Census. GPO, Washington, DC.

U.S. Bureau of Reclamation (USBR) (1962). *Studies of Tractive Forces of Cohesive Soils in Earth Canals*. Hydraulics Br. Rep. Hyd-504. Denver, CO.

———(1963). *Linings for Irrigation Canals*. GPO, Washington, DC.

———(1978). *Drainage Manual*. GPO, Washington, DC.

U.S. Department of Agriculture (USDA) (1962). *Basic Statistics of the National Inventory of Soil and Water Conservation Needs*. Stat. Bull. 317, August. GPO, Washington, DC.

U.S. Soil Conservation Service (SCS) (1973). *Drainage of Agricultural Lands*. Water Information Center, Inc., Port Washington, NY.

PROBLEMS

13.1 Determine the design runoff for an open ditch to provide good drainage in your area for flat, cultivated watersheds of 40 ha (100 ac) and 2600 ha (10 mi^2).

13.2 Compute the most efficient bottom width for an open channel with a flow depth of 2.5 m (8 ft) in silt loam soil. What are the velocity and the channel capacity if the hydraulic gradient is 0.09 percent and $n = 0.035$?

13.3 A farmer in your area desires to drain 260 ha (640 ac) of alluvial sandy loam land. A topographic survey indicates that the maximum slope is 0.19 percent (10 ft/mi). Assuming cultivated land, design the ditch cross section including the spoil bank.

13.4 Design an open ditch to carry the runoff from a 3600-ha (14-mi^2) watershed in your area. The slope of the land along the route of the ditch is 0.27 percent, the average slope of the watershed is 0.5 percent, and the soil is heavy clay. Assume that the ditch should provide sufficient depth for pipe drains and the land is cultivated.

13.5 Determine the deflection angles for the layout of a 24-degree curve (SI units) if the ditch makes a 45-degree change in direction (see Appendix G). The point of curvature is at Station 1 + 00 m (3 + 28 ft).

13.6 What is the design capacity of an irrigation canal with a total depth of 1.5 m (5.0 ft) (including freeboard), bottom width 1.2 m (4.0 ft), 2:1 sideslopes, and hydraulic gradient of 0.09 percent? Assume $n = 0.04$ and the freeboard is 20 percent.

13.7 If the critical tractive force on the bottom of a channel in cohesive soil is 10 Pa (0.21 psf), what is the maximum average velocity of flow at maximum slope for a trapezoidal channel where the depth of flow is 1.2 m (4 ft), bottom width 1.5 m (5 ft), $n = 0.03$, and sideslopes 1:1?

CHAPTER 14

Subsurface Drainage Design

The design of a subsurface drainage system includes the layout and arrangement of the drain lines, selection of a suitable outlet, proper depth and spacing of laterals, determination of the length and size of drains, selection of good-quality materials of adequate strength, and design of such accessories as surface inlets and outlet structures.

14.1 Benefits of Subsurface Drainage

Subsurface drainage is an important conservation practice. Poorly drained lands are usually topographically situated so that when drained they may be farmed with little or no erosion hazard. Many soils having poor natural drainage are, when properly drained, rated among the most productive soils in the world.

Specific benefits of subsurface drainage are (1) aeration of the soil for maximum development of plant roots and desirable soil microorganisms; (2) increased length of growing season because of earlier possible planting dates; (3) decreased possibility of adversely affecting soil tilth through tillage at excessive soil water levels; (4) improvement of soil water conditions in relation to the operation of tillage, planting, and harvesting machines; (5) removal of toxic substances, such as salts, that in some soils retard plant growth; and (6) greater storage capacity for water, resulting in less runoff and a lower initial water table following rains. Through these benefits drainage enhances farm productivity by (1) adding productive land without extending farm boundaries, (2) increasing yield and quality of crops, (3) permitting good soil management, (4) ensuring that crops may be planted and harvested at optimum dates, and (5) eliminating inefficient machine operation caused by small wet areas in fields. In arid regions irrigation and drainage are complementary practices. In some areas leaching of soluble salts through a drainage system is essential before the land can be developed. Drainage is often a necessity as a result of excess water that accumulates from low efficiencies in the conveyance and application of water for irrigation. Although these losses can be reduced, they cannot be entirely eliminated. The benefits of drainage can be realized only when the soil is potentially productive if drained. Government regulations may require leaving soil undrained as range land or as a recreation and wildlife area.

The benefits and costs from drainage in a humid area are estimated from measured corn yields in Table 14.1. Subsurface drainage gave more than twice the increase in yields compared with surface drainage; however, surface drainage resulted in a higher benefit–cost ratio because of much lower investment cost than for subsurface drainage. Costs were based on 1990 prices, and they will vary greatly from field to field, especially for surface drainage. Costs also vary from region to region.

14.2 Drainage Requirements and Plant Growth

Excess water affects plant growth by reducing soil aeration, for plants are not adversely affected even in total water culture if air is provided. Although for design purposes the water table is a convenient term for reference, it is not satisfactory for explaining the water relationships with respect to plant root environment. Because of the capillary fringe above the water table, saturation may occur for some distance above it, especially in a clay soil. A better criterion for drainage depth would be the "upper surface of saturation," or a depth based on the oxygen diffusion rate of the soil. Several investigators have found that roots do not generally penetrate to a static water table but stop growing a short distance above it. Soils that have a relatively small volume of noncapillary pore space may provide an unfavorable environment for root growth even though not saturated.

Drainage requirements for optimum plant growth are determined largely by the volume and content of the soil air. These needs depend on type of crop, soil, availability of plant nutrients, climatic conditions, biological activity, and soil and crop management practices. The rooting depth and the tolerance to excess water

Table 14.1 Drainage Benefits and Costs for Corn in a Silty Clay Soil in Northern Ohio (1990 prices)

	Drainage System		
	Surface Only	*Subsurface Only*	*Surface Plus Subsurface*
Corn yields (kg/ha) (13-year average)	5750	7250	7600
Increased yields (undrained 4400 kg/ha)	1350	2850	3200
Benefits at $0.12/kg	$162	$342	$384
Annual costs/ha[a]	$47	$182	$229
Benefit–cost ratio	3.5	1.9	1.7
Benefits less costs	$115	$160	$155

[a]Estimates do not include fertilizer and other production costs as these would be similar for all drainage systems. Costs were computed assuming 5 percent depreciation (20-yr economic life) per year, 10 percent interest on average investment, and $2.00/ha per year for maintenance of subsurface drains; and 10 percent annual interest on investment and $10.00/ha per year for maintenance of surface drainage, only. Initial investment was $370/ha for surface drains and $1800/ha for subsurface drains (12-m spacing, 1 m deep).
Source: Schwab et al. (1985).

are the most important crop characteristics. For example, rice is tolerant because it has special internal mechanisms for obtaining oxygen from the air, whereas tobacco is sensitive to excess water. Plants have widely varying oxygen requirements, sometimes even greater than the concentration in the atmosphere.

As shown in Fig. 14.1, a sharp reduction in yield with an increase or decrease in water table depth from an optimum of 0.3 m may be expected for a peat soil. As the soil texture becomes heavier, maximum yields are obtained at greater water table depths.

Under field conditions soil water content changes continually with time during the growing season, varying from the wilting point to complete saturation as below the water table. The stress day index (SDI) has been developed to quantify this variable as a single numerical value for a short period or for the entire growing season. It is useful to characterize the effects of water stress on crop yields for both irrigation and drainage requirements of crops. Mathematically,

$$\text{SDI} = \sum_{i=1}^{N} (\text{SD}_i \times \text{CS}_i) \tag{14.1}$$

where SDI = stress day index in cm-days,
SD$_i$ = stress day factor for period i in cm-days,
CS$_i$ = crop susceptibility factor for period i,
N = number of growth periods.

The SD factor for high fluctuating water tables was developed in The Netherlands in the early 1960s. It was referred to as SEW and expressed as

$$\text{SEW}_d = \text{SD}_i = \sum_{i=1}^{n} (d - x_i) \tag{14.2}$$

where SEW$_d$ = sum of excess water above depth d in cm-days,
d = critical depth of the water table below the soil surface in cm,
x = daily measured water level in cm,
n = number of days during the growing season.

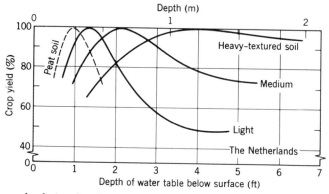

Fig. 14.1   General relationship between crop yield and constant water table depth during the growing season. (Redrawn from Visser, 1959.)

Negative values of SEW from the equation are neglected because water table depths below d are assumed to have no effect on crop yields. The critical depth d is often taken as 30 cm. If $d = 30$ cm and the average water table depth in a field is 20 cm for 10 days, the SEW is 100 cm-days.

The crop susceptibility factor CS in Eq. 14.1 depends on the species and the stage of growth of the plant. As many as five stages of growth have been recognized, but the vegetative, flowering, and maturing stages are generally the most important for annual crops, such as cereals, corn, and soybeans. During the early vegetative stage, research has shown CS values as high as 0.5 for corn and 0.3 for soybeans. Small plants shortly after emergence are particularly susceptible to flooding. During the maturing stage CS values are low; some values are near zero, indicating little effect on crop yield. Studies have shown a good correlation of SDI and crop yields. The SDI procedure has been widely accepted and is adaptable to computer modeling.

The relationship between drainage and nitrogen availability is illustrated in Fig. 14.2. A major effect of drainage is the increased nitrogen supply that can be obtained from the soil. These amounts, which are estimated by projecting the dashed curves to the abscissa, are about 56 and 157 kg/ha of nitrogen at water table depths of 0.4 and 1.5 m, respectively. The practical implication is that nitrogen applications increase crop yields more at high than at low water tables. With a shallow water table nitrogen gives a good response because the crop cannot obtain it from the soil. The data in Fig. 14.2 were obtained in The Netherlands where rainfall rates are low and a constant water table can be maintained. Because a constant water table prevents root development below it, these results may not be applicable where widely fluctuating water table conditions prevail, as in the humid area of the United States.

Soil and air temperatures affect oxygen diffusion rates as well as soil organisms and the biological processes in the plant. Increasing the temperature usually causes a decrease in oxygen and an increase in carbon dioxide concentration because of the increase in respiration rates of roots and organisms. Kramer and Jackson (1954) found that tobacco plants grown at a water temperature of 20°C recovered from flooding if drained within one day, whereas plants grown at 34°C died under the same conditions. Normally, during the dormant season plants are not affected by flooding because at low temperatures the biological processes are slow. Drainage

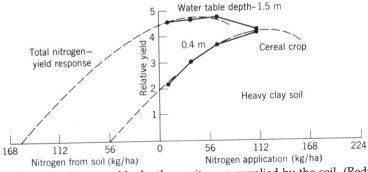

Fig. 14.2 Influence of water table depth on nitrogen supplied by the soil. (Redrawn from Van Hoorn, 1958.)

influences soil temperature because of the differences between wet and dry soil in specific heat, thermal conductivity, and evaporation. The specific heat of soil particles is about 0.2 for most dry soils as compared with 1.0 for water. The thermal conductivity of dry soil is one third to one half that of water. Evaporation from the surface requires solar energy that would otherwise be available to warm the soil; however, in the field, differences of only a few degrees in surface soil temperatures have been measured between points directly over a drain line and midway between lines where the slower drainage occurs. This small difference is probably due to the large mass of the earth and equalization of temperatures by conduction.

Flooding of plant roots causes a rapid reduction in transpiration, reduced absorption of oxygen and other plant nutrients with a corresponding increase in the carbon dioxide content of the soil, a disturbance of microbiological activity, and a reduced mass of soil from which nutrients can be obtained. Kramer (1949) found that excess carbon dioxide reduced absorption much sooner than a deficiency of oxygen. Death of the plant may be caused by toxic substances moving up from the roots, but the effect is complex and not fully understood. When aeration is reduced, iron and manganese may become high enough in concentration to be toxic to the roots. Decomposition of organic matter may produce hydrogen sulfide which is also toxic. Changes in the oxygen and carbon dioxide balance may affect the growth of disease organisms. With prolonged flooding and reduced biological activity, soil structure may be destroyed, which in turn reduces aeration. In heavy-textured soils wetting and drying can have a beneficial effect. In sandy and organic soils, such as peat and muck, a high water table is less critical than in heavy-textured soils.

The relationship of subsurface drainage to root development is illustrated in Fig. 14.3. A high water table in the spring inhibits root development, leaving a plant with an inadequate root system during the summer months, but a low water table in the spring allows maximum root development.

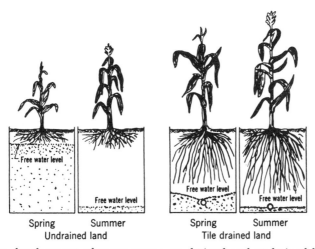

Fig. 14.3 Root development of crops grown on drained and undrained land. (Redrawn from Manson and Rost, 1951.)

14.3 Environmental Impacts of Drainage

In recent years the major impact of drainage has been related to the preservation of wetlands for wildlife habitat and nonpoint pollution of drainage water for downstream users. Federal legislation, such as the swampbuster program, limits the development of wetlands for agricultural purposes. Effects of drainage on water quality are of concern in both arid and humid regions. Compared with surface runoff in humid areas, subsurface drainage water generally contains less sediment and smaller amounts of phosphorus, potassium, and pesticides, and has a higher pH. Nitrogen is normally greater in subsurface water because it moves through the soil with the water. This problem is of great concern for city water supplies. In irrigated areas sodium has long been known to have adverse effects on water quality. In more recent years trace elements, such as selenium, boron, molybdenum, and arsenic, have been found to have potentially serious impacts on the environment, including death of some aquatic organisms and waterfowl, such as occurred at the Kesterson National Wildlife Refuge in California in 1982 (Hoffman, 1990).

The primary concern with the drainage of wetlands is the loss of habitat for wildlife. Federal legislation generally supports the concept that there should be no net loss of wetlands. The greatest controversy is not that wetlands should be drained, but in what constitutes a wetland and how is a farmer to be compensated for wetlands that should be preserved.

The hydrologic impact of subsurface drainage to reduce flooding is often not realized. In Ohio, studies showed that peak runoff rates from small, flat plots (0.2 ha) were reduced 7 percent by installing subsurface drains and that the number of floods was reduced by 46 percent. Similar results were obtained from adjacent 32-ha watersheds in North Carolina. These reductions in peak flow are attributed to the increased capacity of the soil to store water, but the effect may not always be consistent for all soil and hydrologic conditions.

14.4 Pipe Drains

Pipe drains include concrete and clay tile, corrugated plastic tubing, or other perforated conduit. Corrugated steel pipe with a high structural strength is suitable to withstand high soil loads, to cross unstable soils that require the rigidity of a long pipe, and to provide a stable outlet into open ditches.

Tile for lateral drains (drains intended to receive water directly from the soil as contrasted with main lines, which receive most of their flow from laterals) are usually 0.3-m lengths and 75- to 150-mm nominal diameter, and have a wall thickness about one twelfth their diameter. The tile are laid end to end in the bottom of a trench that is then backfilled. Some tile are also perforated. Water enters the tile line through perforations and cracks between the ends of adjoining tile.

Corrugated plastic tubing (CPT) in the United States was first commercially produced in 1967, and by 1973 it had largely replaced concrete and clay tile for the smaller-size drain laterals. High-density polyethylene (HDPE) tubing is widely used in the United States; polyvinylchloride (PVC) is more prevalent in Europe. Corrugated tubing is light in weight (about 1/25 the weight of concrete or clay tile), durable, resistant to soil chemicals, extruded in long lengths, and easy to join and handle in the field. It is especially suitable for installation with a trenchless plow, and less labor is required than for tile. CPT is subject to damage by rodents and its hydraulic roughness is higher than that of tile. It will float in water and tend to stay curved as in the shipping coil. CPT is perforated or slotted with three or more rows of openings and may be covered with a fabricated porous envelope to prevent inflow of fine particles.

14.5 Mole Drains

Mole drains as shown in Fig. 14.4 are cylindrical channels artificially produced in the subsoil without digging a trench from the surface. They are similar to pipe drains except that they are not lined with tile or other stabilizing material.

Moling is a temporary method of drainage. Where soil conditions are suitable, moles function efficiently for the first few years and then gradually deteriorate. Mole drains have been successful in England, New Zealand, and several European countries. Their maximum life is 10 to 30 years. Except in some soils in Louisiana, Florida, and California, mole drainage is generally not practiced in the United States.

Mole drains fail principally because the soil is not sufficiently stable to maintain a channel. Although high-clay soils are generally the most suitable, the clay content is not necessarily a good index for moling. In Iowa a study of three soils showed that the greater the clay content, the more rapid the failure of the channel. The water content at the time of moling should be as high as possible, provided the soil can support the tractor. Mole channels usually range in diameter from 50 to 100 mm. The small sizes are more stable than the larger ones.

Where mole drains are suitable, they are generally pulled across and over pipe drains, using the pipe drains as outlets. Moles may be directly connected to the pipe, or gravel may be placed in the pipe trench. The depth varies from 0.5 to 1.2 m depending on the depth of a stable soil layer. Spacings will range from 1 to 50 m. Length of drain is usually less than 500 m depending on the grade, which may range from nearly level to 5 percent.

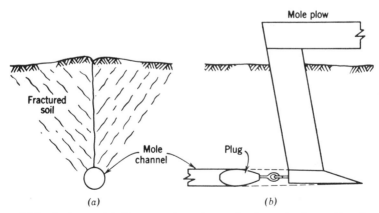

Fig. 14.4 Mole drainage. (*a*) Cross section of a mole channel. (*b*) Profile of a plow for forming a mole drain.

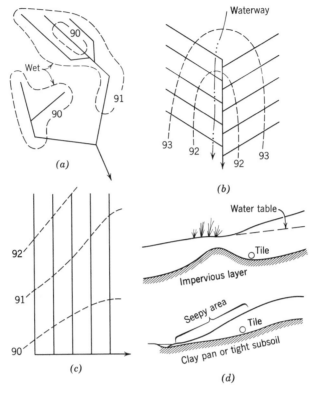

Fig. 14.5 Common types of pipe drainage systems. (a) Natural or random. (b) Herringbone. (c) Gridiron. (d) Cutoff or interceptor.

TYPES OF SYSTEMS

Pipe drainage systems in general use are the natural, herringbone, gridiron, and interceptor types shown in Fig. 14.5. Combinations of two or more of these types are frequently required for the complete drainage of an area.

14.6 Natural or Random

The natural or random system is widely adopted in fields that do not require complete drainage with equally spaced laterals. This system is flexible as well as economical since the drain lines follow natural draws or other low depressions. Such a system is particularly adapted to the drainage of small or isolated wet areas.

14.7 Herringbone

The herringbone system is adapted to areas that have a concave surface or a narrow draw, with the land sloping to it from either direction. The main line is laid out nearly normal to the slope and follows the low area. Despite the large amount of double drainage (land drained both by the laterals and by the main or submain), the herringbone system is particularly suitable where the laterals are long and the waterway requires thorough drainage.

14.8 Gridiron

The gridiron system is similar to the herringbone system except that the laterals enter the main from only one side. The gridiron system is the most common. The gridiron pattern is more economical than the herringbone system because the number of junctions and the double-drained area are reduced. Where the waterway is of considerable width, a main is placed on both sides of the waterway. This system, known as the double-main system, is essentially two separate gridiron patterns. Since the mains need not cross the waterway at critical points, serious erosion problems may be prevented. Although right-angle junctions are shown in Fig. 14.5c, main and laterals may intersect at angles less than 90 degrees.

14.9 Cutoff or Interceptor

The cutoff or interceptor drain is normally placed near the upper edge of a wet area as shown in Fig. 14.5d. The usual cause for such wet conditions is the outcropping of impermeable strata on or near the surface. Frequently, this condition exists along waterways. To drain such areas the interceptor drain is frequently installed on both sides of the waterway.

OUTLETS

14.10 Requirements of an Outlet

The importance of a good outlet is indicated by the fact that a high percentage of failures of drainage systems are due to faulty outlets. The requirements of a good outlet are to (1) provide a free outlet with minimum maintenance, (2) discharge the outflow without serious erosion or damage to the pipe, (3) keep out rodents and other small animals, (4) protect the end of the drain against damage from the tramping of livestock as well as excessive freezing and thawing, and (5) prevent the entrance of flood water where the outlet is submerged for several hours.

14.11 Types of Outlets

The two principal types of outlets for pipe drains are gravity and pump. Pump outlets (see Chapter 16) may be considered where the water level at the outlet is higher than the bottom of the pipe outlet for any extended period.

Gravity outlets, by far the most common, include other pipe drains, constructed waterways, natural channels, and wells. Outlet ditches should have sufficient capacity to carry surface runoff and drain flow. Where the drainage system is connected to other pipe drains, the outlet should have sufficient capacity to carry the additional discharge. As shown in Fig. 14.6, corrugated metal or other rigid pipe is recommended. Some type of grille or flap gate over the end is desirable to prevent entry of rodents. If there is danger of flood water backing up into the drain, an automatic flood or tide gate may be installed in place of the flap gate. The end of the outlet pipe should be 0.3 m or more above the normal water level in the ditch. To prevent damage caused by high velocities in the ditch or failure from snow loads, the exposed end of the pipe should not extend beyond the bank more than one third its total length. The minimum total length should be 5 m, and the

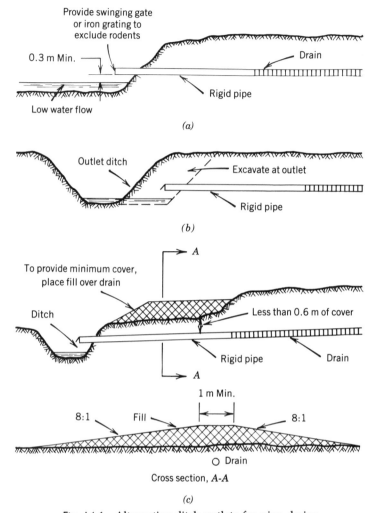

Fig. 14.6   Alternative ditch outlets for pipe drains.

diameter should be the same or larger than the drain pipe size. In connecting the metal pipe to the drain, a concrete collar may be installed. Where available, drop inlets and other permanent structures described in Chapter 9 are suitable for stabilizing the outlet.

Vertical drainage outlets are essentially wells extending into a porous soil layer or open rock formation in the lower horizons. The existence of a substratum that can continually take in large quantities of water and that can be reached without prohibitive cost is the exception rather than the rule. Since there is no positive method for locating or for predicting the permanent capacity of such outlets, their use involves considerable risk. Also, if drain outflow is contaminated, there is danger of polluting the ground water. For these reasons, vertical outlets are not generally recommended.

DEPTH AND SPACING

A definite relationship exists between depth and spacing of drains. For soils of uniform permeability, the deeper the drains the wider the spacing; hence fewer drains are required and the pipe cost is lower. The primary consideration in drainage design is to provide adequate root depth above the saturated zone.

14.12 Depth

The depth of pipe drains should be such as to provide the desired water table depth midway between drain lines. Pipe depth is affected by soil permeability, outlet depth, spacing of laterals, depth to the impermeable layer in the subsoil, and limitations of trenching equipment. Pipe depth as considered here applies only to laterals and not to the depth of mains, since the depth of the main is governed primarily by outlet conditions and topography.

Pipe depth, defined as the distance from the surface to the bottom of the pipe, varies in different soils. Under no conditions should the amount of cover over the top of the pipe be less than 0.6 m. This minimum is necessary to protect the pipe from heavy surface loads and to prevent shifting of the pipe. In uniformly permeable mineral soils the depth of laterals usually varies from 0.8 to 2.5 m. Unless limited by an impermeable layer, one should design for the maximum depth as this will permit a wide spacing. In deep organic soils after initial settlement has taken place, the minimum depth should not be less than 1.2 m.

Where the subsoil is relatively impermeable, the pipe should be placed on or above the impermeable layer. If pipe must be placed below the impermeable layer, the trench should be backfilled with permeable soil or gravel.

In humid regions where the water table will rise to near the surface during heavy rainfall, the rate of drop is the important factor; however, in organic soils the water table may be maintained at a nearly uniform depth, which may be above the pipe. A few empirical criteria for drainage depths are presented in Table 14.2.

In arid regions under irrigation the drainage design criteria are determined more by the minimum depth of the water table for optimum crop growth than by the rate of drop. Depths of 2 to 3 m are common. Water table depths approved by irrigation authorities and by financial institutions as a basis for long-time loans for improving irrigated land (Hansen et al., 1980) are as follows:

Classification	Range in Water Table Depth
Good	Static water table below 2.10 m; up to 1.80 m for about 30 days per year
Fair	Water table at 1.80 m; up to about 1.20 m for 30 days; no general rise
Poor	Some alkali on surface; water table 1.20 to 1.80 m; up to 0.90 m for 30 days
Bad	Water table less than 1.20 m and rising

Selection of salt-tolerant crops, careful application of water, and proper soil management may permit successful cropping with shallower depths than indicated.

Table 14.2 Drainage Depth Requirements for Humid Areas

Crops	Depth and Rate of Drop of the Water Table	Location	Reference
Mineral Soils			
Field	Initial depth, 0.15 m minimum; 0.3 m/day through second 0.15 m; 0.2 m/day through third 0.15 m	Minnesota	Neal (1934)
Field	Drop from surface to 0.3 m in 24 h and 0.5 m in 48 h	Illinois	Kidder and Lytle (1949)
Field	Drop 0.2 m/day	Virginia	Walker (1952)
Grass	Constant depth 0.5 m or less	The Netherlands	Wesseling et al. (1957)
Arable	Constant depth 0.9 to >1.3 m	The Netherlands	Wesseling et al. (1957)
Organic Soils (controlled drainage)			
Grasses	Maximum depth, 0.5 m	United States	SCS (1973)
Vegetables (shallow rooted)	Maximum depth, 0.6 m	United States	SCS (1973)
Field (deep rooted)	Maximum depth, 0.8 m	United States	SCS (1973)
Cereals, short grass, sugar beets	Optimum depth, 0.8–0.9 m	England	Nicholson and Firth (1953)
Truck crops, grass, and sugar cane	Optimum depth, 0.3–0.6 m	Florida	Clayton et al. (1942)

14.13 Spacing

In humid regions most drain lines are spaced 10 to 50 m apart and up to 90 m in very permeable soils. Where high-value crops are grown or under special conditions, spacings of 10 to 15 m are sometimes necessary. In irrigated areas of the West spacings ranging from 50 to 200 m are possible.

14.14 Depth and Spacing Design Criteria

In general, the depth and spacing of drains vary largely with soil permeability, crop and soil management practices, kind of crop, and extent of surface drainage. With good crop and soil management practices, the depth and spacing for normal conditions vary within the limits given in Table 14.3.

Many depth and spacing formulas have been proposed that are approximate mathematical solutions. With all of these the procedure is briefly (1) to determine a drainage rate or a water table height, (2) to estimate or measure the hydraulic

Table 14.3   Average Depth and Spacing for Pipe Drains

Soil	Hydraulic Class	Conductivity (m/day)	Spacing (m)	Depth (m)
Clay	Very slow	0.001	9–15	0.9–1.1
Clay loam	Slow	0.001–0.005	12–21	0.9–1.1
Average loam	Moderately slow	0.005–0.02	18–30	1.1–1.2
Fine sandy loam	Moderate	0.02–0.06	30–37	1.2–1.4
Sandy loam	Moderately rapid	0.06–0.13	30–60	1.2–1.5
Peat and muck	Rapid	0.13–0.25	30–90	1.2–1.5
Irrigated soils	Variable	0.025–25	45–180	1.5–3.0

conductivity and other required soil characteristics, (3) select a suitable depth for the drains, and then (4) to compute the spacing.

Steady State.   For conditions where the rainfall rate or irrigation rate is constant, the drain discharge will also be constant if the water table does not change with time. The ellipse equation, which was one of the first mathematical solutions for drain spacing, will be derived by using the symbols and geometry shown in Fig. 14.7. Assuming that a constant rate of rainfall is removed equally well at all distances from the drain,

$$q_x = \left[\frac{(S/2) - x}{S/2} \right] \frac{q}{2} \tag{14.3}$$

where q_x = rate of flow toward the drain across a vertical plane of unit width at any distance x (L^3/T),

S = spacing between drains (L),

q = total flow rate into a drain from both directions per unit length of drain (L^3/T).

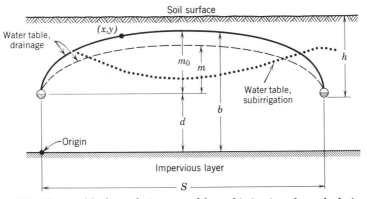

Fig. 14.7   Water table from drainage and by subirrigation through drains.

From Darcy's law (see Chapter 3) and the assumption that the flow velocity is proportional to the water table slope,

$$q_x = -y v_x = -Ky \left(\frac{dy}{dx} \right) \tag{14.4}$$

where v_x = velocity at x (L/T),
K = hydraulic conductivity (L/T).

By equating Eqs. 14.3 and 14.4, the differential equation is

$$y \, dy = \left(\frac{q}{SK} \right) \left(\frac{S}{2} - x \right) dx \tag{14.5}$$

Integrating from $x = 0$ and $y = d$ to $x = x$ and $y = y$, where d is the height of the water table in the drain above the impervious layer, gives

$$y^2 - d^2 = \frac{q}{SK} (Sx - x^2) \tag{14.6}$$

which is the equation of an ellipse. Substituting $x = S/2$ and $y = b$ for the midpoint yields

$$S = \frac{4K(b^2 - d^2)}{q} = \left[\frac{4K(b^2 - d_e^2)}{i} \right]^{1/2} \tag{14.7}$$

The flow q may also be expressed as Si, where i is the drainage rate in meters per day. The basic ellipse equation was first developed by Colding in 1872.

The principal limitation of the ellipse equation is that the resistance resulting from convergence of flow lines near the drains is ignored. Thus, it is suitable only where the spacing is large compared with the depth to the impervious layer. The accuracy of the spacing can be improved by substituting the equivalent depth d_e given in Fig. 14.8 for d in the equation. The depth b is equal to $d_e + m$. Moody (1966) has developed equations for d_e for other drain diameters.

□ Example 14.1

For an irrigated area compute the drain spacing assuming the depth to the center of the drain is 1.8 m (6 ft) and the minimum depth to the water table, 1.5 m (5 ft). Subsurface explorations indicate a hydraulic conductivity of 0.5 m/day (1.6 fpd) above an impervious layer at a depth of 6.7 m (22 ft). The excess irrigation rate is equivalent to a drainage coefficient of 1.2 mm/day (0.048 ipd) (see Section 14.16).

Solution.   Assume a spacing of 55 m (180 ft) and, for $d = 6.7 - 1.8 = 4.9$ m (16 ft), read from Fig. 14.8, $d_e = 2.9$ m (9.5 ft). With tile flowing half full as in Fig. 14.7, $m = 1.8 - 1.5 = 0.3$ m (1.0 ft). Therefore, $b = 2.9 + 0.3 = 3.2$ m (10.5 ft). Substituting in Eq. 14.7,

$$S = \left[\frac{4 \times 0.5(3.2^2 - 2.9^2)}{0.0012} \right]^{1/2} = 55.2 \text{ m (181 ft)}$$

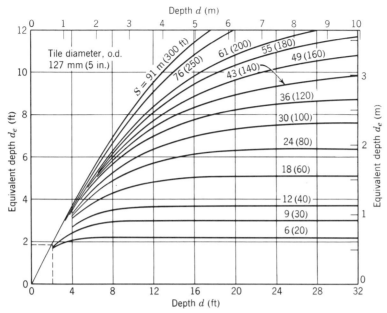

Fig. 14.8 Equivalent depth for the water-conducting layer below the drain. (Compiled by van Schilfgaarde, 1963).

Since 55.2 m (181 ft) is close to the assumed spacing of 55 m (180 ft), the solution is satisfactory. If the computed spacing is not close to the assumed value, assume a new spacing, which will change d_e, and recompute the spacing. Continue until the assumed and computed spacings are in close agreement. □

Nonsteady State. Mathematical analysis of the falling water table case is far more difficult than for the steady state. The Dupuit–Forchheimer theory, which assumes horizontal flow with the velocity proportional to the slope of the free-water surface, has been applied to the nonsteady state by adjusting the continuity equation. This equation after appropriate substitutions for soil porosity f and time t may be written

$$\frac{\partial^2 y}{\partial x^2} = \frac{f}{K(d + m_0/2)} \frac{\partial y}{\partial t} \tag{14.8}$$

It is often referred to as the heat flow equation since it is identical in form to the differential equation for solving heat flow problems. By assigning appropriate boundary conditions Eq. 14.8 can be solved in terms of a sine series, but only the first term is retained. The spacing equation as derived by Glover and reported by Dumm (1954) using symbols in Fig. 14.7 is

$$S = \pi \left[\frac{Kt(d + m_0/2)}{f \ln(4/\pi)(m_0/m)} \right]^{1/2} \tag{14.9}$$

Because of the assumptions made in its derivation, Eq. 14.9 also ignores the resistance resulting from the convergence of the flow lines near the drain and is based on a thickness of the water-conducting zone equal to $d + m_0/2$. Van Schilf-gaarde (1963) modified the Glover equation to take these factors into consideration. This equation is

$$S = \left[\frac{9Ktd_e}{f[\ln m_0(2d_e + m) - \ln m(2d_e + m_0)]} \right]^{1/2} \tag{14.10}$$

where S = drain spacing (L),
  K = soil hydraulic conductivity (L/T),
  d_e = equivalent depth from Fig. 14.8 (L),
  m = height of water table above the center of the drain at midplane after time t (L),
  m_0 = initial height of water table (L),
  t = time for water table to drop from m_0 to m (T),
  f = drainable porosity of the water-conducting soil expressed as a fraction (voids drained at 0.6-m tension).

This equation is recommended for design purposes (see Example 14.2). It does not yield a practical solution where $d_e = 0$. For this case another form of the equation is available.

The soil hydraulic conductivity K should be the effective value for the water-conducting zone, particularly for heterogeneous soils that may have horizontal strata of varying permeability and for soils with different permeability in the horizontal (K_h) and vertical (K_v) directions. In such soils $K = (K_h K_v)^{1/2}$. Auger hole, piezometer, core sample, and other methods have been proposed for measuring soil hydraulic conductivity, but these are time consuming and costly. Skaggs (1975) developed a procedure that permits evaluation of the ratio K/f, where f is the drainable porosity. The water table drawdown may be taken from a single drain or from two adjacent drains. With this procedure neither K nor f need be determined individually, as only the ratio is required in the equation.

☐ *Example 14.2*

Calculate the drain spacing for a falling water table assuming $K = 0.3$ m/day (1 fpd), $f = 0.02$, drain depth to center line $= 0.9$ m (3 ft), $m_0 = 0.9$ m (3.0 ft), $m = 0.6$ m (2.0 ft), $t = 1$ day for the water to drop from the soil surface to 0.3 m (1.0 ft) below, and $d = 0.6$-m (2.0-ft) depth of impervious layer below drains.

Solution.   For purposes of evaluating d_e, assume $S = 20$ m (65 ft). From Fig. 14.8 read $d_e = 0.6$ m (2.0 ft). Substituting in Eq. 14.10,

$$S = \left\{ \frac{9 \times 0.3 \times 1 \times 0.6}{0.02[\ln 0.9(2 \times 0.6 + 0.6) - \ln 0.6(2 \times 0.6 + 0.9)]} \right\}^{1/2} = 18 \text{ m (59 ft)}$$

Since 18 m is close to the estimated spacing of 20 m, d_e would not change and the spacing of 18 m is correct. Since the water table initially was at the soil surface, the drain depth h for a 100-mm (4-in.) drain is $0.9 + 0.06 = 0.96$ m (3.2 ft), the depth to the bottom of the drain. The extra depth, 0.06 m (0.2 ft), is one half the outside diameter of the 100-mm pipe.   ☐

Table 14.4   Drain Spacings With and Without Surface Drainage

Water Table Drawdown (mm/day)	Drain Spacing (m)	
	With Surface Drainage	*Without Surface Drainage*
100	22	13
200	14	8
300	10	6

Note: Spacings computed from drain flow measurements for silty clay soil from drains with 12-m spacing, 0.9-m depth, and initial water table near the surface. *Source:* Hoffman and Schwab (1964).

Drain spacing is influenced by the amount of surface water to be removed by subsurface drains, although spacing equations do not take this factor into consideration. Computed spacings (Table 14.4) from experimental plots in northern Ohio with and without surface drainage demonstrate this effect. For a given rate of drawdown, spacings were increased nearly 70 percent where surface drainage was provided. On plots having both surface and subsurface drainage about half of the flow from 10-year return period rainfall was removed in surface runoff and half from the tile.

14.15 Interceptor Drain Location

Typical interceptor drains in humid areas, shown in Fig. 14.5*d*, are located above the seepage area where seepage is caused by the shallow depth or outcropping of the impervious layer on the surface. In irrigated regions seepage from an unlined canal, shown in Fig. 14.9, often causes wetness and damage to crops below. The distance of the first pipe drain below the ditch or canal can be determined by procedures described by SCS (1973) and USBR (1978). Spacing of drains below the first one are computed from previous equations, where the land slope is less than 10 percent. Problems of this type with special boundary conditions can be solved with computers.

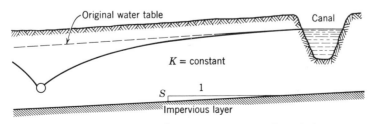

Fig. 14.9 Water table drawdown from an interceptor drain below a canal.

SIZE OF PIPE DRAINS

14.16 Design Flow

The design flows for pipe drains for humid and irrigated conditions are based on entirely different criteria. In either case the "drainage coefficient" is a convenient term for expressing the flow rate. It is defined as the depth of water to be removed from the drainage area in 24 h.

Humid Conditions. Normally, the drainage area is computed from the length and spacing of the drains. Where surface inlets are installed, the contributing watershed is the drainage area rather than the pipe-drained area. For example, if a surface inlet is located in a pothole and if the contributing watershed area is 5 ha, the outlet should be designed using the appropriate drainage coefficient for surface inlets with a drainage area of 5 ha; however, if no surface inlets are installed, the outlet drain should be designed using only the area actually drained by the pipe and the normal drainage coefficient as indicated in Table 14.5. The area drained by the pipe and the contributing watershed represent minimum and maximum areas, respectively (ASAE, 1988).

In humid areas the drainage coefficient depends largely on rainfall. It is difficult to correlate rainfall with the drainage coefficient since the distribution of rainfall during the growing season and its intensity must be considered along with evaporation and other losses. For example, when the soil is dry, an intense storm of short duration produces rapid surface runoff and little infiltration; however, rains of low intensity over a long period may produce high rates of outflow from the drains.

The selection of a drainage coefficient is based primarily on experience and judgment. Recommended drainage coefficients are shown in Table 14.5. The drainage coefficient should be such as to remove excess water rapidly enough to prevent serious damage to the crop. Where the underlying stratum is sand or other porous material normal coefficients may be reduced. Where the design rate of drop

Table 14.5 Drainage Coefficients for Pipe Drains in Humid Regions

Crops and Degree of Surface Drainage	Drainage Coefficient[a]	
	Mineral Soil (clay and silt) (mm/day)	Organic Soil (mm/day)
Field crops		
Normal[b]	10–13	13–19
With blind inlets	13–19	19–25
With surface inlets	13–25	25–38
Truck crops		
Normal[b]	13–19	19–38
With blind inlets	19–25	38–51
With surface inlets	25–38	51–102

[a]These values may vary depending on special soil and crop conditions. Where available, local recommendations should be followed.
[b]Adequate surface drainage with outlets to other drains or ditches to be provided.
Source: ASAE (1988).

of the water table is known, the drainage coefficient should be computed from the drainable porosity. For example, for a 0.3-m drop per day of the water table and 3 percent porosity, the drainage coefficient is 9 mm/day.

Irrigated Conditions.   In irrigated areas the discharge from drain lines may be expected to vary from about 10 to 50 percent of the water applied. In these regions the drainage coefficient may vary with the size of the area contributing to the flow. Since not all of the area is irrigated at the same time, the design drainage area is not the same as the entire area but is estimated from the area being irrigated. Seepage water should also be considered in selecting the drainage coefficient, or it may be included by modifying the drainage area. The drainage coefficient based on seepage and canal losses may be estimated from Eq. 13.6 (ASAE, 1984).

SCS (1973) reported that from surveys throughout the western states pipe discharge varied from about 0.2 to 70 L/s per 100 m of drain, with an average of about 9.3 L/s. Because of such wide variation, local recommendations must be developed. Pillsbury et al. (1965) suggest the following discharge equation for the Coachella Valley of California:

$$q = 1.56A^{0.75} \tag{14.11}$$

where q = maximum flow in L/s,
A = drained area in ha.

14.17 Grades

Maximum grades are limiting only where pipes are designed for near-maximum capacity or where pipes are embedded in unstable soil. Tile embedded in fine sand or other unstable material may become undermined and settle out of alignment unless special care is taken to provide joints that fit snugly against one another. Under extreme conditions it may be necessary to install bell and spigot tile, tongue and groove concrete tile, metal pipe, or plastic tubing. On mains, steep grades up to 2 or 3 percent are acceptable, provided the capacity at all points nearer the outlet is equal to or greater than that of the drain above.

A desirable minimum working grade is 0.2 percent. Where sufficient slope is not available, the grade may be reduced to that indicated in Table 14.6. The minimum grade should be sufficient to remove any sediment from the drain. The grade

Table 14.6   Minimum Grades for Pipe Drains

Nominal Diameter (mm)	Fine Sand or Silt Not Present[a]		Fine Sand or Silt May Enter the Drain[b]	
	Tile	Tubing	Tile	Tubing
	Grade (%)			
75	0.08	0.10	0.60	0.81
100	0.05	0.07	0.41	0.55
125	0.04	0.05	0.30	0.41
150	0.03	0.04	0.24	0.32

[a]Minimum cleaning velocity of 0.15 m/s.
[b]Minimum cleaning velocity of 0.42 m/s.
Source: Adapted from ASAE (1988).

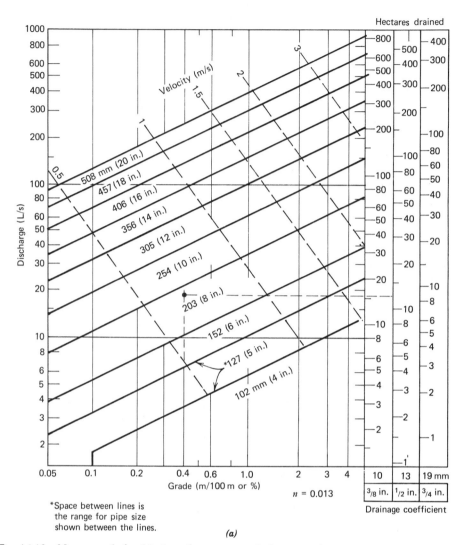

Fig. 14.10 Nomograph for (*a*) size of concrete and clay pipe drains for $n = 0.013$ and (*b*) size of corrugated plastic tubing. (Revised from ASAE, 1988.)

should not be less than the minimum, unless special precautions are taken during construction. Reverse grades are undesirable.

14.18 Hydraulic Design — Drain Size

The hydraulic capacity of drains can be determined from the Manning velocity equation. By equating the design flow to the hydraulic capacity at full flow, the required diameter is

$$d = 51.7(D_c \times A \times n)^{0.375} \, s^{-0.1875} \tag{14.12}$$

where d = inside drain diameter in mm,
D_c = drainage coefficient in mm/day,

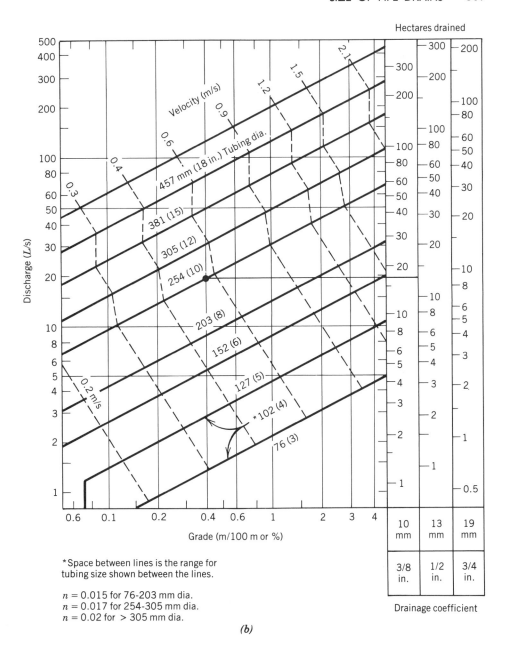

*Space between lines is the range for
tubing size shown between the lines.

$n = 0.015$ for 76-203 mm dia.
$n = 0.017$ for 254-305 mm dia.
$n = 0.02$ for > 305 mm dia.

(b)

A = drainage area in ha,
n = roughness coefficient in Manning's equation,
s = drain slope in m/m.

After computation of the required drain size, the next larger commercial size is
selected. Graphical solutions of this equation are given in Fig. 14.10.

The roughness coefficient in the Manning equation will increase with misalign-
ment at tile joints and with irregularities of the drain surface, such as roughness of
the walls and joints and corrugations in tubing. Design coefficients should be
0.013 for clay and concrete and 0.015 to 0.02 for corrugated plastic tubing. For

drains in irrigated lands of the West a roughness coefficient of 0.015 for concrete and clay tile is often recommended (see Appendix B).

For practical reasons a minimum pipe size is usually specified. If the capacity were to match the design flow exactly, the drain would be gradually enlarged starting from the upper end of the line. Local custom, availability of pipe, accuracy of installation, grade, and possible failure from sedimentation largely determine the minimum size. In most humid regions 75 or 100 mm is the minimum diameter recommended, but 125 and 150 mm are considered the minimum in organic soils and in irrigated regions. On flat slopes the minimum grade shown in Table 14.6 may determine the minimum size.

☐ *Example 14.3*
___

Determine the diameter of corrugated plastic tubing and the flow rate where the slope is 0.4 percent, the drainage area is 12 ha (30 ac), and the drainage coefficient is 13 mm (½ in.).

Solution.   Substituting in Eq. 14.12 using $n = 0.017$ (see Appendix B),

$$d = 51.7(13 \times 12 \times 0.017)^{0.375}(0.004)^{-0.1875} = 210 \text{ mm (8.3 in.)}$$

Select the next largest size, 254 mm from Fig. 14.10*b* and read the flow rate of 18 L/s from the left scale. Converting 13 mm (½ in.) per day from 12 ha (30 ac),

$$q = \frac{12 \times 13 \times 10^4 \times 10^3}{1000 \times 60 \times 60 \times 24} = 18.1 \text{ L/s (0.64 cfs)} \qquad ☐$$

☐ *Example 14.4*
___

Determine the clay tile size to carry the design flow from 910 m (3000 ft) of tile spaced 30 m (100 ft) apart if the soil-drainable porosity is 4 percent and the drainage requirement for optimum plant growth is a water table drop of 0.24 m/day (0.8 fpd). The grade for the tile is 0.15 percent.

Solution.   Assume a uniform drop of the water table over the entire drained area. The design flow is

$$q = \frac{910 \times 30 \times 0.24 \times 0.04 \times 1000}{60 \times 60 \times 24} = 3.0 \text{ L/s (0.11 cfs)}$$

Using this flow rate enter Fig. 14.10*a* on the left side, move to the right to a slope of 0.15 percent, and read 127 mm (5 in.) as the tile size. Size may also be computed from Eq. 14.12 where $D_c = 0.24 \times 1000 \times 0.04 = 9.6$ mm/day. Computed diameter is 117 mm, but select 127 mm, nominal commercial size.   ☐

The size of the main is determined from the drainage area of all connected drains. It should not be computed from the maximum capacity of all laterals. The capacity at any point in the system should be adequate to carry the discharge of the drainage system above, allowance being made for surface inlets and future additions to the system.

14.19 Drain Openings

Adequate inflow area can be provided by crack or joint spacings between 0.3-m lengths of clay and concrete tile if the opening area is equal to about 1 percent of the outside (o.d.) surface area of the drain (about 3500 mm²/m length for 100-mm drains). Experience has shown that such openings do not seriously restrict drainage. For 100-mm-diameter tubing about 88 slots 1.6 × 25-mm or 125 perforations 6-mm-diameter per meter length are adequate. Doubling the area or number of uniformly distributed openings will increase inflow about 20 percent. With CPT, soil in the valley of the corrugation has a large effect on inflow. Inflow is slightly greater if the opening is located on the ridge rather than in the valley. Shape of openings is relatively unimportant if they are uniformly distributed. The restriction of openings is greatly affected by the presence of soil within the opening itself and the permeability of the soil within about 25 mm of it. Within this distance about 50 percent of the head is lost as a result of convergence and friction. If a drain is enclosed in a gravel envelope or surrounded by a permeable cover, the loss in head is usually less than 10 percent.

ACCESSORIES

Accessories for pipe drainage systems include special facilities, such as surface inlets, sedimentation basins, and blind inlets.

14.20 Surface Inlets

As shown in Fig. 14.11, a surface inlet, sometimes called an open inlet, is an intake structure for the removal of surface water from potholes, road ditches, other

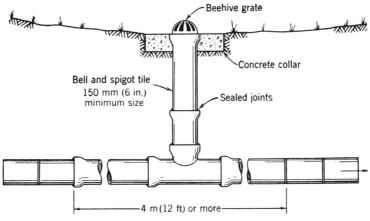

Fig. 14.11 Surface inlet for a pipe drain.

depressions, and farmsteads. Whenever practicable, surface water should be removed with surface drains (see Chapter 12) rather than surface inlets.

Surface inlets should be properly located and constructed. They should be placed at the lowest point along fence rows or in land that is in permanent vegetation. Where the inlet is in a cultivated field, the area immediately around the intake should be kept in grass. The surface inlet should be constructed of sealed bell and spigot tile and extended at least 2 m as a short lateral on either side of the main line. Galvanized metal pipe (asphalt coated) or a manhole constructed of brick or concrete is also satisfactory. At the surface of the ground a concrete collar should extend around the intake to prevent the growth of vegetation and to hold it in place. On top of the riser a beehive cover or other suitable grate is necessary to prevent trash from entering the line.

14.21 Blind Inlet or French Drain

Where the quantity of surface water to be removed is small or the amount of sediment is too great to permit surface inlets to be installed, blind inlets may, at least temporarily, improve drainage. Though these inlets often do not function satisfactorily for more than a few years, they are economical to install and do not interfere with farming operations.

As shown in Fig. 14.12, a blind inlet is constructed by backfilling the trench with various gradations of material. The coarsest material is placed immediately over the line, and the size is gradually decreased toward the surface. These inlets seldom meet the filter design criteria given in Appendix H and for that reason are not permanently effective. Since the soil surface has a tendency to seal, the area should be kept in grass or permanent vegetation. Other materials, such as corncobs, sawdust, and straw, are considered less dependable than more durable materials.

14.22 Sedimentation Basin

Soils containing large quantities of fine sand frequently cause sedimentation since the particles enter the drains through the openings. A sedimentation basin is any type of structure that provides for sediment accumulation, thus reducing deposition in the drain.

A sediment basin may be desirable where the slope downstream is greatly

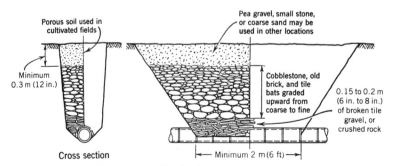

Fig. 14.12 Blind inlet or French drain.

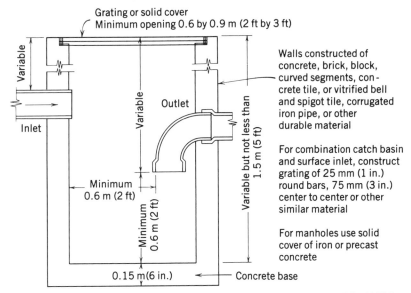

Fig. 14.13   Sedimentation basin or manhole. (Redrawn from SCS, 1973.)

reduced, where several laterals join the main at one point, or where surface water enters. The structure shown in Fig. 14.13 has a turned-down elbow on the outlet for retarding the outflow when the basin becomes filled with sediment. Where structures do not have this feature, the tendency is to neglect the clean-out of the basin.

Junction boxes may be installed where several lines join at different elevations. Except for the catch basin, they are similar to the structure shown in Fig. 14.13. To facilitate farming operations, the top of the manhole should be placed 0.3 m or more below the surface.

14.23   Controlled Drainage Structures

Control structures in drain lines for maintaining the ground water table at a specified level are similar to sedimentation basins except that crestboards are placed in the structure to keep the water at the desired level. The functioning of such structures is similar to that for control dams in open ditches described in Chapter 13. Controlled drainage is essential in the management of organic soils because of its effectiveness in controlling subsidence. Commercial valves for pipe drains are available for subirrigation or controlled drainage (see Chapter 9 and Section 14.33).

14.24   Relief Pipes and Breathers

Relief pipes and breathers are small-size vertical risers extending from the drain line to the surface. The riser should be made of steel pipe or sealed bell and spigot tile and should be located at fence lines where they are not likely to be damaged. Breathers are installed on long lines to prevent the development of a vacuum. In model studies Lembke et al. (1963) found that for slopes over 1 percent, negative pressures could develop in a tile running full. Quantitative values could not be predicted in the field nor were the effects of these negative pressures on flow determined. Many engineers do not feel that breathers are justified.

Relief pipes serve to relieve the excess water pressure in the drain during periods of high outflow, thus preventing blowouts. A relief pipe should be installed where a steep section of a main changes to a flat section, unless the capacity of the flat section exceeds the capacity of the steep section by 25 percent.

14.25 Envelope Filters

For drainage of irrigated land in the West gravel envelopes are placed completely around the drain; however, in humid areas this practice is not normally recommended. These envelopes are installed (1) to prevent the inflow of soil particles into the drain, which may cause failure, and (2) to increase the effective drain diameter, which increases the inflow rate. Design criteria for these filters are given in Appendix H and Chapter 15. Because of practical considerations and cost, the envelope is usually limited to one gradation of material, which is selected from local naturally occurring deposits.

Several types of sheet filters, called geotextiles, are available commercially. Strips can be placed above and below the drain during installation, and some filter materials are encased around corrugated plastic tubing at the extruding plant. Filters are made from nylon, polypropylene, and other materials. The mesh size should allow some of the fine soil particles to pass through so as not to plug the filter, yet retain the larger particles that would tend to deposit in the drain. One recommendation is that the 50 percent particle size of the soil should be greater than or equal to the average diameter of openings in the filter.

14.26 Artesian Relief Wells

Where artesian aquifers are the cause of drainage problems, wells may be drilled into the aquifer to reduce the pressure and to lower the water table. Sometimes these wells are connected directly into a drain, permitting the water to flow by gravity rather than by pumping. Water table contour maps or piezometric surface contours are often required prior to designing the drainage system.

DRAIN PIPE QUALITY

Only high-quality pipe should be installed in a drainage system. It is false economy to install second-grade or poor-quality materials.

14.27 Characteristics of Good Drain Tile

Clay or concrete tile should have the following characteristics: (1) resistance to weathering and deterioration in the soil; (2) sufficient strength to support static and impact loads under conditions for which they are designed (see Chapter 15); (3) low water absorption, that is, a high density; (4) resistance to alternate freezing and thawing; (5) relative freedom from defects, such as cracks and ragged ends; and (6) uniformity in wall thickness and shape. Drain tile that meet current specifications of the American Society for Testing Materials (ASTM) have the essential qualities listed above (see Appendix D). Specifications have been prescribed for three classes of drain tile: standard, extra-quality, and special-quality (concrete only) or heavy-duty (clay only). Standard-quality tile are satisfactory for drains of moderate size and depths found in most farm drainage work.

14.28 Concrete Tile

Concrete tile should be made with high-quality materials and be properly cured. Good-quality concrete tile are resistant to freezing and thawing but may be subject to deterioration in acid and alkaline soils. Where concrete tile are to be placed in acid or alkali soils, the tile should be extra-quality and made with cements having specific chemical characteristics. Curing methods will also depend on the degree of acidity or alkalinity of the soil (Miller and Manson, 1948).

14.29 Clay Tile

Clay tile should be well burned, with no checks or cracks, and should have a distinct ring when tapped with a metal object. Ordinary drain tile are not burned as hard as vitrified sewer tile. Clay tile made from shale are more durable and usually have less absorption than those made from surface clays. Clay tile are not generally affected by acid or alkaline soils. When subjected to frequent alternate freezing and thawing conditions, it is safer to use concrete tile, although most clay tile are resistant to frost damage. Where clay tile are laid with less than 0.7 m of cover, they should be extra-quality. In most mineral soils the kind of tile (clay or concrete) is not as important as the quality.

14.30 Corrugated Plastic Tubing

CPT is not damaged by soil chemicals, is light in weight, and is shipped in long lengths. Tubing should be uniform in color and density and free from visible defects. Parallel plate stiffness when deflected at 127 mm per minute should not be less than 0.17 N/mm per millimeter of length at 5 percent deflection and 0.13 N/mm per millimeter of length at 10 percent deflection for diameters up to 200 mm. In addition to the above standards polyethylene CPT should meet other criteria specified by ASTM F405 and F667 (see Appendix D). Standards for PVC tubing (more common in Europe) are covered in ASTM F800.

SYSTEM DESIGN PROCEDURE

The subsurface drainage system should be coordinated with existing and proposed surface ditches and other drains.

14.31 Preliminary Design Survey

The first step in design is to make a preliminary survey of the area, much as described for open ditches in Chapter 13. Soils should be investigated to determine whether subsurface drainage is practicable and economical. If so, sufficient information should be obtained to permit the selection of an adequate depth and spacing for the drains. The extent and depth of impermeable strata and ground water pressure and the source of seepage should be determined. On flat land, topographic maps may be prepared, especially where an extensive drainage system is planned. If the drainage system is not extensive, the preliminary survey and the location survey may be made at the same time. See Chapter 15 for layout procedure.

14.32 System Design

The procedure for designing a simple subsurface drainage system, shown in Fig. 14.14, is described in Example 14.5. Example 14.6 shows the procedure where part of the system includes a pothole from which surface water is to be removed through the pipe drain.

Pipe drains like open ditches are arbitrarily designated in decreasing order of importance as mains, submains, and laterals. As shown in Fig. 14.14, the first lateral above the outlet to be connected to main A may be designated A1, the next A2, and so on. The lines entering A1 may be indicated as A1.1, A1.2, and so on.

□ *Example 14.5*

Determine the quantity of each size of pipe for the drainage system shown in Fig. 14.14. The spacing of all laterals is 30 m (100 ft.). The slope of Main A and A1 is

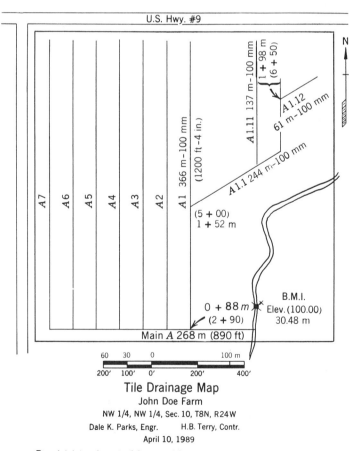

Fig. 14.14 A suitable map for a pipe drainage system.

0.4 percent and all other laterals are 0.15 percent. The soil is fine sandy loam in a humid area. Field crops are to be grown. Minimum pipe size is 100 mm (4 in.), which is available only in CPT. Larger sizes are to be clay tile.

Solution. From Table 14.5 select a drainage coefficient of 10 mm (⅜ in.) per day. Calculate drainage area (D.A.) from the length and spacing (30 m) of each line, but include the area drained by a lateral and the main only once. Determine pipe size from Fig. 14.10 or Eq. 14.12 using the appropriate n value. Lengths required: 3004 m (9850 ft), 100 mm (4 in.) CPT; 180 m (600 ft), 150 mm; (6 in.); and 88 m (290 ft), 200 mm (8 in.) clay tile plus 3 percent breakage allowance. Appropriate size and type junctions required as shown by plan. SI units are rounded.

	Slope (%)	D.A. [ha (ac)]	Nominal Pipe Diameter (mm)	(in.)	Length (m)	(ft)
Laterals						
A2 to A7	0.15	1.1 (2.7)	100	(4)	2196	(7200)
A1.1, A1.11,						
A1.12	0.15	1.3 (3.2)	100	(4)	442	(1450)
A1	0.4	2.4 (5.9)	100	(4)	366	(1200)
Main A						
A1 to A7	0.4	6.9 (17.0)	150[a]	(6)	180	(600)
0 + 00 to A1	0.4	9.6 (23.7)	200	(8)	88	(290)

[a]Size could be reduced to correspond to D.A. upstream from A1. □

□ Example 14.6

Determine the pipe size in Example 14.5 if laterals A1.1, A1.11, and A1.12 are all located within a 2.4 ha (5.9 ac) pothole and all of the surface water from this area is removed through a surface inlet at Sta. 1 + 98 m (6 + 50) on A1.1.

Solution. From Table 14.5 select a D_c of 19 mm (¾ in.) that applies to the entire pothole area. Since a flow rate for surface water is more than adequate for the three laterals within the pothole, the area drained by the three laterals need not be considered in computing outlet size.

	Slope (%)	D.A. [ha (ac)]	D_c (mm)	(in.)	Nominal Pipe Diameter (mm)	(in.)
A1.1 below inlet	0.15	2.4 (5.9)	19	(3/4)	150	(6)
A1 at Main A	0.4	5.7 (14.1)[a]	10	(3/8)	150	(6)
Main A						
A1 to A7	0.4	6.9 (17.0)	10	(3/8)	150	(6)
0 + 00 to A1	0.4	12.9 (31.8)	10	(3/8)	200	(8)

[a]D.A. for A1 = 1.1 + 4.6 = 5.7 ha. The equivalent of 2.4 ha at 19 mm is 4.6 ha at 10 mm. Conversion to a common D_c is necessary to read size from Fig. 14.10. The alternate procedure is to convert to flow rates in L/s (cfs).

Tile size on Main A is the same as in Example 14.5, but sizes should be increased on A1 and A1.1 to 150 mm (6 in.) to carry the increased flow from the pothole.

□

14.33 Controlled Drainage and Subirrigation Systems

Controlled drainage is the regulation of the water table by means of pumps, control structures (weirs), or a combination of these, to maintain the water level at depths that will produce maximum crop production. The water may be distributed through pipe drains or a system of small ditches. Some existing pipe systems may be retrofitted by installing control structures (see Chapter 9) and pumping facilities (see Chapter 16) to provide subirrigation. In Michigan, field crop yields have been increased as much as 15 to 30 percent by subirrigation (water added to raise the water table). Because drain spacings for subirrigation should be as much as 30 percent closer than those for drainage only, these combination systems should best be designed as new systems. Some cost savings may be achieved, compared with having a separate surface or sprinkler irrigation system. During the nongrowing seasons, the combination system is operated in the drainage mode and then switched to the irrigation mode when the crop needs water. The water table levels between the pipe drains for the two modes of operation are shown in Fig. 14.7. See ASAE (1990) and Fouss et al. (1990) for further design criteria.

Controlled drainage and subirrigation systems are limited to special soil and climatic conditions. The farmer must also have an adequate understanding of their operation and management. The soil must be nearly impermeable below the drain depth, but have a good horizontal permeability to permit a reasonable drain spacing. Land slopes should generally be less than 1 percent to reduce the variation in water table levels, and/or the elevation difference between adjacent control structures should be less than 0.5 m. Automatic controls are desirable, especially when unexpected rainfall occurs and the water level is high. Water supplies should be adequate to meet irrigation requirements.

14.34 Computer Modeling for Drainage Design

Modern computers simplify the design of a subsurface drainage system and make possible the incorporation of other variables, such as climate and plant growth factors, so as to predict relative crop yields. Several models have been developed, one of which is DRAINMOD (Skaggs, 1980). This model considers hourly and daily weather data, soil properties, crop characteristics, soil water distribution, and other factors. DRAINMOD is widely adopted in the eastern United States and a user's manual and program are available (see Appendix I). This model is accessible in state SCS offices and is accepted by many extension and research engineers in the humid states. The program generates the number of working days for tillage operations, a quantitative evaluation of excess wetness conditions, the number of dry days with deficient soil water, and the yield effects of these stresses. The model can also evaluate drainage system design for wastewater treatment. With some modification it can be adapted to irrigation design. DRAINMOD simulates the performance of a given drainage system for as many years of climatic record as desired. It will predict relative crop yields for the period of record from which economic probability analyses can be made. Surface drainage inputs can be evaluated as can the effect of subirrigation through the existing drainage system. The model has been validated from field data at a number of locations in the humid area. One of its greatest merits is the ability to predict the crop response to changes in drainage or subirrigation system design. One or more design parameters can be changed without affecting others.

REFERENCES

American Society of Agricultural Engineers (ASAE) (1984). *Design, Construction and Maintenance of Subsurface Drains in Arid and Semiarid Areas.* EP463. ASAE, St. Joseph, MI.

——— (1988). *Design, Construction of Subsurface Drains in Humid Areas.* EP260.4. ASAE, St. Joseph, MI.

——— (1990). *Design, Installation and Operation of Water Table Management Systems for Subirrigation/Controlled Drainage in Humid Regions.* EP479. ASAE, St. Joseph, MI.

Clayton, B. S., R. R. Neller, and R. V. Addison (1942). *Water Control in the Peat and Muck Soils of the Florida Everglades.* Florida Agr. Expt. Sta. Bull. 378. University of Florida, Gainesville, FL.

Dumm, L. D. (1954). "Drain-Spacing Formula." *Agr. Eng.* **35**, 726–730.

Fouss, J. L., R. W. Skaggs, J. E. Ayars, and H. W. Belcher (1990). "Watertable Control and Shallow Groundwater Utilization." In *Management of Farm Irrigation Systems*, G. J. Hoffman, T. A. Howell, and K. A. Solomon (eds.). ASAE Monograph, Chap. 21. ASAE, St. Joseph, MI.

Hansen, V. E., O. W. Israelsen, and G. E. Stringham (1980). *Irrigation Principles and Practices*, 4th ed. Wiley, New York.

Hoffman, G. J. (1990). *Environmental Impacts of Subsurface Drainage.* Proceedings, 4th International Workshop on Land Drainage, Feb. 1990, Cairo, Egypt. International Commission on Irrigation and Drainage, New Delhi, India.

Hoffman, G. J., and G. O. Schwab (1964). "Spacing Prediction Based on Drain Outflow." *ASAE Trans.* **7**, 444–447.

International Institute for Land Reclamation and Improvement (1972–1974). *Drainage Principles and Applications*, Vols. 1–4. The Institute, Wageningen, The Netherlands.

Kidder, E. H., and W. F. Lytle (1949). "Drainage Investigations in the Plastic Till Soils of Northeastern Illinois." *Agr. Eng.* **39**, 384–386, 389.

Kramer, P. J. (1949). *Plant and Soil Water Relationships.* McGraw-Hill, New York.

Kramer, P. J., and W. T. Jackson (1954). "Cause of Injury to Flooded Tobacco Plants." *Plant Physiol.* **29**, 241–245.

Lembke, W. D., J. W. Delleur, and E. J. Monke (1963). "Model Study of Flow in Steep Tile Drains." *ASAE Trans.* **6**, 142–144.

Manson, P. W., and C. O. Rost (1951). "Farm Drainage—An Important Conservation Practice." *Agr. Eng.* **32**, 325–377.

Miller, D. G., and P. W. Manson (1948). "Essential Characteristics of Durable Concrete Drain Tile for Acid Soils." *Agr. Eng.* **29**, 437–441.

Moody, W. T. (1966). "Nonlinear Differential Equation of Drain Spacing." *Proc. ASCE* **92** (IR2), 1–9.

Neal, J. H. (1934). *Proper Spacing and Depth of Tile Drains Determined by the Physical Properties of the Soil.* Minnesota Agr. Expt. Sta. Tech. Bull. 101. University of Minnesota, St. Paul, MN.

Nicholson, H. H., and D. H. Firth (1953). "The Effect of Ground Water Levels on the Performance and Yield of Some Common Crops." *J. Agr. Sci.* **43**, 95–105.

Pavelis, G. A. (ed.) (1987). *Farm Drainage in the United States: History, Status, and Prospects.* USDA Misc. Publ. 1455. GPO, Washington, DC.

Pillsbury, A. F., J. R. Spencer, W. R. Johnston, and L. O. Weeks (1965). "The

Drainage Performance in Coachella Valley, California." *Proc. ASCE* **91** (IR2), 1–10.

Schwab, G. O., N. R. Fausey, E. D. Desmond, and J. R. Holman (1985). *Tile and Surface Drainage of Clay Soils.* Res. Bull. 1166. Ohio Agr. Res. & Dev. Center, Wooster, OH.

Skaggs, R. W. (1975). "Determination of the Hydraulic Conductivity-Drainable Porosity Ratio from Water Table Measurements." *ASAE Trans.* **19**, 73–80.

——(1980). *A Water Management Model for Artificially Drained Soils.* North Carolina Agr. Res. Serv. Tech. Bull. 267. North Carolina State University, Raleigh, NC.

Smedema, L. K., and D. W. Rycroft (1983). *Land Drainage.* Cornell University Press, Ithaca, NY.

U.S. Bureau of Reclamation (USBR) (1978). *Drainage Manual.* GPO, Washington, DC.

U.S. Soil Conservation Service (SCS) (1973). *Drainage of Agricultural Lands.* Water Information Center Inc., Port Washington, NY.

Van Hoorn, J. W. (1958). "Results of a Ground Water Level Experimental Field with Arable Crops on Clay Soil." *Neth. J. Agr. Sci.* **6** (1), 1–10.

Van Schilfgaarde, J. (1963). "Tile Drainage Design Procedure for Falling Water Tables." *Proc. ASCE* **89** (IR2), June, and discussion Dec. 1963, Mar. and Dec. 1964.

Van Schilfgaarde, J. (ed.) (1974). *Drainage for Agriculture.* Monograph 17. Am. Soc. Agron., Madison, WI.

Visser, W. C. (1959). "Crop Growth and Availability of Moisture." Tech. Bull. 6. Inst. for Land and Water Mgt. Res., Wageningen, The Netherlands.

Walker, P. (1952). "Depth and Spacing for Drain Laterals as Computed from Core-Sample Permeability Measurements." *Agr. Eng.* **33**, 71–73.

Wesseling, J., W. R. vanWijk, M. Fireman, B. D. van't Woudt, and R. M. Hagan (1957). "Land Drainage in Relation to Soils and Crops." In *Drainage of Agricultural Lands*, J. N. Luthin (ed.). Monograph 7, Am. Soc. Agron., Madison, WI.

PROBLEMS

14.1 Calculate the discharge of a 150-mm (6-in.)-diameter CPT flowing full if the slope is 0.4 percent.

14.2 How many hectares (acres) will a 254-mm (10-in.) clay tile drain with a slope of 0.3 percent if the tile are in mineral soil in your area and field crops are to be grown?

14.3 Determine the quantity of each size of CPT required for a gridiron system, as shown in Fig. 14.5c, if the length of each of the five laterals is 400 m (1310 ft) and the main is 300 m (1000 ft). Slope in the laterals is 0.12 percent and in the main, 0.3 percent. Spacing of the laterals is 30 m (100 ft). Design the system for field crops in your area.

14.4 What size CPT is required to remove the surface water with a surface inlet if the runoff accumulates from 8.0 ha (20 ac) and the slope in the drain is 0.4 percent? Design for field crops in your area.

14.5   If a surface inlet draining 8.0 ha (20 ac) is added at the upper end of the main in Problem 14.3, what size CPT is required from the surface inlet to the outlet?

14.6   What slope is required to provide a velocity of 0.5 m/s (1.5 fps) at full flow in a 100-mm (4-in.) CPT? In a 300-mm (12-in.) CPT?

14.7   If a tile drainage system draining 12.0 ha (30 ac) flows full for 3 days following a period of heavy rainfall, determine the effluent volume during this period if the system was designed for a drainage coefficient of 13 mm (½ in.) per day.

14.8   Determine the depth of flow as a percentage of the drain diameter to obtain maximum discharge. Note maximum discharge does not occur at full flow.

14.9   Determine the design flow from 3600 m (9840 ft) of pipe at a spacing of 60 m (200 ft) following local recommendations for irrigated conditions. What size of CPT is required for the main at the outlet if the drain slope is 0.25 percent and $n = 0.016$?

14.10   Determine the depth and spacing of drains to maintain a constant water table 1.5 m (5 ft) below the surface assuming that the height of the water table at the midplane is 0.5 m (1.5 ft) above the center of 150-mm (6-in.) drains. The average hydraulic conductivity of the soil is 0.6 m/day (2 fpd), the depth d to the impervious layer is 3 m (10 ft), and the excess irrigation rate is 3.0 mm/day per meter of length (0.01 fpd per foot of length). Assume a spacing of 40 m to select the equivalent depth.

14.11   Compute the drain spacing for a clay loam soil having a drainable porosity of 0.06, a hydraulic conductivity of 0.6 m/day (2 fpd), and an impervious layer 2.4 m (8 ft) below the drains. Drainage requirements for the crop are such that the initial water table depth below the surface is 0.15 m (6 in.), with a rate of drop of 0.21 m (0.7 ft) for the first day. Because of outlet depth, the maximum depth of the drain is limited to 1.2 m (3.9 ft) measured to the bottom of the 125-mm (5-in.) drains.

14.12   From a contour map supplied by your instructor design and layout the most efficient pipe drainage system for the soil and cropping practices specified for the field. Determine the length of each size of pipe and a list of all other materials required.

14.13   Determine the size at the outlet of 3600 m (11 800 ft) of CPT if the spacing is 18 m (60 ft) and the grade is 0.1 percent. The soil porosity is 3 percent and the optimum rate of drop of the water table is 0.15 m (0.5 ft) in the first 12 h following heavy rainfall.

14.14   Determine the drainage coefficient for an irrigated field where the seepage from a canal is 10 percent of the water applied and the deep percolation from irrigation is 25 percent. Assume an irrigation depth of 75 mm (3 in.) every 8 days. Compute the flow rate in L/s from 3 ha (7.4 ac).

CHAPTER 15

Location, Installation, and Maintenance of Subsurface Drains

Although a subsurface drainage system is adequately designed and is staked out according to plan, it will not function satisfactorily unless properly installed and maintained. The trench should be dug to the specified grade, the bedding conditions and width of the trench should be such as to prevent overloading of the pipe, good workmanship should be secured in laying the pipe and in making junctions, the cost of installation should be reasonable, and the drainage system should be mapped and properly recorded.

LAYOUT PROCEDURES

Experience is desirable in the proper location of pipe drains, but there are a few general rules: (1) place the outlet at the best possible location; (2) provide as few outlets as possible; (3) lay out the system with short mains and long laterals; (4) use the available slope to best advantage, especially on flat land; (5) follow the general direction of natural waterways, particularly with mains and submains on land with considerable slope; (6) avoid routes that result in excessive cuts; (7) avoid crossing waterways except at an angle of 45 degrees or greater; and (8) avoid soil conditions that increase installation and maintenance costs.

15.1 Alignment

Where a change in direction of pipe drains is necessary, a junction box, a fitted tile, or a T- or Y-manufactured junction is required. The alignment of pipe at junctions is shown in Fig. 15.1. Where sufficient slope is available, the grade line of the

320

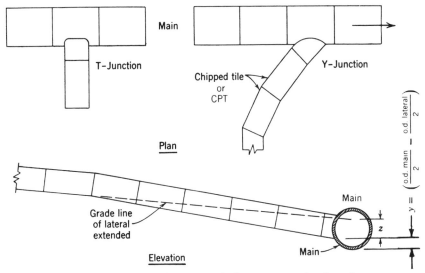

Fig. 15.1   Vertical and horizontal alignment at pipe junctions.

lateral should intersect the main near the top of the main; however, if sufficient slope is not available, the grade line of the lateral may intersect the main at a lower elevation but should always be high enough (y in Fig. 15.1) to permit the center line of the lateral to intersect the center line of the main. A difference in elevation (z in Fig. 15.1) between the extended grade line of the lateral and the main is adjusted by increasing the slope in the last few meters on the lateral.

15.2 Staking

The drain line is staked by placing hub and guard stakes at 15- or 30-m stations, starting at the outlet (Station 0 + 00). To prevent these stakes from being removed during construction, the stake line must be offset about 2 m from the center line of the trench. With the laser beam grade control system described in Section 15.4, staking is necessary only for location of the line and of points where grade changes are required.

INSTALLATION PRACTICES

Installation should always begin at the outlet and progress upstream so that seepage water is free to move to the outlet. The installation includes digging to an established grade, laying the pipe, and backfilling.

15.3 Trenching Methods

Pipe is usually installed with a trenching machine or trenchless plow: Trenching machines may be divided into three general classes: (1) wheel excavators, (2) endless-chain excavators, and (3) hoe excavators. The first two types are shown in Fig. 15.2 and the hoe in Fig. 13.8. Type of machine will vary with geographic location, soil, and other local conditions. In the 1980s in Ontario, Canada, of 388

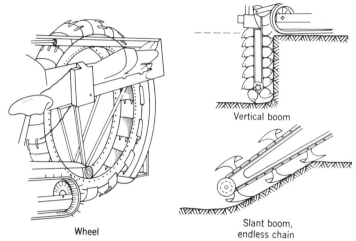

Fig. 15.2 Wheel and endless-chain trenching machines.

machines licensed by the government, 53 percent were wheel-type, 37 percent trenchless plows, and 10 percent endless-chain excavators.

With wheel excavators, soil is carried to the top of the wheel and then dropped onto a conveyor belt. Endless-chain excavators may be subdivided into two groups: the vertical-boom type and the slant-boom type. The vertical-boom type reduces the amount of handwork where obstructions, such as pipelines and cables, are encountered. On many hoe-type machines, described in Chapter 13, the digging bucket is simply a concave blade of the desired width. This class of excavator is suitable for making deep cuts, removing large stones, and making junctions. They also work satisfactorily in wet subsoil. Draglines are suitable for deep, wide trenches.

A trenching machine should dig to a uniform grade under a variety of soil conditions. A crumber (shoe) following the digging mechanism should make a curved or V-shaped bottom in the trench to ensure good bedding conditions and to align the pipe.

The factors that influence the rate of installation for a trenching machine are (1) soil water; (2) soil characteristics, such as hardness, stickiness, stones, and submerged stumps; (3) depth of trench; (4) condition of the trenching machine; (5) skill of the operator; (6) width of the trench; and (7) delays resulting from interruptions during operation. Increasing the depth from 0.9 to 1.5 m decreased the digging speed by 56 percent under Iowa and Minnesota conditions. Although the width of the trench may not be important for machines with sufficient power, the rate of installation may be reduced especially for trenches wider than 0.4 m.

The average output for a trenching machine at depths from 0.9 to 1.5 m may be as much as 700 to 300 m per day, respectively. A ditching machine operated by a contractor in northern states can be expected to install pipe only about 150 working days during the year. Studies showed that a total of 66 percent of the available working hours were lost. This time loss is accounted for as follows: weather, 19 percent; repairs, 14 percent; making junctions, 10 percent; miscellaneous, 9 percent; moving from job to job, 8 percent; and servicing of the machine, 6 percent.

The rate of installation is much faster with trenchless plows equipped with automatic laser grade-control systems than with a trencher. They are best suited for large installations and for smaller-size tubing (200 mm in diameter or less) at shallow depths (1.2 m or less). The plow with a vertical blade similar to a mole plow (Fig. 14.4) is generally mounted on the rear of large crawler tractors. Plastic tubing is fed through a chute at the rear of the blade to guide it into the channel formed by the plow. High investment cost of the equipment and limitations to special field conditions have restricted widespread use.

15.4 Methods of Establishing Grade

Grade may be established by laser or manual methods. A laser system consists of a rotating laser beam that can give either a level or a sloping plane of reference. A detector on the trenching machine picks up the beam and automatically keeps the machine on the same grade as the plane produced by the laser beam. A grade modification device with a distance-measuring wheel mounted on the trencher, called a grade breaker, will change the drain grade from that made by the laser beam. The laser system has been adapted to many other earth-moving machines, such as the blade grader, scraper, and bulldozer.

For hand trenching or for trenching machines not equipped with an automatic grade-control system a sight or string line is a simple grade-control method. The string line and the sight methods are essentially the same except that string is stretched between stations in the line method and sighting targets are set at each station in the sight method. As shown in Fig. 15.3, the depth of the trench is established by sighting over two targets. The line of sight or string line must be parallel to the grade line of the drain. The distance from the line of sight to the grade line is known as the *base distance*, which in Fig. 15.3 is 3.0 m. The target is set near the hub stake, and the cut is subtracted from the base distance to obtain the difference in elevation between the top of the hub and the line of sight. For example, at the first station the target is 1.93 m (3.00 − 1.07) above the hub stake. At least three consecutive targets should be set so that errors in design or calculation may be detected.

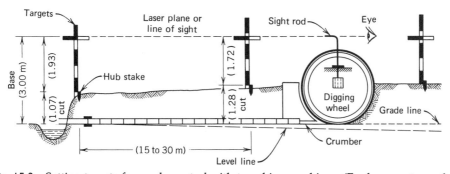

Fig. 15.3 Setting targets for grade control with trenching machines. (For laser systems, the laser plane would be parallel to the line of sight as shown.)

15.5 Laying the Drain

Pipe drains should be laid on true grade, with the bottom of the trench shaped with a 90-degree groove angle to provide good alignment and bottom support. The crack spacing between individual tiles should be about 3 mm, unless the soil is sandy (Chapter 14). In unstable soils the crack spacing should be small to prevent inflow of fine sand. The upper ends of all drain lines should be closed tightly with flat slabs of concrete, stones, or prefabricated ends.

If for some reason the trench is excavated below grade, it should be refilled to the desired grade with well-graded gravel. If the trench does not have water in it, moist soil may be suitable if thoroughly compacted under the drain.

15.6 Junctions

A properly made junction is an essential part of the drain system. Either a T- or a Y-manufactured junction is satisfactory. With the Y-junction the flow should be directed downstream. Extensive tests conducted by Manson and Blaisdell (1956) showed that the energy loss for a 90-degree junction angle is insignificant for all practical purposes. Because of the ease and convenience in installation and the resulting improvement of workmanship, the 90-degree junction, or T-junction, is preferred.

Manufactured junctions are more satisfactory than those fabricated in the field. A homemade junction can be fitted together by chipping and breaking ordinary tile and by using concrete mortar completely around the connection. Plastic tubing can be connected in a similar manner using special fittings, usually of the snap-on type furnished by each manufacturer. Often they are not interchangeable.

15.7 Checking Grade

Immediately after installation and before backfilling, the drains should be checked for grade, alignment, and other design specifications. Allowable variation from true grade should be within reasonable limits. Some engineers allow a variation of 3 mm per 25 mm of drain diameter (about 10 percent) for sizes up to 250 mm. In no circumstances should there be backfall in the drain. Somewhat greater variation may be allowed where slopes are greater than minimum (Chapter 14).

15.8 Blinding and Envelope Filters

As shown in Fig. 15.4, blinding is accomplished by placing loose, mellow topsoil or soil from another permeable layer in the trench immediately over the pipe. The purpose of blinding is to prevent the tubing or tile from being moved out of alignment, to prevent tile breakage or tubing deflection during the backfilling operation, and to provide permeable soil in contact with the pipe. If stones are present in the soil, the minimum depth of blinding should be about 0.15 m. In heavy soils it may be desirable to backfill the trench with corncobs, cinders, well-graded gravel, and other porous materials; however, organic materials may not be effective for more than a few years.

In irrigated areas of the West, where gravel-envelope filters are recommended, trenching machines are often equipped with two gravel chutes as shown in Fig. 15.5. The first one places the gravel in the bottom of the trench and the pipe is laid

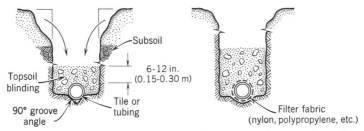

Fig. 15.4 Methods of protecting and stabilizing pipe drains in humid areas.

on this layer. The second chute places the gravel on the sides and top of the drain to the desired thickness, usually a minimum of 0.1 m. Tar paper or filter mats may be placed on the drain in the manner shown. Factory-wrapped tubing with thin fabric completely around the drain may be a substitute for a gravel envelope. These fabrics, called geotextiles, may be made of nylon, polypropylene, fiberglass, and other materials. The mesh size should allow some of the fine soil particles to pass through so as not to plug the filter, yet retain the larger particles that tend to deposit in the drain. Fabrics may be easier and more economical to install than gravel, but they are more subject to soil sealing. Some plastic tubing is perforated with small pinholes that restrict the inflow of sand. Tests show that the maximum opening size should be not more than about four times the diameter of sand at the 60 percent passing size.

Gravel envelopes should be well graded and free of vegetation, clays, and other materials that could reduce their hydraulic conductivity. All envelope material should pass the 38-mm-square screen opening and not more than 5 percent should pass the 0.3-mm sieve (USBR, 1978). Since few pit run sands and gravels meet these requirements, most envelope material must be machine sorted. Washing is required when the only source is pits containing silt or clay-coated particles. Upper and lower size limits in relation to the base material (existing soil) are shown in Table 15.1. These criteria are not as restrictive as those for toe drains in dams given

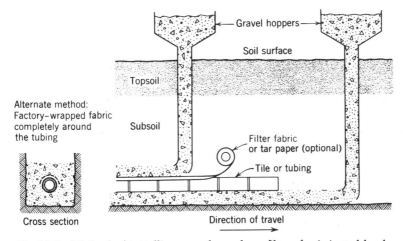

Fig. 15.5 Method of installing gravel envelope filters for irrigated land.

Table 15.1 Gradation Relationships Between Base Material and Graded Envelope Material

Diameters of Base Material, 60% Passing (mm)	*Percent Passing*				
	100	60	30	10	0
	Diameter of Particles for Envelopes (mm)— Lower Limits and Upper Limits				
0.02–0.05	9.5–38	2–10	0.8–8.7	0.3–2.5	0.07–0.6
0.05–0.10	9.5–38	3–12	1.1–10.4	0.4–3.0	0.07–0.6
0.10–0.25	9.5–38	4–15	1.3–13.1	0.4–3.8	0.07–0.6
0.25–1.00	9.5–38	5–20	1.5–17.3	0.4–5.0	0.07–0.6

Example: If the base material has a 60% size of 0.075 mm (second line in the table) and envelope sizes of 30, 7, 2, 1, and 0.3 mm for 100, 60, 30, 10, and 0 percent passing, respectively, the envelope material is satisfactory.
Source: Adapted from USBR (1978).

in Appendix H. In Europe, cinders, processed expanded clay aggregates, and coconut fiber mats are installed as envelope filters.

Properly designed gravel envelopes normally prevent the inflow of soil and increase inflow rates. Poor construction techniques may cause failures. Gravel is normally deposited in the chute hoppers with motorized scoop loaders. In California tests, envelope failure was attributed to contamination by topsoil picked up with the gravel.

15.9 Backfilling

Backfilling of the trench can be accomplished with many machines, such as bulldozers, hoes, motor graders, manure loaders, and blades on small tractors. All soil should be replaced and heaped up so that after settlement the trench can be crossed with tillage equipment. If the soil is dry or contains stones, extreme care should be taken to prevent breakage of the tile or crushing of the plastic tubing.

15.10 Installation in Peat and Muck

Before drains are installed in newly developed peat and muck soils, the land should be drained with open ditches for several years to allow initial subsidence. The pipe should be laid on the underlying mineral soil, provided the mineral soil is not too deep and not impermeable. In peat or muck soils, 150-mm drains are often recommended as the minimum size because of differential subsidence. Perforated tile several feet in length or plastic tubing is recommended.

15.11 Installation in Quicksand

Quicksand is not a type of sand but rather a soil condition in which fine sand in a saturated, fluid condition is buoyed up by hydrostatic pressure from below. Some of the following recommendations may be helpful in planning and installing drains under quicksand conditions: (1) install pipe only during the driest season; (2) use plastic tubing with a suitable prefabricated covering; (3) install pipes that are at least 125 or 150 mm in diameter; (4) use grades of not less than 0.4 percent; (5) protect all cracks between individual tile with durable materials, covering the upper two thirds of the tile circumference; (6) install each line as rapidly as possible and without interruptions to prevent settlement of the crumber on the machine; (7) blind the pipe with coarse material, such as gravel, sawdust, or sod; (8) prevent the caving-in of the ditch bank prior to and during backfilling operations; and (9) provide careful maintenance of the line, particularly if sinks or blowholes develop. A plastic stabilizing strip may be placed under the pipe to maintain alignment and grade. In some situations, reducing the hydrostatic pressure by pump drainage may be necessary.

LOADS ON CONDUITS

Loads on underground conduits include those caused by the weight of the soil and by concentrated loads resulting from the passage of equipment or vehicles. At shallow depths concentrated loads largely determine the strength requirements of conduits; at greater depths the load due to the soil is the most significant. Loads for both conditions should be determined, particularly where the depth of the controlling load is not known. For pipe drains the weight of the soil usually determines the load. Concentrated loads may be calculated by methods described in Appendix E.

15.12 Types of Conduit Loading

For purposes of analyzing loads, the two types of underground conduit are ditch conduits and projecting conduits, as shown in Fig. 15.6. Ditch conduit conditions apply to narrow trenches, and projecting conduit conditions occur in trenches wider than about two or three times the outside diameter of the pipe; however, the type of loading can be determined from Eqs. 15.1 and 15.2. Projecting conditions generally exist when the settlement of soil prisms A and C shown in Fig. 15.6b is

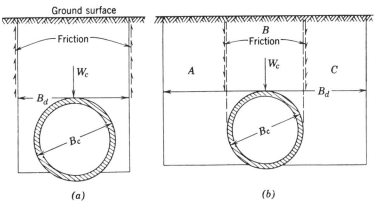

Fig. 15.6 Frictional forces for (a) ditch-type and (b) projecting-type rigid conduit loading.

greater than that of prism B. Since this condition applies to conduits placed under an embankment on undisturbed soil as in road fills or pond dams, the conduit projects above a relatively solid surface; hence its name.

15.13 Load Factors Based on Bedding Conditions

The load factor (L.F.) is the ratio of the strength of a rigid conduit under given bedding conditions to its strength as determined by the three-edge bearing test. As shown in Fig. 15.7, four classes of bedding conditions are generally recognized.

For nonpermissible bedding conditions, shown in Fig. 15.7a, no attempt is made to shape the foundation or to compact the soil under and around the conduit. Such a bedding has a low load factor (1.1) and is not suitable for drainage work.

For ordinary bedding conditions (L.F. 1.5) the bottom of the trench must be shaped to the conduit for at least half its width. The conduit should be surrounded with fine, granular soil extending at least 0.15 m above the top of the pipe.

For first-class bedding (Fig. 15.7c) the conduit must be placed in fine, granular material for 0.6 the conduit width and should be entirely surrounded with this material extending 0.3 m or more above the top. In addition, the blinding material should be placed by hand and thoroughly tamped in thin layers on the sides and above the conduit. Although the load factor for first-class bedding is 1.9, it is seldom necessary in drainage work.

The concrete cradle bearing is constructed by placing the lower part of the conduit in concrete. Such construction is not practical in conservation work except for earth dams or high embankments where excessive loads are encountered. This type of bedding provides the best conditions, with load factors varying from 2.2 to 3.4.

Based on a study of trench bottoms made by several ditching machines, load factors were found to vary from 0.9 to 2.5. With a keel on the bottom of a curved trencher shoe or a V-shaped bottom, a load factor of 1.5 or greater may be expected.

15.14 Soil Loads on Rigid Pipe

Loads on both ditch conduits and projecting conduits are based on studies by Marston (1930), Spangler (1947), and Schlick (1932). For loads on ditch conduits it

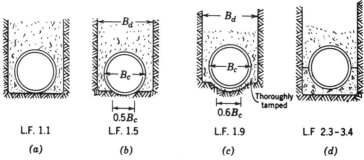

Fig. 15.7 Typical bedding specifications and load factors (L.F.) for rigid ditch conduits. (a) Not permissible. (b) Ordinary. (c) First class. (d) Concrete cradle. (Redrawn from Spangler, 1947.)

is assumed that the density of the fill material is less than that of the original soil. As settlement takes place in the backfill, the sides of the ditch resist such movement (Fig. 15.6a). Because of the upward frictional forces acting on the fill material, the load on the conduit is less than the weight of the soil directly above it. In its simplest form the ditch conduit load formula is

$$W_c = C_d w B_d^2 \qquad (15.1)$$

where W_c = total load on the conduit per unit length (F/L),
$\quad C_d$ = load coefficient for ditch conduits,
$\quad w$ = unit weight of fill material (F/L³),
$\quad B_d$ = width of ditch at top of conduit (L).

The backfill directly over a projecting conduit (prism B in Fig. 15.6b) will settle less than the soil to the sides of the conduit (prisms A and C). For projecting conditions the load on the conduit is greater than the weight of the soil directly above it, because shearing forces resulting from greater settlement of soil on both sides are downward rather than upward. The projecting conduit formula for wide ditches is

$$W_c = C_c w B_c^2 \qquad (15.2)$$

where C_c = load coefficient for projecting conduits,
$\quad B_c$ = outside diameter of the conduit (L).

The load coefficients C_d and C_c are functions of the frictional coefficient of the soil, the height of fill, and, respectively, the width of the ditch or the width of the conduit. Since these coefficients are rather complex, C_d and C_c curves, shown in Fig. 15.8, have been developed. The load on the conduit is the smaller value as computed from Eq. 15.1 or 15.2 (Example 15.1). The width of trench for a given conduit that results in the same load when computed by both equations is known as the transition width (Fig. 15.9).

Drain pipe should be installed so that the load does not exceed the required average minimum crushing strength of the pipe, as given in Appendix D. Whenever practicable, ditch conduit conditions should be provided rather than projecting conduit conditions. To prevent overloading in wide trenches, a narrow sub-ditch can be excavated in the bottom of the main trench. The width of the subditch measured at the top of the pipe determines the load, regardless of the shape of the trench above this point. The subditch need not extend above the top of the pipe. A common method of construction in deep cuts is to remove the excess soil with a bulldozer and to dig the subditch with a trenching machine. For trench widths of 0.4 and 0.46 m or less, standard-quality and extra-quality concrete or clay tile, respectively, may be installed at any depth.

☐ Example 15.1

Determine the static load on 250-mm (10-in.)-diameter pipe (B_c = 300 mm or 1 ft) installed 2.1 m (7.0 ft) deep in a trench 0.45 m (18 in.) wide if the backfill is ordinary clay weighing 20 000 N/m³ (120 pcf). What quality of concrete pipe is required, assuming ordinary bedding conditions?

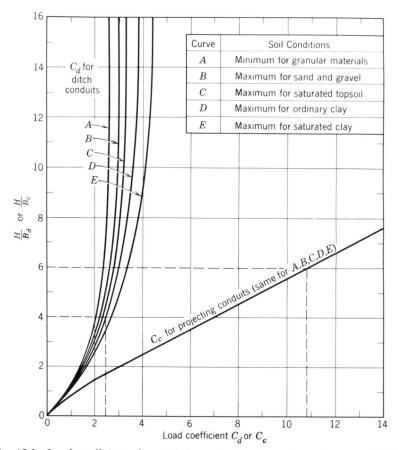

Fig. 15.8 Load coefficients for rigid drain tile. (Redrawn from Spangler, 1947.)

Solution. $H = d - B_c$, where H = depth to top of the pipe and d = depth to the bottom of the pipe. $H = 2.1 - 0.3 = 1.8$ m (6.0 ft). Read from Fig. 15.8 for $H/B_d = 1.8/0.45 = 4.0$, $C_d = 2.4$. Substitute in Eq. 15.1 to obtain the load for ditch conditions:

$$W_c = 2.4 \times 20\ 000 \times (0.45)^2 = 9720 \text{ N/m (648 lb/lin ft)}$$

Read from Fig. 15.8 for H/B_c of 6.0, $C_c = 10.8$. Substitute in Eq. 15.2 using pipe o.d. = 0.3 m (with 25-mm wall thickness) to obtain load for projecting conditions:

$$W_c = 10.8 \times 20\ 000 \times (0.3)^2 = 19\ 440 \text{ N/m (1296 lb/lin ft)}$$

The lower value, 9720 N/m, is the actual load; hence, ditch conditions apply (Fig. 15.9). To allow for variations in pipe strength and bedding conditions include a safety factor of 1.5, which results in a design load of $9720 \times 1.5 = 14\ 580$ N/m (972 lb/lin ft).

From Fig. 15.7 the load factor for ordinary bedding is 1.5, and from Appendix D the supporting strength of standard-quality concrete pipe is 11 900 N/m (800 lb/lin ft), based on the three-edge bearing test method. Strength for ordinary

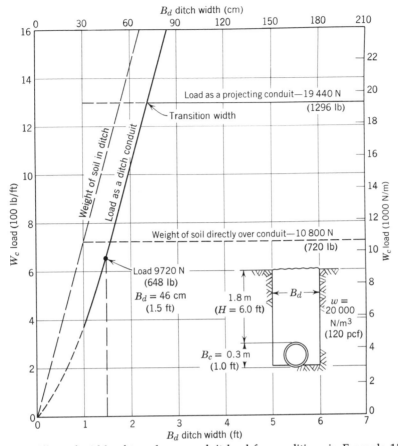

Fig. 15.9 Effect of width of trench on conduit load for conditions in Example 15.1.

bedding conditions is $11\ 900 \times 1.5 = 17\ 850$ N/m (1200 lb/lin ft), which is sufficient to support a design load of 14 580 N/m (972 lb/lin ft).   □

15.15 Soil Loads on Flexible Pipe

For loads on flexible pipe Spangler and Handy (1981) multiplied Eq. 15.1 for ditch conditions by B_c/B_d to obtain

$$W_c = C_d w B_c B_d \tag{15.3}$$

This equation is valid only when thoroughly tamped soil on the sides of the tubing has essentially the same stiffness as the tubing itself. By knowing the load-carrying capacity of the tubing the deflection can be computed from the modified Iowa formula (Spangler and Handy, 1981)

$$\Delta x = \frac{D_1 K W_c}{(EI/r^3) + 0.061E'} \tag{15.4}$$

where Δx = horizontal deflection ($\Delta x = 0.91 \Delta y$) (L),
D_1 = deflection lag factor (usually 1.5),
K = bedding angle factor (0.096 for 90-degree angle),
W_c = load-carrying capacity per unit length (F/L),
r = mean radius of tubing (L),
E = modulus of elasticity of tubing (F/L^2)
I = moment of inertia of tubing (L^4),
E' = modulus of soil reaction (F/L^2).

The two additive terms in the denominator represent the contributions of pipe stiffness and soil side support, respectively. The tubing stiffness term, EI/r^3, can be more easily computed from the parallel plate stiffness test, in which

$$EI/r^3 = 0.149C(F/\Delta y) \tag{15.5}$$

where $F/\Delta y$ = parallel plate stiffness for tubing as specified in ASTM F405 or F667,
C = correction factor for radius of curvature with deflection.

Field tests by Schwab and Drablos (1977) on tubing varying in diameter from 102 to 381 mm showed that the load, when computed by the flexible pipe equation, was much too low, probably because soil compaction and the modulus of soil reaction did not conform to the assumptions in Eq. 15.3. They found that tubing reaches a maximum deflection about 4 years after installation and that measured maximum deflections were about 17 percent. Deflections up to 20 percent in the field usually do not cause failure of corrugated high-density polyethylene (HDPE) tubing, but at 30 percent, collapse from buckling may occur.

Fenemor et al. (1979) found that the Iowa formula using $D_1 = 3.4$, $E' = 345$ kN/m^2, and pipe stiffness at 20 percent deflection gave reasonable results that agreed with field-measured deflections. Recommended maximum depths for tubing buried in loose, fine-grained soil are given in Appendix E. Howard (1977) developed E' values for a wide range of soils.

In recent years the manufacture of large-diameter corrugated HDPE pipe has made it an alternative for concrete or steel pipe for culverts and storm sewers. Allowable fill heights for diameters from 380 to 760 mm have been developed by Katona (1988) for several soil compaction and backfill types. From the longitudinal cross-sectional area per unit length of pipe, which varies with the manufacturer, the maximum fill height can be determined. Minimum depth of cover for 300- to 910-mm-diameter HDPE pipe culverts varies from 0.3 to 0.6 m depending on the soil, diameter, and truck type (Katona, 1990).

ESTIMATING COST

The three principal items of cost for a pipe drainage system are installation, engineering, and material costs, such as tile, tubing, outlet structures, and surface inlets. In the Midwest, the total cost for small laterals about 1 m in depth is approximately 45 percent for materials, 45 percent for installation, and 10 percent for engineering. In California, the total cost for a depth of 2 m is about three times the cost of 150-mm pipe; that is, labor and installation costs constitute about two thirds of the total, including up to 16 percent for the gravel envelope. The cost of installation depends primarily on the method of installation; depth of cut; size of pipe; presence of unusual soil conditions, such as stones and quicksand; and number of junctions. The relative cost of drain pipe is given in Table 15.2. Sewer tile or metal pipe are suitable for outlet structures, surface inlets, and relief wells. Manufactured junctions, either T- or Y-shape, usually cost about five to ten times as much as 0.3 m of ordinary pipe. Engineering and supervision costs normally vary from 5 to 10 percent of the total cost.

MAPPING THE DRAINAGE SYSTEM

A suitable map should be made of the drainage system and filed with the deed to the property. Much time, effort, and money have been wasted because the location of old lines was not known. A record of the drainage system is also of considerable value to present and future owners for planning new lines, maintenance, and repair. The essential features of a drainage map are shown in Fig. 14.14. The drainage system may be mapped by aerial photography, which shows greater detail than sketch maps. Aerial photos should be taken shortly after installation, but the lines may show later because of drier soil or differences in vegetation over the drains.

MAINTENANCE

15.16 Causes for Pipe Drain Failures

Pipe drain failures can be classified into four categories: (1) lack of inspection or maintenance, (2) improper design, (3) improper construction, and (4) manufacturing processes and materials used. As an indication of the relative importance of these classes of failures, a survey in Ohio showed that the percentages of failure from each of these causes were (1) 29, (2) 28, (3) 23, and (4) 20, respectively. The principal causes of concrete and clay tile failures are the lack of resistance to freezing and thawing and inadequate strength. In design the major causes of

Table 15.2 Relative Cost of Drain Pipe by Size

Nominal Diameter (mm)	Relative Cost		
	Corrugated Plastic Tubing	Clay or Concrete Tile	
		Standard Quality	Extra Quality
75	82	90	95
100	100[a]	100[a]	105
125	158	152	158
150	225	192	209
200	440	326	336
250	750	586	595
300	1320	788	820

[a]Arbitrarily taken as 100%, which was about $0.80 per meter in Ohio in 1990.

Fig. 15.10   Tile outlet failure resulting from lack of maintenance. (Courtesy SCS.)

failure are insufficient capacity, pipe placed too near the surface, and lack of auxiliary structures, such as surface inlets. As shown in Fig. 15.10, failure to provide a cantilevered pipe or other structure at the outlet is a common cause of failure. Improper construction, which in the survey accounted for 23 percent of the failures, results from too-wide crack spacings between the tile, improper bedding conditions, poor junctions, nonuniform grade, careless backfilling, and poor alignment. The lack of inspection and maintenance in the above survey was the major cause for failure, representing 29 percent of the total. Such failures are due mainly to the washout of surface inlets and outlet structures, piping over the lines, clogging from root growth, and failure to keep the outlet in good condition.

15.17   Preventive Maintenance

In comparison with open ditches pipe drains require relatively little maintenance. Open ditch outlets should be kept free of weed and tree growth, and the end of the outlet pipe should be covered with a flood gate, a screen, or a flap gate. The outlet ditch must not fill with sediment so as to obstruct the flow of the water from the pipe. Surface water should not be diverted into the ditch at or near the pipe outlet. Where shallow depths are required, drains should be protected from tramping of livestock and should not be crossed with heavy machinery, particularly during wet periods.

Pipe drains should be kept free of sediment and other obstructions. Roots of trees and certain other plant roots may grow into the drain and obstruct the flow, especially if the drain is fed by springs supplying water far into the dry season. Brush and trees, particularly willow, elm, cottonwood, soft maple, and eucalyptus, must be removed if within 30 m of the drain. Where it is impractical to remove this vegetation, nonperforated plastic tubing, sealed bell and spigot tile, or metal pipe should be installed. Sugarbeet and alfalfa roots have been known to enter drain lines, but these plants usually do not cause trouble because the roots die, decay, and are washed away.

During the first year after installation drain lines should be carefully watched to detect evidence of failure. Sinkholes over the line indicate a broken tile or too wide a crack or opening. Surface water should be diverted across the trench since this water may enter the drain or erode the backfill and wash out the pipe. Sediment basins should be cleaned at regular intervals. Surface inlets must be kept free of weed growth and sediment around the entrance.

15.18 Corrective Maintenance

Pipe drains that become filled with sediment or plant roots may be cleaned with a suitable plug, swab, or sewer rod. Where sufficient water and grade are available, sediment can be washed out. Water can be forced at high pressure through a hose inserted from the outlet end.

In Europe and in California and Florida, mineral deposits of predominantly iron oxide and manganese oxide have formed in pipe drains. MacKenzie (1962) reported that these minerals can be satisfactorily removed by injecting sulfur dioxide gas for a period of 24 h. To obtain the desired chemical action water must also be in the lines during this time.

Especially in peat and muck soils with air in the pipe, iron ochre may form a gelatinous mass that seals the drain openings. Ochre is generally a combination of bacterial and chemical sludge. Raising the water level above the pipe as with subirrigation may reduce oxidation.

REFERENCES

American Society of Agricultural Engineers (ASAE) (1988). *Design and Construction of Subsurface Drains in Humid Areas*, EP260.4. ASAE, St. Joseph, MI.

Fenemor, A. D., B. R. Bevier, and G. O. Schwab (1979). "Predictions of Deflection for Corrugated Plastic Tubing." *ASAE Trans.* **22**, 1338–1342.

Howard, A. K. (1977). "Modulus of Soil Reaction Values for Buried Flexible Pipe." *J. Geotech. Eng. Div. ASCE* **103** (GTI), 33–43.

Katona, M. G. (1988). *Allowable Fill Heights for Corrugated Polyethylene Pipe.* Transportation Res. Record No. 1191. Transportation Res. Board, National Res. Council, Washington, DC.

——— (1990). *Minimum Cover Heights for Corrugated Plastic Pipe Under Vehicle Loading,* Transportation Res. Record No. 1288. Transportation Res. Board, National Res. Council, Washington, DC.

MacKenzie, A. J. (1962). "Chemical Treatment of Mineral Deposits in Drain Tile." *J. Soil Water Cons.* **17**, 124–125.

Manson, P. W., and F. W. Blaisdell (1956). "Energy Losses at Draintile Junctions." *Agr. Eng.* **37**, 249–252, 257.

Marston, A. (1930). *The Theory of External Loads on Closed Conduits in the Light of the Latest Experiments,* Iowa Eng. Expt. Sta. Bull. 96.

Schlick, W. J. (1932). *Loads on Pipe in Wide Ditches,* Iowa Eng. Expt. Sta. Bull. 108.

Schwab, G. O., and C. J. W. Drablos (1977). "Deflection — Stiffness Characteristics of Corrugated Plastic Tubing." *ASAE Trans.* **20**, 1058–1061, 1066.

Spangler, M. G. (1947). "Underground Conduits — An Appraisal of Modern Research." *Proc. ASCE* **73**, 855–884.

Spangler, M. G., and R. L. Handy (1981). *Soil Engineering*, 4th ed. Harper & Row, New York.

U.S. Bureau of Reclamation (USBR) (1978). *Drainage Manual*. GPO, Washington, DC.

PROBLEMS

15.1 Select the grade and compute cuts at each station for a 180-m (600-ft) drain line outletting into an open ditch so that cuts do not exceed 1.3 m (4.2 ft). Design for an average depth of 1.2 m (4 ft). Elevation of the water surface in the ditch is 25.15 m (82.50 ft) and hub elevations of successive 30-m (100-ft) stations are 26.46 m (86.81 ft), 26.61 (87.29), 26.80 (87.92), 26.87 (88.15), 26.94 (88.40), 27.01 (88.61), and 27.06 (88.79).

15.2 Determine the soil load on 300-mm (12-in.) drain tile installed at a depth of 3.0 m (10 ft) in a trench 0.6 m (24 in.) wide, assuming that the maximum weight of saturated topsoil in the backfill is 17 kN/m³ (110 pcf). Calculate the design load, using a safety factor of 1.5.

15.3 What quality of drain tile is required to withstand the design load as determined in Problem 15.2 if ordinary bedding conditions are provided?

15.4 For the tile in Problem 15.3, what width of trench should be dug to permit the installation of standard-quality tile with ordinary bedding?

15.5 Based on the design load, what bedding conditions are necessary if only standard-quality tile were available for the installation described in Problem 15.2?

15.6 Estimate the cost of the drainage system in Fig. 14.14. The average cut in the main is 1.4 m (4.5 ft) and for the laterals, 1.2 m (4.0 ft). *Installation Cost:* 1.2 m (4.0 ft) or less, $30 per 30 m (100 ft); overcut, $0.60 per 0.03 m (0.1 ft) per 30 m (100 ft). *Materials:* Tile price based on Table 15.2, using $300 per 305 m (1000 ft) for 100-mm (4-in.) pipe and adding 5 percent for breakage; corrugated metal outlet pipe at $13 per m ($4 per ft); junctions at five times cost for 0.3-m length. *Engineering:* 5 percent of total materials and installation costs.

15.7 Determine the slope of a pipe drain in percentage from the following field notes:

Station [m (ft)]	Hub Elevation [m (ft)]	Cut [m (ft)]
0 + 00 (0 + 00)	30.00 (98.40)	1.22 (4.00)
0 + 30 (1 + 00)	30.08 (98.70)	1.22 (4.00)
0 + 60 (2 + 00)	30.14 (98.90)	1.19 (3.90)
0 + 90 (3 + 00)	30.36 (99.60)	1.36 (4.45)

15.8 If the sight bar on a trenching machine is set 2.74 m (9.0 ft) above the bottom of the trench, compute the height above the hub stakes to set the target at each station along the drain line given in Problem 15.7.

15.9 At the junction of a 300-mm (12-in.) main and a 100-mm (4-in.)-diameter lateral, a manufactured T-junction fitting is to be installed. If the two openings are aligned on their center lines, compute the minimum elevation rise for the lateral (y in Fig. 15.1). Assume the outside diameter is 10 percent greater than the inside diameter of the pipe.

CHAPTER 16

Pumps and Pumping

Engineers should be able to determine the proper type of pump, the number and size of pumps required, and the size of power units. In addition, they may be called on to design the plant installation, estimate the cost of operation, and supervise construction and operation of the plant.

Pumping plant installations are encountered principally in drainage and irrigation enterprises. Pumping plants for drainage provide outlets for open ditches and pipe drains or may lower the water level by pumping from shallow wells. In irrigation, pumping from wells and storage reservoirs into irrigation canals or pipes, or into other reservoirs, is common practice.

TYPES OF PUMPS

The most common pumps use rotating impellers or reciprocating pistons to transfer energy to the fluid. Reciprocating pumps, sometimes called piston or displacement pumps, are capable of developing high pressures, but their capacity is relatively small. They are not ordinarily suitable for drainage and irrigation, especially if sediment is present; however, small piston pumps may inject chemicals into pressurized irrigation systems.

Flow-through impeller pumps may be classified as radial, axial, or mixed (Fig. 16.1). In radial-flow pumps, the fluid moves through the pump impeller perpendicular to the axis of rotation of the impeller. In axial-flow pumps, the fluid moves through the pump parallel to the axis of rotation of the impeller. Pumps that discharge flow from their impellers on vectors that lie between radial and axial are mixed-flow pumps.

Impeller pumps are commonly known as centrifugal, propeller, and turbine pumps. In centrifugal pumps the flow from the impeller is radial; in propeller pumps the flow from the impeller is axial. Most turbine pumps have mixed flow. Pumps may be selected from these three types for a wide range of discharge and head characteristics.

An index of pump type is specific speed. Specific speed is calculated from the

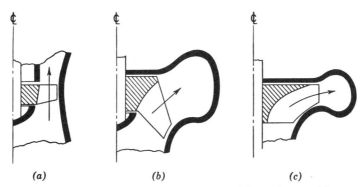

Fig. 16.1   Flow direction through impellers with (*a*) axial flow, (*b*) mixed flow, and (*c*) radial flow.

following formula (Church, 1944), with discharge and head at optimal pump efficiency:

$$n_s = CNq^{1/2}/h^{3/4} \tag{16.1}$$

where n_s = specific speed,
N = impeller speed in rpm,
q = pump discharge (L³/T),
h = total head (L),
C = 51.65 with q in m³/s and h in m, 1.63 with q in L/s and h in m, 1.0 with q in gpm and h in ft.

Figure 16.2 shows the efficiency of impeller pumps as a function of specific speed. Centrifugal pumps have low specific speeds, low discharges, and high heads. Propeller pumps are characterized by high specific speeds, high discharges, and low heads. Turbine pumps have intermediate specific speeds, discharge, and head; however, most turbine pumps have operational characteristics closer to those of centrifugal than propeller pumps.

CENTRIFUGAL AND TURBINE PUMPS

Centrifugal and turbine pumps are economical and simple in construction, yet they produce a smooth, steady discharge. They are small compared with their capacity, easy to operate, and capable of handling sediment and other foreign material. Since turbine pumps are frequently used in wells, they are also known as deep-well turbines.

16.1 Principles of Operation

Centrifugal and turbine pumps consist of two main parts: (1) the impeller or rotor that adds energy to the water in the form of increased velocity and pressure, and (2) the casing that guides the water to and from the impeller. As shown in Fig. 16.3, the water enters the pump at the center or eye of the impeller and passes outward through the rotor to the discharge opening. By changing the speed of the pump, the discharge can be varied. Theoretically, the so-called pump laws for a

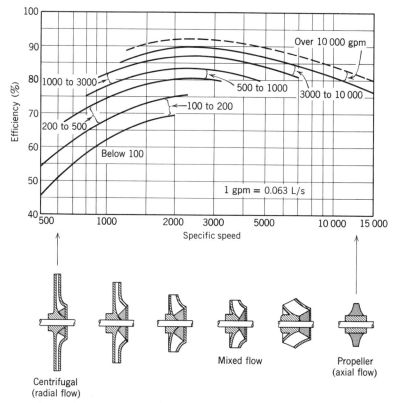

Fig. 16.2 Relationship among specific speed, impeller shape, efficiency, and pump type. (Courtesy Worthington Corporation.)

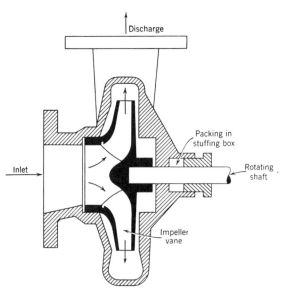

Fig. 16.3 Cross section of a horizontal centrifugal pump (single-suction enclosed impeller).

centrifugal pump state that (1) the discharge is directly proportional to the speed of the impeller, (2) the head varies as the square of the speed, and (3) the power varies as the cube of the speed. For a given pump, changing the diameter of the impeller, and thereby the peripheral velocity, has the same effect as changing the speed (Fig. 16.6).

16.2   Classification

With regard to the construction of the casing around the impeller, pumps are classified as volute or diffuser, as shown in Fig. 16.4. The volute-type pump is so named because the casing is in the form of a spiral with a cross-sectional area increasing toward the discharge opening. Diffuser-type pumps, sometimes called turbine-type, have stationary guide vanes surrounding the impeller. As the water leaves the rotor, the vanes gradually enlarge and guide the water to the casing, resulting in a reduction in velocity with the kinetic energy converted to pressure. The vanes provide a more uniform distribution of the pressure.

Centrifugal pumps are built with horizontal or vertical drive shafts and with different numbers of impellers and suction inlets. The suction inlet may be either single or double acting, depending on whether the water enters from both sides or one side of the impeller. Single-suction, horizontal centrifugal pumps are frequently used where the suction lift does not exceed 4 to 6 m. Practically all turbine pumps are of the vertical type. Pumps with more than one impeller are known as multiple-stage pumps. Both centrifugal and turbine pumps may have multiple impellers, but they are more common in the turbine type.

16.3   Centrifugal-Type Impellers

The design of the impellers greatly influences the efficiency and operating characteristics of the pump. Centrifugal-type impellers shown in Fig. 16.5 are classified as (a) open, (b) semienclosed, and (c) enclosed. The open-type impeller has exposed blades that are open on all sides except where attached to the rotor. The semienclosed impeller has a shroud (plate) on one side; the enclosed impeller has shrouds on both sides, thus enclosing the blades completely. The open and semienclosed impellers are most suitable for pumping suspended material or

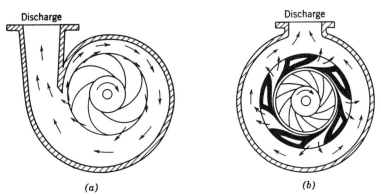

(a)   *(b)*

Fig. 16.4   (*a*) Volute-type and (*b*) diffuser-type centrifugal pumps.

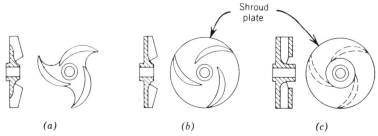

Fig. 16.5 (*a*) Open, (*b*) semienclosed, and (*c*) enclosed centrifugal-type impellers.

trashy water. Enclosed impellers are generally not suitable where suspended sediment is carried in the water as this material greatly increases the wear on the impeller.

16.4 Performance Characteristics

In selecting a pump for a particular job the relationship between head and capacity at different speeds should be known as should pump efficiency. Curves that provide these data are called *characteristic* curves, as shown in Fig. 16.6. The head-capacity curve shows the total head developed by the pump for different rates of discharge. At zero flow the head developed is known as the shut-off head. Head losses in pumps are caused by friction and turbulence in the moving water, shock losses resulting from sudden changes in momentum, leakage past the impeller, and mechanical friction. For a given speed the efficiency can be deter-

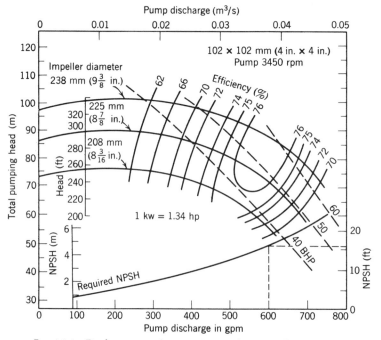

Fig. 16.6 Performance characteristics of a centrifugal pump.

mined for any discharge from these characteristic curves. A pump should be selected that will have a high efficiency for a wide range of discharges. For example, from the upper curve in Fig. 16.6, 71 percent or greater efficiency can be obtained for capacities varying from 0.027 to 0.046 m^3/s at heads of 99 to 75 m, respectively.

Characteristic curves can be obtained from the pump manufacturer. Such curves will vary in shape and magnitude, depending on size of the pump, type of impeller, and overall design. Performance characteristics are normally obtained by tests of a representative production line pump rather than by tests of each pump manufactured.

For operation of a centrifugal pump without cavitation, the suction lift plus all other losses must be less than the theoretical atmospheric pressure. The maximum practical suction lift can be computed with the equation

$$H_s = H_t - H_f - e_s - \text{NPSH} - F_s \qquad (16.2)$$

where H_s = maximum practical suction lift or elevation of the pump center line minus the elevation of the water surface (L),
H_t = atmospheric pressure at the water surface (L),
H_f = friction losses in the strainer, pipe, fittings, and valves of the suction line (L),
e_s = saturated vapor pressure of the water (L),
NPSH = net positive suction head, which is the head required to move water into the eye of the impeller (L),
F_s = safety factor, which may be taken as 0.6 m with all other units in m.

The approximate correction of H_t for altitude is a reduction of about 1.2 m/1000 m of altitude. Friction losses and suction lift should be kept as low as possible. For this reason the suction line is usually larger than the discharge pipe, and the pump is placed as close as possible to the water supply. The head loss equivalent resulting from the vapor pressure of the water must be considered to prevent cavitation, but it does not add to the total suction head when the pump is operating. NPSH is a characteristic of the pump and should be furnished with the characteristic curves as shown in Fig. 16.6. The following example illustrates the calculation of suction lift.

☐ Example 16.1

Determine the maximum practical suction lift for a pump having the characteristics shown in Fig. 16.6 if the discharge is 0.038 m^3/s (600 gpm), the water temperature is 20°C (68°F), the total friction loss in the 102-mm (4-in.) suction line and fittings is 1.58 m (5.2 ft), and the pump is to be operated at an altitude of 300 m (1000 ft) above sea level. (Sea level pressure is 10.36 m of water.)

Solution. Substituting in Eq. 16.2 after reading NPSH = 4.8 m (15.8 ft) from Fig. 16.6, obtaining e_s = 0.24 m (0.8 ft) from Chapter 3, assuming F_s = 0.6 m (2.0 ft), and correcting for altitude H_t = 10.36 − 0.36 m (34.0 − 1.2 ft) = 10.00 m,

$$H_s = 10.00 - 1.58 - 0.24 - 4.8 - 0.6 = 2.78 \text{ m (9.0 ft)} \qquad \square$$

Pump operation at the maximum practical lift should be avoided in design since the collection of debris on the screen and increase of pipe friction with age may cause the pump to lose prime. To prime the pump the suction line and pump should be nearly full of water. This can be accomplished by filling the pump manually with water or by removing the air with a suction pump or an internal combustion engine exhaust primer. A gate valve on the discharge side of the pump and a check valve at the lower end of the suction line are essential for priming, except at very low suction heads.

PROPELLER PUMPS

16.5 Principle of Operation

As distinguished from centrifugal pumps, the flow through the impeller of a propeller-type pump is parallel to the axis of the driveshaft rather than radial. These pumps are also referred to as axial-flow or screw-type pumps. The principle of operation is similar to that of a boat propeller except that the rotor is enclosed in a housing. Many propeller pumps have diffuser vanes mounted in the casing, similar to the diffuser-type pump.

16.6 Performance Characteristics

Propeller pumps are designed principally for low heads and large capacities. The discharge of these pumps varies from 0.04 to 4.4 m³/s, with speeds ranging from 450 to 1760 rpm and heads usually not more than 9 m. Although most propeller pumps are of the vertical type, the rotor may be mounted on a horizontal shaft.

Performance curves for a large propeller pump are shown in Fig. 16.7. In comparison with centrifugal pumps, the efficiency curve is much flatter, heads are considerably less, and the power curve is continually decreasing with greater discharge. With propeller pumps the power unit may be overloaded by increasing

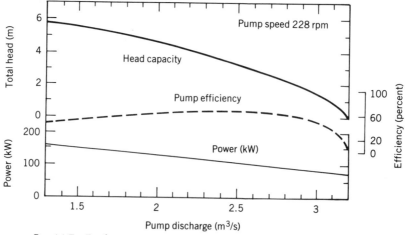

Fig. 16.7 Performance curves for an axial-flow (propeller) pump.

the pumping head. These pumps are not suitable where the discharge must be throttled to reduce the rate of flow.

PUMPING

Although in many respects pumping water for irrigation is similar to that for drainage, design requirements differ. For example, in pump drainage, heads are generally 6 m or less, whereas for irrigation heads up to 90 m are common.

16.7   Power Requirements and Efficiency

For a given discharge the power requirements for pumping are proportional to the total pumping head including velocity head; friction losses; static head, which is the difference between the free-water surface at the inlet and at the discharge point; and the pressure head at the point of discharge. Where the end of the discharge pipe is higher than the water at the outlet, the head is measured to its center. Power requirements for pumping are computed with the formula

$$kW = 9.8qh/E_p \qquad (16.3)$$

where kW = (input) power delivered to pump,
q = discharge rate in m³/s,
h = total head in m,
E_p = pump efficiency as a decimal fraction.

Considerable energy is used in overcoming friction losses in the pump and for velocity head losses. Because of such losses pumping plant efficiencies range from about 75 percent under favorable conditions to as low as 20 percent or less under unfavorable conditions. A well-designed pump should have an efficiency of 70 percent or greater over a wide range of operating heads. Such efficiencies in the field are difficult to maintain because of wear in the pump and other factors. It is good practice to check the pump in the field to be sure that satisfactory efficiencies are obtained. If efficiencies are low, the source of difficulty should be located and corrected. The overall efficiency of the installation, which includes the efficiency of the power plant, the pipe system, and the pump, should also be checked.

16.8   Power Plants and Drives

Power plants for pumping should deliver sufficient power at the specified speed with maximum operating efficiency. Internal combustion engines and electric motors are by far the most common types of power units. The selection of the type of unit depends on (1) the amount of power required, (2) initial cost, (3) availability and cost of fuel or electricity, (4) annual use, and (5) duration and frequency of pumping.

Internal combustion engines operate on a wide variety of fuels, such as gasoline, diesel oil, natural gas, and butane. Where the annual use is more than 800 to 1000 h, the diesel engine may be justified. Otherwise, it may be more practical to use a gasoline engine. For continuous operation water-cooled gasoline engines may be expected to deliver 70 percent of their rated horsepower, diesel engines 80 percent, and air-cooled gasoline engines 60 percent. Where a vertical centrifugal or a deep-well turbine pump is to be driven with an internal combustion engine, right-angle gear drives are usually employed. In some instances, internal combustion engines are mounted with vertical crankshafts so they may be directly coupled to a vertical driveshaft. Belt drives may use either flat or V-belts suitable for driving vertical shafts or gear drives. In comparison with direct drives, which have a transmission efficiency of nearly 100 percent, the efficiency for gear drives ranges from 94 to 96 percent, for V-belts from 90 to 95 percent, and for flat belts from 80 to 95 percent.

Electric motors operate best at full-load capacity. They have many advantages over internal combustion engines, such as ease of starting, low initial cost, low upkeep, and suitability for mounting on horizontal or vertical shafts; however, obtaining electricity at a remote site is often expensive. Direct drive motors are preferred because gears and belts are eliminated. Deep-well pumps are available with watertight vertical motors. The motor is submerged in the well near the impellers, thus eliminating the long shaft otherwise required.

16.9 Selection of Pump

Performance curves serve as a basis for selecting a pump to provide the required head and capacity for the range of expected operating conditions at or near maximum efficiency. The factors that should be considered include the head-capacity relationship of the well or sump from which the water is removed, space requirements of the pump, initial cost, type of power plant, and pump characteristics, as well as other possible uses for the pump. Storage capacity, rate of replenishment, and well diameter may limit the pump size and type. For example, a large drainage ditch provides a nearly continuous source of water, whereas a well usually has a small storage capacity.

The initial cost that can be justified depends largely on the annual use and other economic considerations. The size of the power plant and type of drive should be adapted to the pump.

Figure 16.2 may aid in selecting a pump type that provides a high efficiency for the discharge and head required. The centrifugal pump is suitable for low to high capacities at heads up to several hundred meters, the propeller pump for high capacities at low heads, and the mixed-flow pump for intermediate heads and capacities. Since impellers of these pumps may be placed near or below the water level, the suction head developed may not be critical. Horizontal centrifugal pumps are best suited for pumping from surface water supplies, such as ponds and streams, provided the water surface does not fluctuate excessively. Where the water level varies considerably, a vertical centrifugal or a deep-well turbine pump is more satisfactory.

16.10 Pumping Costs

The cost of pumping consists of fixed and operating costs. Fixed costs include interest on investment, depreciation, taxes, and insurance. Included in the investment cost is the construction and development of the well, pump, power plant, pump house, and water storage facilities. Construction costs for wells depend largely on size and depth of the well, construction methods, nature of the material

Table 16.1 Service Life for Pumping Plant Components

Item	Estimated Service Life
Well and casing	20 years
Plant housing	20 years
Pump turbine	
Bowl (about 50% of cost of pump unit)	16 000 h or 8 years
Column, etc.	32 000 h or 16 years
Pump, centrifugal	32 000 h or 16 years
Power transmission	
Gear head	30 000 h or 15 years
V-belt	6000 h or 3 years
Flat-belt, rubber and fabric	10 000 h or 5 years
Flat-belt, leather	20 000 h or 10 years
Electric motor	50 000 h or 25 years
Diesel engine	28 000 h or 14 years
Gasoline engine	
Air-cooled	8000 h or 4 years
Water-cooled	18 000 h or 9 years
Propane engine	28 000 h or 14 years

Source: From SCS, (1959).

through which the well is drilled or dug, type of casing, length and type of well screen, and time required to develop and test the well.

Table 16.1 gives the estimated service life of various components of pumping plants for estimating depreciation. Taxes and insurance are approximately 1.5 percent of the total investment. Operating costs include fuel or electricity, lubricating oil, repairs, and labor for operating the installation. Fuel costs for internal combustion engines are generally proportional to the consumption, whereas the unit cost of electrical energy generally decreases with the amount consumed. A demand charge or minimum is made each month regardless of the amount of energy consumed. This demand charge may or may not include a given amount of energy.

16.11 Design Capacity

The first consideration in selection of a pump and design of a pumping plant must be the determination of the required capacity. Irrigation pumping requirements depend on peak evapotranspiration rates, irrigation efficiencies, and the planned frequency and duration of pumping plant operation (Chapters 3 and 18–21). Design capacity of the pumping plant may also be dictated by characteristics of the water supply. A crop that has a peak consumptive use rate of 5 mm/day will require a pumping capacity of 0.58 L/s per hectare irrigated plus additional capacity to supply conveyance and other irrigation losses if 24-h/day pumping is required.

Drainage coefficients for pumping plant design vary according to soil, topography, rainfall, and cropping conditions. Recommended coefficients for the upper Mississippi Valley range from 7 to 25 mm/day. Requirements in southern Texas and Louisiana are as high as 75 mm, including pumping discharge and reservoir storage. In Florida a drainage coefficient of 25 mm/day is suitable for field crops on organic soils. Truck crops may require and economically justify a drainage coefficient of as much as 75 mm/day. Additional recommendations for drainage coefficients may be obtained from ASAE (1990), Chapters 13 and 14, or state and local guides.

16.12   Drainage Pumping Plants

Pumps offer a means of providing drainage where gravity outlets are not otherwise available. Typical applications of pump drainage include land behind levees, outlets for pipe drains and open ditches, and flat land where the gravity outlet is at considerable distance. Drainage pumping plants are normally required to handle large capacities at low lifts.

The design of small farm pumping plants may be simplified by using the design runoff rate recommended for open ditches plus 20 percent. Where electricity is available, electric motors with automatic controls are recommended for small installations. For drainage areas larger than 40 ha, storage capacity should be obtained by enlarging the open ditches or by excavating a storage basin. Constructed sumps as shown in Fig. 16.8 are generally too expensive where the diameter is greater than 5 m. Where storage is available in open ditches or other reservoirs, sumps about 2 m in diameter are satisfactory. Figure 16.9 illustrates a large pumping plant for pumping water from an open ditch. Most pumps operate only during the wet spring months, but some installations draining extremely low land may operate more or less continuously throughout the year. Usually, these plants operate intermittently and are used only 10 to 20 percent of the time.

For economical performance it is better to operate a small pump for several days than to run a large pump intermittently for short periods. Frequently, drainage installations lack storage capacity and the allowable head variation is small. Under these conditions automatically controlled electric motors are especially suitable. Larson and Manbeck (1961) have found that 10 to 15 cycles per hour are satisfactory for drainage pumping plants with automatic controls. Internal combustion engines are usually started manually, but an automatic shut-off can be provided. The number of stopping and starting cycles should not exceed more than two per day for nonautomatic operation (ASAE, 1989).

Engine-operated plants require a larger storage capacity than those electrically operated. For large installations two pumps with different capacities may be necessary, one for handling surface runoff during wet seasons and the other for seepage or flow from pipe drains.

Sufficient storage must be provided in a sump to avoid excessive starting and stopping. The continuity equation for the total time per cycle is the sum of the running and the stopping time for the pump and may be written

$$\frac{3600}{n} = \frac{S}{q - q_i} + \frac{S}{q_i} \qquad (16.4)$$

where n = number of cycles per hour,
  S = storage volume in m³,
  q = average pumping rate in m³/s,
  q_i = inflow rate in m³/s.

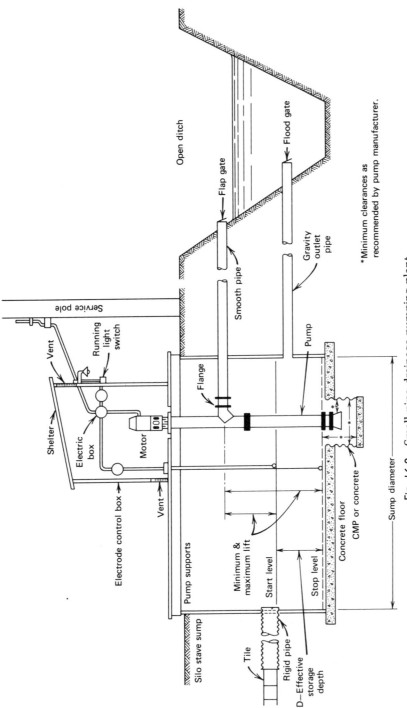

Fig. 16.8 Small pipe-drainage pumping plant.

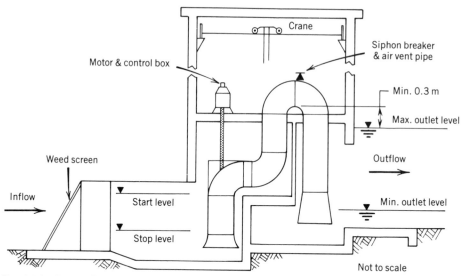

Fig. 16.9 Large drainage pumping plant for an open ditch. (Courtesy APE Allen Ltd., England.)

By solving for Sn, assuming q constant, differentiating S with respect to q_i, and setting $dS/dq_i = 0$ to obtain maximum storage,

$$q_i = \frac{q}{2} \tag{16.5}$$

Substituting q_i above in Eq. 16.4 results in

$$S = \frac{900q}{n} \tag{16.6}$$

Thus, maximum storage occurs when the inflow rate is one half the discharge of the pump, and the storage requirements depend on the discharge rate and the frequency of cycling. When the inflow rate exceeds the pumping rate, the pump operates continuously. When the inflow rate is less than the pumping rate, cycling occurs.

☐ Example 16.2

A drainage pump has a design capacity of 0.04 m³/s (635 gpm) and is to be installed in a 2.5-m (8-ft)-diameter sump with a maximum cycling frequency of 10 cycles/h. What is the minimum difference in elevation between the start and stop levels in the sump?

Solution. Substituting in Eq. 16.6,

$$S = \frac{900 \times 0.04}{10} = 3.6 \text{ m}^3 \ (127 \text{ ft}^3)$$

$$\text{Depth of storage} = \frac{3.6 \times 4}{3.14 \times 2.5 \times 2.5} = 0.73 \text{ m (2.4 ft)}$$

Other combinations of depth and sump diameter are possible, provided the storage volume and other design requirements of the installation are met. □

The pumping plant should be installed to provide minimum lift and located so that the pump house will not be flooded. A farm pumping plant for pipe drain or open ditch drainage of small areas is shown in Fig. 16.8. A circular sump is recommended for such an installation since less reinforcement is required and it is easier to construct. The electric motor is usually operated automatically by means of an electrode switch or start and stop collars on a float rod. Where open ditches provide storage, the sump can be reduced in size. The discharge pipe should outlet below the minimum water level in the outlet ditch, where practicable, to reduce the pumping head to a minimum. The gravity outlet pipe should be installed only when the water level in the ditch is lower than the pipe invert for at least 10 percent of the flow period. The savings in pumping cost should justify the cost of the gravity outlet pipe and flood gate.

16.13 Irrigation Pumping Plants

Pumping requirements for irrigation vary considerably, depending on such factors as source of water, method of irrigation, and size of area irrigated. Water may be obtained from wells, rivers, canals, pits, ponds, and lakes. Pumping from surface supplies is similar to pump drainage except where pressure irrigation is practiced. Much higher heads are normally required where water is pumped from wells or pumped to pressure irrigation systems. Water pumped from wells may be discharged into open ditches or into pipe lines for distribution to irrigated areas. Wells located in the vicinity of the area to be irrigated have many advantages as compared with canals supplying water from distant sources. A typical deep-well turbine pump installation discharging into an underground pipeline is shown in Fig. 16.10.

Because of excessive pressure (water hammer), which may develop from starting or stopping the pump when pumping into pipelines, surge tanks should be installed in the discharge line, as illustrated in Fig. 16.10. However, they are usually not required for portable sprinkler irrigation systems. Surge tanks that are open to atmospheric pressure are suitable for pumping into pipelines made of thin metal, concrete, or vitrified clay pipe. The surge tank must be high enough to allow for any increase in elevation of the pipeline, for friction head, and for a reasonable freeboard. When the total head exceeds 6 m, an enclosed metal tank having an air chamber is usually more practical than a surge tank. The maximum head at which irrigation pumping is practical varies with locality, value of the crop, and other economic factors.

The capacity and head requirements of the irrigation system (Chapters 18–21) must be known before selecting a pump. The pump should be selected to match the head-capacity characteristics of the irrigation system and the water supply. In addition, the pump should operate at optimal efficiency. Plotting the head-capacity curve for the pump and the head-discharge curve for a well (Chapter 17) aids in the selection. Such a set of curves for a typical field installation is shown in Fig.

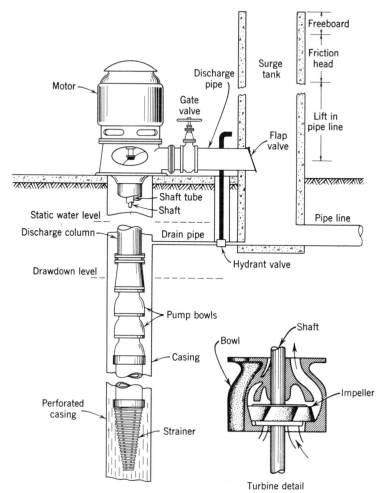

Fig. 16.10 Two-stage deep-well turbine pump installation with a surge tank for low heads. (Modified from Rohwer, 1943, and Wood, 1950.)

16.11. The most desirable conditions for operating the pump occur at the intersection of the two curves when the capacity is 0.045 m³/s and the head is 13.4 m. This intersection should also be the desired head and capacity of the irrigation system and may be shown by adding the irrigation head-capacity curve to such a plot. At this discharge and head the pump also operates at its maximum efficiency (65 percent) and the power unit at nearly its maximum power. If the discharge is changed, there is no danger of overloading the power unit since at other capacities the power requirement is always less. When maximum efficiency does not occur at the desired capacity, changing the pump speed may shift the efficiency curve to a higher value. If different speeds are not practicable, another size or type of pump must be selected. The above procedure for selecting a pump emphasizes the importance of test-pumping the well to obtain the head-capacity curve before purchasing the pump.

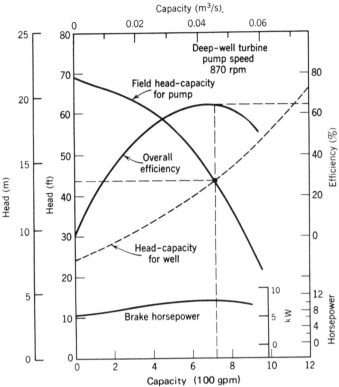

Fig. 16.11 Fitting the pump to the well using head-capacity curves. (Redrawn from Code, 1936.)

REFERENCES

American Society of Agricultural Engineers (ASAE) (1989). *Design of Agricultural Drainage Pumping Plants.* EP369.1. ASAE, St. Joseph, MI.

———(1990a). *Design and Construction of Subsurface Drains in Humid Areas.* EP260.4. ASAE, St. Joseph, MI.

Church, A. H. (1944). *Centrifugal Pumps and Blowers.* Wiley, New York.

Code, W. E. (1936). *Equipping a Small Irrigation Pumping Plant.* Colorado Agr. Expt. Sta. Bull. 433.

Driscoll, F. G. (1986). *Groundwater and Wells.* Johnson Division, St. Paul, MN.

Haman, D. Z., F. T. Izuno, and A. G. Smajstrla (1989). *Pumps for Florida Irrigation and Drainage Systems.* Cir. 832, Inst. of Food and Agr. Sci., University of Florida. Gainesville.

Larson, C. L., and D. M. Manbeck (1961). "Factors in Drainage Pumping Efficiency." *Agr. Eng.* **42**, 296–297, 305.

Lobanoff, V. S., and R. R. Ross (1985). *Centrifugal Pumps: Design & Application.* Gulf Publishing Co., Houston, TX.

Longenbaugh, R. A., and H. R. Duke (1983). "Farm Pumps." In *Design and Operation of Farm Irrigation Systems,* M. E. Jensen (ed.). Monograph No. 3, pp. 347–391. ASAE, St. Joseph, MI.

Rohwer, C. (1943). *Design and Operation of Small Irrigation Pumping Plants.* USDA Cir. 678. GPO, Washington DC.

U.S. Soil Conservation Service (SCS) (1959). "Irrigation Pumping Plants." In *Irrigation,* Sect. 15, Chap. 8 (litho.). Washington, DC.

Wood, I. D. (1950). *Pumping for Irrigation.* SCS-TP-89 SCS, Washington, DC.

PROBLEMS

16.1 A centrifugal pump discharging 0.03 m³/s (475 gpm) against a 25-m (82-ft) head and operating with an efficiency of 70 percent requires 10.5 kW (14.1 hp) at 1800 rpm. What is the estimated discharge if the speed is decreased to 1500 rpm, assuming the efficiency remains constant? What are the estimated head and power at 1500 rpm?

16.2 From the performance curves for the centrifugal pump with a 238-mm (9⅜-in.) impeller shown in Fig. 16.6, what is the pump efficiency at a head of 90 m (295 ft)? What is the discharge? What are the efficiency and discharge if the head is reduced to 75 m (246 ft)?

16.3 What is the power requirement for pumping 0.07 m³/s (1110 gpm) against a head of 50 m (165 ft) assuming a pump efficiency of 65 percent? What size electric motor is required assuming an efficiency of 90 percent?

16.4 If the average discharge of a propeller pump is 0.06 m³/s (950 gpm), what is the required depth between start and stop levels where the water is pumped from a sump 3.0 m (9.8 ft) in diameter? The maximum cycling rate is 12 cycles per hour. At what flow rate would maximum cycling occur?

16.5 Determine the maximum practical suction lift for a centrifugal pump if the discharge is 0.07 m³/s (1110 gpm), the water temperature is 25°C (77°F), friction losses in pipe and fittings are 1.5 m (4.9 ft), the NPSH of the pump is 3.0 m (9.8 ft), and the altitude for operation is 1500 m (4920 ft) above sea level.

16.6 For the centrifugal pump in Fig. 16.6 and an impeller diameter of 208 mm (8³⁄₁₆ in.), determine the head, efficiency, and power for a discharge of 0.03 m³/s (475 gpm). Should this pump be recommended for these flow conditions?

CHAPTER 17

Water Supply
and Quality

Precipitation is presently our only practical source of renewable fresh-water supply for all agricultural, industrial, and domestic uses. Increasing attention being given to large-scale desalinization of brackish or salty waters may eventually result in reasonable supplies of water for high-value uses in some locations, but precipitation will remain the dominant source of water. Nearly 1.4×10^9 ha-m of water falls on the North American continent annually. Developed water supplies in the United States use only 4 percent of the precipitation, which is only 13 percent of the residual precipitation after allowing for evaporation and transpiration from natural plants and unirrigated crops. Thus, there is actually ample water for our needs; however, it is often not available at the time and place of need. The development of water resources involves storage and conveyance of water from the time and place of natural occurrence to the time and place of beneficial use.

This chapter emphasizes the development of water resources for agricultural use without intending in any way to minimize the importance of water for other uses. More detailed discussion is presented by Linsley and Franzini (1979).

17.1 Classification of Waters

Water in one or more of its three physical states — solid, liquid, or gas — is present in greater or lesser quantities in or on virtually all the earth, its atmosphere, and all things living or dead. Water, important from the standpoint of water-resource development, falls into the categories of atmospheric water, surface waters, and subsurface waters. Atmospheric water and resultant precipitation have been discussed in Chapter 2 and are the source of replenishment of surface and subsurface waters. Surface and subsurface waters are the direct source of our developable water resources.

SURFACE WATER

17.2 Surface Water

Surface waters exist in natural basins and stream channels. Where minimum flows in streams or rivers are large in relation to water demands of adjacent lands, towns, and cities, development of surface waters is accomplished by direct withdrawal from the flow. On many streams and rivers, however, flow fluctuates widely from season to season and from year to year. Further, peak demands from many major rivers occur at seasons of minimum flow and, in fact, require that as much of the annual flow as possible be conserved and diverted for beneficial use.

This situation requires construction of reservoirs to hold flow during seasons or years of high runoff for later release to beneficial use. These reservoirs frequently incorporate hydroelectric, flood control, and recreational features in addition to their water supply function. Frequently, the income from electric power development is used to subsidize the water supply function.

Reservoirs range in size from several million hectare-meters for large multiple-purpose reservoirs in the West to small ponds with less than a hectare-meter of storage. As examples of large reservoirs, the Grand Coulee Dam and the Hoover Dam are 168 and 221 m high and have storage capacities of 0.74 and 3.46 million ha-m, respectively.

Surface water reservoirs are usually constructed to permit effective use of water downstream from the reservoir site. An interesting exception to this is the Glen Canyon Dam on the Colorado River in northern Arizona. This dam stores water from the upper basin during wet years for release to downstream users during dry years. This allows more water to be diverted in the upper basin since the reservoir ensures that yearly release of water from the upper basin will be adequate to meet lower basin appropriations.

Surface water supplies may be increased by water harvesting, which is any watershed manipulation carried out to increase surface runoff. In many arid areas, the majority of precipitation that infiltrates into the soil is lost for use either through direct evaporation or through transpiration by natural vegetation. For example, in the Colorado River Basin less than 6 percent of the precipitation appears as streamflow. Vegetation management has been shown to be effective as a means of increasing streamflow.

More immediate in application is the practice of water harvesting by catchments as described by Lauritzen and Thayer (1966). Catchments are areas of concrete, sheet metal, asphalt, or otherwise treated or waterproofed soil specifically constructed to catch and collect precipitation. Successful water harvesting requires attention not only to collection of water but to the conveyance and storage of the water collected. Catchment treatment may include soil smoothing and removal of vegetation, application of salts or other chemicals that disperse the soil aggregates and greatly reduce infiltration, or application of plastic film or wax seals. Simple soil smoothing and vegetation removal have increased runoff by a factor of 3 in some tests. Complete sealing will, of course, increase runoff to essentially 100 percent. Water harvesting is being applied to develop water supplies for wildlife, livestock and occasional domestic use.

The study of the source and fate of water on the soil surface is known as surface water hydrology. This involves the processes of precipitation, evaporation, infiltration, and runoff which were discussed in Chapters 2 to 4.

GROUND WATER

17.3 Ground Water

Subsurface water available for development is normally referred to as ground water. Ground water results predominantly from precipitation that has reached the zone of saturation in the earth through infiltration and percolation. Ground water is developed for use through wells, springs, or dugout ponds.

In many areas where ground water is an important source of water supply it is being withdrawn much faster than it is being replenished. Figure 17.1 shows the severity of ground water lowering in a basin where the entire water supply is from ground water.

17.4 Ground Water Hydrology

Ground water hydrology is the science of water in the unsaturated and saturated zones below the earth's surface. The classification of the earth's crust as a reservoir for water storage and movement of water is shown in Fig. 17.2. Although other portions of this text are primarily concerned with the soil water zone, the earth's surface must be considered at much greater depths for the study of ground water. The profile of the earth is divided into two primary rock zones: the zone of rock fracture and the zone of rock flowage. Formations below the zone of rock fracture are of no concern in ground water hydrology. In the rock flowage zone the water is in a chemically combined state and is not available. In the zone of rock fracture

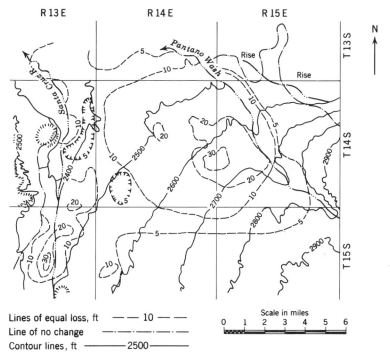

Fig. 17.1 Ground water changes in the Tucson basin of the Santa Cruz valley of Arizona, spring 1956–1961. (Adapted from Schwalen and Shaw, 1961.)

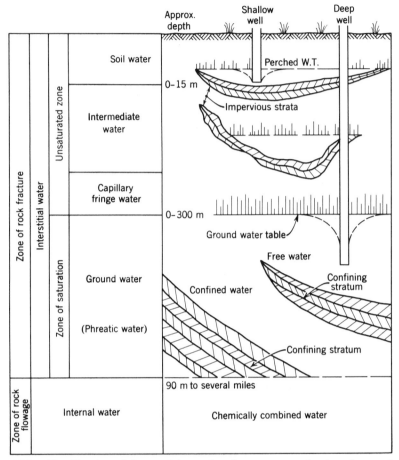

Fig. 17.2 Classification of the earth's crust and occurrence of subsurface water.

interstitial water is contained in the pores of the soil or in the interstices of gravel and rock formations. This zone containing interstitial water is divided into the unsaturated and saturated zones. As here considered, ground water occurs only in the zone of saturation. Perched water tables shown in Fig. 17.2 are often encountered in the unsaturated zone. Good examples of perched and ground water tables can be observed in shallow and deep wells, respectively. Since the depth of perched water is shallow, ground water, which extends continuously downward, is of greater importance. Although the ground water table is at the soil surface near lakes, swamps, and continuously flowing streams, it may be several hundred meters deep in drier regions. Ground water is often referred to as phreatic (a Greek term meaning "well") water. Capillary fringe water and intermediate water exist above the ground water table and are present above perched water tables as well.

Formations from which ground water is derived in the zone of saturation have considerably different characteristics than the soil near the surface. The various types of deposits that furnish water supplies are shown in Fig. 17.3. A good water-bearing formation should have high hydraulic conductivity and high drainable porosity, called specific yield. As shown in Table 17.1, sand and gravel have

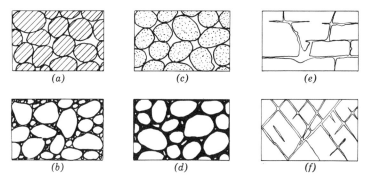

Fig. 17.3 Several types of rock interstices and the relationship of rock texture to porosity. (*a*) Well-sorted sedimentary deposit having high porosity. (*b*) Poorly sorted sedimentary deposit having low porosity. (*c*) Well-sorted sedimentary deposit with pebbles that are porous so that the deposit as a whole has a very high porosity. (*d*) Well-sorted sedimentary deposit whose porosity has been diminished by the deposition of mineral matter in the interstices. (*e*) Rock rendered porous by solution. (*f*) Rock rendered porous by fracturing. (Redrawn from Meinzer, 1923.)

these characteristics although fractured limestone and rock formations are also good aquifers.

Wells may be classified as gravity, artesian, or a combination of artesian and gravity, depending on the type of aquifer supplying the water. Gravity wells are those that penetrate the water table where water is not confined under pressure. Frequently, maps are prepared that show contour lines (depth) of the water table and indicate the type of formation.

Gravity water may be obtained from wells, springs, and dugout ponds. Wells, by far the most common, are either shallow or deep as shown in Fig. 17.2. Where the ground water table reaches the soil surface because the underlying strata are impervious, springs or seeps may develop. In areas where the ground water table is near the surface, dugout ponds or open pits dug below the water level are practical for water storage.

Whenever the piezometric surface, as measured in a vertical hole drilled into the aquifer, rises above the ground water table, artesian conditions are present. According to this definition, an artesian well is not necessarily a flowing well. The

Table 17.1 Approximate Characteristics of Ground Water Aquifers

Soil Material	*Total Porosity (%)*	*Specific Yield (%)*	*Relative Hydraulic Conductivity*
Dense limestone or shale	5	2	1
Sandstone	15	8	700
Gravel	25	22	5000
Sand	35	25	800
Clay	45	3	1

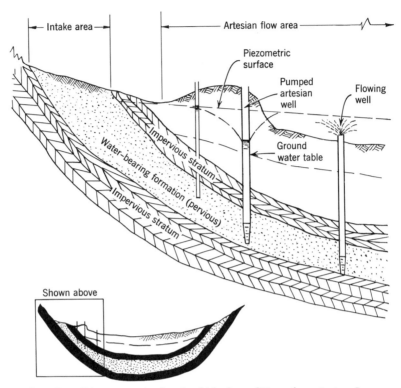

Fig. 17.4 Diagrammatic sketch of ideal conditions for artesian flow.

conditions under which artesian flow takes place are shown in Fig. 17.4. For artesian flow the following conditions must be present: (1) pervious stratum with an intake area, (2) impervious strata below and above the water-bearing formation, (3) inclination of the strata, (4) source of water for recharge, and (5) absence of a free outlet for the water-bearing formation at a lower elevation. The level to which water rises in a pipe placed in the water-bearing formation is known as the piezometric head or, in three dimensions, the piezometric surface. Where the pervious stratum outcrops at the surface, artesian springs may develop.

17.5 Hydraulics of Wells

A cross section of a well installed in homogeneous soil overlying an impervious formation is shown in Fig. 17.5. Under static conditions the water level will rise to the water table. When pumping begins, the water level in the well is lowered, thus removing free water from the surrounding aquifer. The distance from the well to where the static water table is not lowered significantly by drawdown is known as the radius of influence. The water level at the edge of the well will be slightly higher than in the well because of friction losses through the perforated casing. The effect of the number and size of perforations on inflow has been investigated by Muskat (1942) (also see Driscoll, 1986). For a given rate of pumping the water table surrounding a well in time reaches a stable condition. The shape of the drawdown curve is similar to an inverted cone. The base of the cone is at the water table or, for artesian conditions, the base is on the piezometric surface (Fig. 17.4).

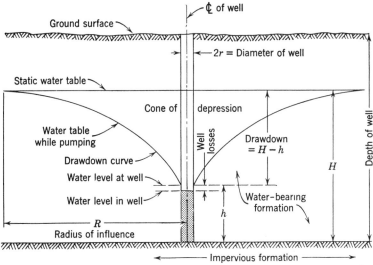

Fig. 17.5   Cross section of a typical gravity well in homogeneous soil.

The rate of flow into a gravity well, illustrated in Fig. 17.5, is

$$q = \frac{\pi K(H^2 - h^2)}{\log_e R/r} \tag{17.1}$$

where q = rate of flow (L^3/T),
$\quad\quad K$ = hydraulic conductivity (L/T),
$\quad\quad H$ = height of the static water level above the bottom of the water-bearing
$\quad\quad\quad$ formation (L),
$\quad\quad h$ = height of the water level at the well, measured from the bottom of the
$\quad\quad\quad$ water-bearing formation (L),
$\quad\quad R$ = radius of influence (L),
$\quad\quad r$ = radius of the well (L).

The radius of influence given in Table 17.2 may be estimated from the texture
and other characteristics of the aquifer. Since the discharge varies inversely as the
logarithm of the radius of influence, an error in estimating this radius results in a
much smaller error in the discharge. Equation 17.1 also indicates that the discharge
is proportional to the logarithm of the well diameter. Assuming an R of 300 m and
other conditions constant, a well 600 mm in diameter will yield about 10 percent
more than a well 300 mm in diameter.

The flow into an artesian well completely penetrating an extensive confined
aquifer (Fig. 17.6), developed from Darcy's law, is

$$q = \frac{2\pi K d(H - h)}{\log_e R/r} \tag{17.2}$$

where d = thickness of the confined layer (L),
$\quad\quad H$ = height of the static piezometric surface above the top of the water-
$\quad\quad\quad$ bearing formation (L),
$\quad\quad h$ = height of the water in the well above the top of the water-bearing
$\quad\quad\quad$ formation (L).

Table 17.2 Radius of Influence of Wells

Soil Formation and Texture	Radius of Influence (m)
Fine sand formations with some clay and silt	30–90
Fine to medium sand formations fairly clean and free from clay and silt	90–180
Coarse sand and fine gravel formations free from clay and silt	180–300
Coarse sand and gravel, no clay or silt	300–600

Source: Bennison (1947).

If the aquifer is level, the flow into the well is radial and horizontal.

For a given well there is a definite relationship between drawdown and discharge. For thick water-bearing aquifers or artesian formations, the discharge–drawdown relationship is nearly a straight line. As shown in Fig. 17.7, the discharge–drawdown relationship can be obtained by pumping the well at various rates and plotting the drawdown against the discharge. Test-pumping a well should be continued for a considerable length of time. Short pumping tests are often misleading. It has been found that even 24-h tests are not long enough; 30-day tests are more likely to indicate the true capacity of the well.

The theoretical water level for continuous pumping from a well having no recharge is shown in Fig. 17.8. These curves show the fallacy of using short-period tests for determining the discharge of a well. For example, 0.019 m³/s may be pumped for a period of 20 days without exceeding an assumed maximum economical lift of 46 m; however, to maintain the same drawdown, the discharge must be reduced if pumping is to be continued longer than 20 days.

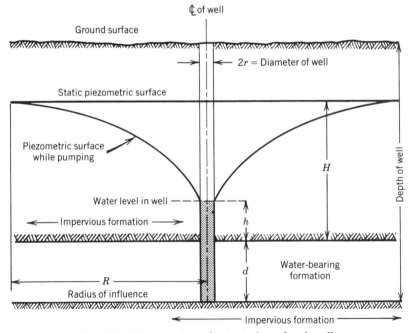

Fig. 17.6 Cross section of a typical confined well.

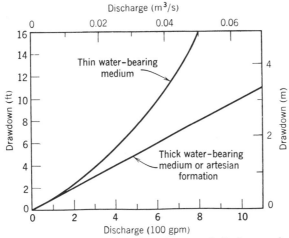

Fig. 17.7 Relationship between drawdown and discharge of a well.

17.6 Construction of Wells

The three general types of wells are dug, driven, and drilled. The dug well consists of a pit dug to the ground water level. Often it is lined with masonry, concrete, or steel to support the excavation. Because of difficulty in digging below the water level, dug wells do not penetrate the ground water to a depth sufficient to produce a high yield.

Wells up to 76 mm in diameter and 18 m deep may be constructed by driving a well point into unconsolidated material. A well point is a section of perforated pipe pointed for driving and connected to sections of plain pipe as it is driven to the desired depth. Sometimes penetration of well points is aided by discharging a high-velocity jet of water at the tip of the point as it is driven.

Deeper and larger-diameter wells are drilled with cable-tool or rotary equipment. With cable tools a heavy bit is repeatedly dropped onto material at the bottom of the well. Crushed material is removed periodically with a bailer. Cable-tool wells have been drilled to depths of 1500 m. Deep wells are also drilled with

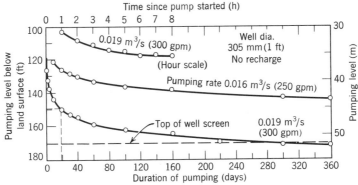

Fig. 17.8 Theoretical pumping level in a well having no recharge. (Redrawn from Ferris, 1959.)

rotary tools consisting of a bit rotated by a drilling pipe. A mud slurry pumped through the drill pipe brings cuttings to the surface as it flows up the outside of the drill pipe. In unconsolidated materials wells may be cased as drilling progresses. Casings may also be installed after drilling is completed.

Well casings are perforated where they pass through water-yielding strata. In some situations perforated casing may be formed in place by ripping or shooting holes through a solid casing. Better results are achieved by placing screens made of brass, bronze, or special alloys to resist corrosion. Screen openings are selected to permit 50 to 70 percent of the particles in the aquifer to pass the screen. The open screen area should keep entrance velocities below 0.15 m/s to minimize head loss. In aquifers of uniformly fine, unconsolidated material a gravel pack may be placed around the screen. Figure 17.9 shows a cross section of a gravel-packed well.

17.7 Development of Wells

After the screen is placed or the casing is perforated, a well should be developed by pumping at a high discharge or surging with a plunger. This practice develops higher velocities through the screen and in the aquifer adjacent to the screen than will be developed in normal pumping from the well. This action brings fine materials into the well where they are removed by pumping or bailing. As a result the aquifer is opened for freer flow of water and a stabilized filter varying from coarse to fine material is developed.

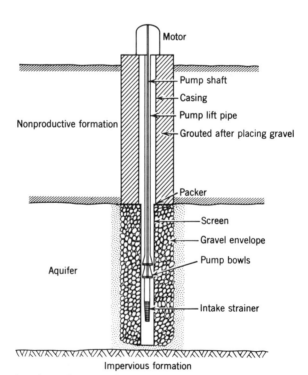

Fig. 17.9 Cross section through a gravel-packed well. (Modified from Linsley and Franzini, 1979.)

WATER CONSERVATION

17.8 Water and Crop Management

Water can be conserved by application of well-designed crop production systems. Many practices discussed in Chapters 5 and 6 on erosion also contribute to water conservation. An example is strip cropping in dryland farming regions to retain snow and rainfall near where it falls to allow more time for infiltration. Well-designed drainage systems (Chapters 13 to 15) or irrigation systems (Chapters 19 to 21) can result in substantial water savings and potentially enhance water quality.

17.9 Utilizing Poor-Quality Waters

Poor-quality waters include drainage water, saline ground water, treated sewage effluent, and other sources that do not meet the quality criteria for some users. In some instances poor-quality waters may be used by agriculture to save high-quality waters for other users. One practice is to mix poor- and good-quality waters to increase water supplies while maintaining an acceptable quality. This practice is more common in humid regions where large quantities of fresh water are available to mix with drainage water. In arid regions saline water can be used to irrigate salt-tolerant crops, mixed with good water, or used late in the growing season when most crops are less susceptible to salinity. Treated wastewater effluent is a potential water source that is increasing in use. This water can be easily diverted to an irrigation distribution system for golf courses, parks, or irrigated crops. Treated effluent use on crops that are consumed without cooking is restricted and local health authorities should be consulted regarding acceptable uses.

17.10 Control of Seepage

Water conveyance losses from canals and ditches can be greatly reduced through reduction or elimination of seepage (Chapter 13). Concrete linings are frequently used in irrigation canals and ditches. Asphalt and plastic linings are also used in canals and ponds. Chemical additives are often successful in canals, ponds, and ditches for seepage reduction.

17.11 Artificial Recharge

Ground water reservoirs are supplied by water percolating to them from the surface. Under natural conditions only a small fraction of rainfall reaches the ground water. Since ground water reservoirs provide evaporation-free storage and since surface runoff waters are often wasted, a logical water conservation measure is to attempt to increase the recharge of ground water from surface runoff.

There are four general methods of artificial recharge: basin, furrow or ditch, flooding, and pit or well. The basin method of spreading water consists of a series of small basins formed by dikes or banks. The dikes often follow contour lines, and they are so arranged that the water flows from one basin to the next. In the furrow or ditch method the water flows along a series of parallel ditches placed closely together. The flooding method consists of ponding a thin layer of water over the land surface. Pits or wells as a method of recharge are used primarily in municipal areas and industrial centers. Regardless of the method, it is desirable to spread water that is relatively free of sediment. It is not unusual to use a combination of several methods.

In artificial recharge basins layers of accumulated sediment are periodically removed and replaced with sand. Soil conditioners such as organic residues, grasses, and chemical treatments are effective in increasing infiltration rates. Some waters require desilting or biological control treatment before they can be recharged without clogging the infiltration area or the aquifers.

17.12   Phreatophyte Control

The term *phreatophyte* includes plants that habitually obtain their water supply from the zone of saturation or from the overlying capillary fringe. Examples are tamarisk, cottonwood, willow, and mesquite. Thompson (1958) reported that phreatophytes covered approximately 7 million ha in the western United States and used an estimated 3 million ha-m of water annually. Phreatophyte growth is concentrated largely along streams and rivers. Consumptive use of water by phreatophytes varies with species, climate, and depth to the ground water table. Under high-water-table conditions, water used by phreatophytes will approach open pan evaporation. Maximum use of water by tamarisk in the Rio Grande Valley of New Mexico has been measured at 3.3 m per year with a 0.6-m depth to the water table.

Control of phreatophytes thus offers a great potential for water conservation. Control can be effected by either chemical or mechanical means; however, the costs of control have limited its application. Thompson (1958) concluded that channelization was the most effective means of salvaging water that would otherwise be lost to phreatophytes. Through channelization and drainage, ground water can be lowered in phreatophyte-infested areas and conveyed to downstream reservoirs. The accompanying lower water table greatly reduces the consumptive use by phreatophytes; however, phreatophyte removal must be balanced against the associated loss of wildlife habitat.

17.13   Evaporation Suppression

Reduction of evaporation from free-water surfaces is an important water conservation measure. Two broad approaches are employed: reduction of the free-water surface area and protection of free-water surfaces.

Reduction of free-water surface is accomplished by minimizing the surface area to volume ratio of reservoirs. Storage of water in natural ground water reservoirs rather than in surface reservoirs also reduces evaporation losses.

Protection of free-water surfaces has been uneconomical except in special situations. Some attention has been given to application of monomolecular films for evaporation suppression. The film inhibits the escape of water molecules, but must be continuously supplied because of breakup by wind action and deterioration by biological processes. Evaporation reductions in excess of 25 percent have been achieved.

WATER MEASUREMENT

Effective use of water requires that flow rates and volumes be measured and expressed quantitatively. Water measurement is based on application of the formula

$$q = av \qquad (17.3)$$

where q = flow rate (L^3/T),
 a = cross-sectional area of flow (L^2),
 v = mean velocity of flow, (L/T).

Some techniques determine mean velocity and area of flow separately and use them directly in Eq. 17.3. In others the calibration of the measurement device gives the flow directly. Measurement of flow volume requires integration of the flow rate q over the time period involved.

17.14   Units of Measurement

In agriculture, the common units of rate of flow in English units are the gallon per minute, cubic foot per second, and Miner's inch, and in SI units, the liter per second and cubic meter per second. The Miner's inch is defined by state legislation and varies from state to state. Conversion factors among the various common units of rate and volumes are given in Appendix F and inside the back cover of the book.

17.15   Float

A crude estimate of the velocity of a stream may be made by determining the velocity of an object floating with the current. A straight uniform section of stream about 100 m long should be selected and marked by stakes or range poles on the bank. The time required for an object floating on the surface to traverse the marked course is measured and the velocity calculated. The average surface velocity is determined by averaging float velocities measured at a number of distances from the bank. Mean velocity of the stream is often taken as 0.8 to 0.9 of the average surface velocity.

Floats consisting of a weight attached to a floating buoy sometimes measure directly mean velocity. The weight is submerged to the depth of mean velocity, and the buoy marks its travel downstream. The float method has the advantage of giving an estimate of velocity with a minimum of equipment; however it lacks precision.

17.16   Impeller Meters

Instruments employing an impeller that rotates at speeds proportional to the velocity of flowing water are often used for velocity determination in open channels. In such applications they are called current meters. Figure 17.10 shows a typical current meter with accessories. The essential part of the meter is a wheel that revolves when suspended in flowing water. An electrical circuit indicates the speed of revolution of the wheel. The meter may be suspended by a cable for deep

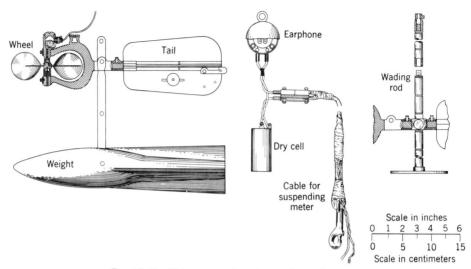

Fig. 17.10   Price current meter and attachments.

streams or attached to a rod in shallow streams. When supported by a cable a streamlined weight holds the meter against the current. A vane attached to the rear of the meter keeps the wheel headed into the stream. Other current meters determine the water velocity by measuring the disturbance in an electrical field.

When the mean velocity of a stream is determined with a current meter, the cross section of flow is divided into a number of subareas. Width of the subareas depends on the size and shape of the stream and the precision desired. This is illustrated in Fig. 17.11. The subareas may be indicated by marks on a tape or cable stretched across the stream or by marks on a bridge railing or other convenient structure. The average velocity at each station across the section is determined with the current meter. It has been found that the average of readings taken at 0.2 and 0.8 of the depth below the surface is an accurate estimate of the average velocity in the subarea. Where the stream is too shallow to allow a reading at 0.8 of the depth, the velocity at 0.6 of the depth below the surface may be taken as the average velocity. The area of the cross section may be determined by sounding with the current meter or other convenient device. Table 17.3 gives the calculation of the discharge for the section shown in Fig. 17.11.

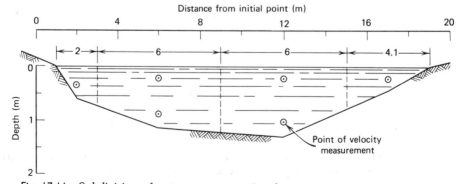

Fig. 17.11   Subdivision of a stream cross section for current meter measurements.

Table 17.3 Calculation of Discharge from Current Meter Measurements

Gaging of Skunk River at Ames, Iowa
Meter No. SC5514394

Date: April 1, 1988
Measurement began at 1:15 P.M.
Measurement ended at 2:30 P.M.

Gaging by DeHart and Storm

Distance from Initial Point (m)	Width (m)	Depth (m)	Observation Depth Ratio	Meter Revolutions	Time (s)	At Point	Mean in Subarea	Area (m²)	Discharge (m³/s)
2	2	0.62	0.6	5	42	0.10	0.10	1.24	0.12
6	6	1.14	0.8	20	41	0.34	0.36	6.84	2.46
			0.2	25	45	0.38			
12	6	1.32	0.8	15	46	0.24	0.27	7.92	2.14
			0.2	20	45	0.30			
17	4.1	0.46	0.6	5	57	0.06	0.06	1.89	0.11
Total	18.1							17.89	4.83

In using impeller meters for velocity measurements in closed conduits, the cross-sectional area of flow remains constant, and the meters are calibrated to read directly in cumulative volume or in flow rate. Figure 17.12 shows typical installations of impeller meters in closed conduits.

17.17 Slope Area

The Manning velocity equation for designing open channels described in Chapter 7 may be applied to streamflow measurement. A nomograph for calculating the velocity from the Manning formula is given in Appendix B. Application of the formula to estimation of flow in open channels requires measurement of the slope of the water surface and measurement of the properties of the cross section of flow. The reach of the channel selected should be uniform and, if possible, as long

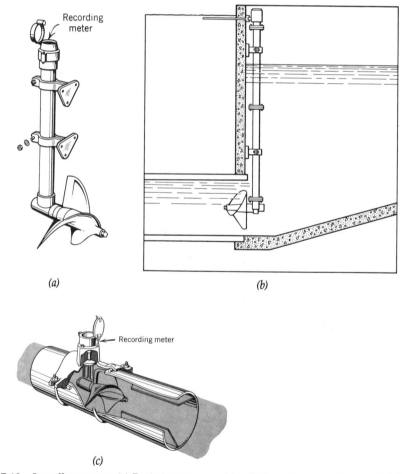

Fig. 17.12 Impeller meters. (*a*) Basic meter assembly. (*b*) Location in an inverted siphon. (*c*) Low-pressure pipe impeller meter. (Courtesy Sparling Division, Hersey-Sparling Meter Company, El Monte, California.)

as 300 m. The value of the roughness coefficient must be estimated and this is difficult to do accurately. Appendix B gives values of n, which are helpful in arriving at such estimates.

The slope–area method is sometimes used in estimating the discharge of past flood peaks. Cross-sectional area and flow gradient are measured from high-water marks along the channel. This method must be regarded as giving only a rough approximation of the peak flow.

17.18   Orifices

An orifice is an opening with a closed perimeter through which water flows. The velocity of flow through an orifice is a function of head as discussed in Chapter 9. Figure 17.13a illustrates an end-cap orifice for measurement of discharge from an irrigation pump. The ratio of orifice to pipe diameter should be between 0.5 and 0.83. The orifice coefficient is dependent on orifice configuration and must be determined for each design. The pipe must be level and the manometer located about 0.6 m from the orifice.

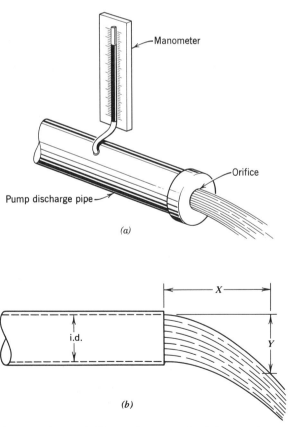

Fig. 17.13 (a) End-cap orifice and (b) coordinate method for measurement of flow from pump discharge.

17.19 Weirs and Flumes

For accurate measurement of flow in open channels, structures of known hydraulic characteristics are required. As discussed in Chapter 9, these structures cause flow to pass through critical depth. They have a consistent relationship between head and discharge. The action of weirs and flumes in open channels is analogous to that of orifices for closed conduits.

Weirs. A weir consists of a barrier placed in a stream to constrict the flow and cause it to fall over a crest. The flow rate through such a structure is given in Chapter 9. Weir openings may be rectangular, trapezoidal, or triangular in cross section, or they may take special shapes to give desired head–discharge relationships. Consult standard hydraulic handbooks and references such as Brater and King (1976). A typical temporary weir for measuring streamflow is illustrated in Fig. 17.14.

Flumes. Specially shaped and stabilized channel sections such as a flume may be used to measure flow. Flumes are generally less inclined to catch floating debris and sediment than are weirs and, for this reason, are particularly suited to measurement of runoff. They also require a very low head loss for operation. Parshall (1950) developed a common measuring flume, which is illustrated in Fig. 17.15. Discharge tables for all sizes of Parshall flumes are available in Parshall (1950) or hydraulics handbooks. Other similar-type flumes, which are simpler and easier to construct, have been developed. These include the cutthroat flume described by Skogerboe (1973) and a critical-depth flume designed by Replogle (1971), which is now known as a long-throated flume (Clemmens and Replogle, 1980; Bos et al., 1984). The long-throated flume is illustrated in Fig. 17.16.

For runoff measurement from small watersheds a special type H flume gives good accuracy at low flows as well as providing high capacity. The opening is V-shaped, with the top of the V sloped toward the upstream side.

Fig. 17.14 Rectangular weir for measurement of flow in a small stream or irrigation ditch.

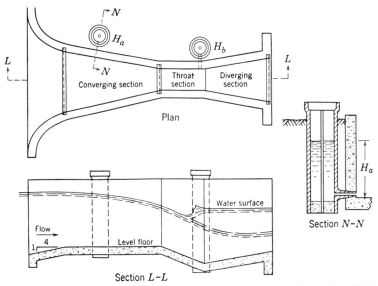

Fig. 17.15   Parshall measuring flume. (Redrawn from Parshall, 1950.)

17.20   Other Methods

Velocity determinations may be made in open channels or closed conduits with pitot tubes (Brater and King, 1976). Determination of mean velocity and calculation of discharge are done by methods similar to those applied to open channels when making current meter measurements. A pitot tube used in a closed conduit may be calibrated to read directly the flow rate from one velocity measurement.

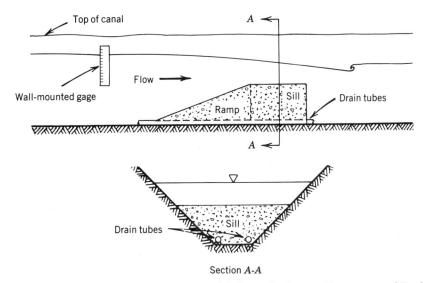

Fig. 17.16   Long-throated flume in a trapezoidal channel. (*Source:* Clemmens and Replogle, 1980.)

One such device, designed for insertion through the wall of a pipe, is known as a Cox meter.

Pipe elbows offer an opportunity for measurement of flow as described by Langsford (1936). Water flowing through an elbow exerts different centrifugal forces at the inside and outside radii of the elbow. The resulting difference in pressure may be measured through a differential manometer. Discharge through the elbow is a function of the square root of the pressure differential, with the coefficient determined by calibration for each size of elbow.

The injection of a fluorescent dye or substance that changes the water conductivity of the stream can be used. The dye is detected with suitable equipment downstream so as to determine the velocity of flow.

The coordinate or trajectory method of measuring pipe flow, illustrated in Fig. 17.13b, requires the measurement of X and Y distances with the pipe level. The pipe should be long enough to produce a smooth flow from the pipe. Flow is estimated from appropriate calibration tables.

WATER QUALITY

Water quality is determined by the concentration of biological, chemical, and physical contaminants. Most water pollution is the result of human activities. Biological contaminants result from human and animal wastes plus some industrial processes. Chemicals enter the water supply from industrial processes and agricultural use of fertilizers and pesticides. Physical contaminants result from erosion and disposal of solid waste. Since all of these sources contribute to degradation of water quality, standards have been developed for drinking water by the U.S. Public Health Service (Table 17.4). These standards strive to prevent health problems by defining the quality of water available for human consumption. Many local, state, and federal regulations have been instituted to prevent contamination of both surface and ground water supplies.

Sources of water pollution are recognized as point or nonpoint in origin. Point sources include animal feedlots, chemical dump sites, storm drain and sewer outlets, acid mine outlets, industrial waste outlets, and other identifiable points of origin. Nonpoint sources include runoff from forest and agricultural land, hillside seepage, small subsurface drain outlets, and other diffuse sources. Nonpoint pollution is often more difficult to identify and to correct.

17.21 Biological Contaminants

In agriculture, biological contaminants are primarily from animal and human waste. Feedlots, dairies, and septic systems are major sources of biological pollution. Bacteria are the most common organisms; however, viruses and other microorganisms may also be present, and all can create serious health problems. Biological contamination can be controlled by proper disposal of wastes; separation of septic systems, feedlots, and other sources from drinking and surface water supplies; and treatment of drinking water before consumption.

Table 17.4 Drinking Water Standards

A.	Chemical	
	1. Maximum Contaminant Level (mg/L)	
	arsenic	0.05
	asbestos[a]	—
	barium	2.0
	cadmium	0.0005
	chromium	0.1
	lead	0.05
	mercury	0.0002
	nitrate (as N)	10.0
	nitrite (as N)	1.0
	selenium	0.05
	alachlor	0.002
	aldicarb	0.003
	atrazine	0.003
	carbofuran	0.04
	chlordane	0.002
	endrin	0.0002
	ethylene dibromide	0.00005
	heptachlor	0.0004
	lindane	0.0002
	methoxychlor	0.04
	toxaphene	0.005
	total trihalomethanes (THMs)	0.1
	2. Secondary Maximum Contaminant Level (mg/L)	
	chloride	250
	copper	1
	fluoride	2.0
	iron	0.3
	manganese	0.05
	pH	6.5 to 8.5
	sulfate	250
	total dissolved solids (TDS)	500
	zinc	5
B.	Physical	
	color	15 color units
	odor	3 odor units
C.	Bacteriological	
	coliform bacteria	none

[a]7 million fibers (longer than 10 mm)/L
Source: EPA (1990); Mancl et al. (1991)

17.22 Physical Contaminants

The most common physical contaminant of water is suspended sediment. Other physical contaminants include organic materials such as plant residues. Most sediment occurs because of soil erosion; however, sand may be obtained during pumping from wells.

Where sediment is deposited on sandy soil, the textural composition and fertility may be improved; however, if the sediment has been derived from eroded areas, it may reduce fertility or decrease soil permeability. Sedimentation in canals or ditches may be serious, resulting in higher maintenance costs. Sediment can greatly decrease the capacity of ponds and reservoirs. Eroded sediments not only indicate an erosion loss, but carry attached chemical ions, such as phosphorus and potassium, which contribute to chemical pollution as well. These chemicals cause eutrophication in lakes and streams, increase the cost of treatment for domestic and industrial supplies, and adversely affect fish and other aquatic life. Mean annual sediment concentration in U.S. streams is shown in Fig. 17.17. These concentrations are highest in the west-central states, but concentration is not necessarily related to total sediment loss because runoff is lower in the western than in the eastern states.

Sediment must be removed from water used in microirrigation systems to prevent plugging. Sands may cause excessive wear to pump impellers and to the nozzles in sprinkler irrigation systems.

17.23 Chemical Contaminants

Chemicals are a major source of water contamination. Some chemicals occur naturally in the water, others are introduced during water movement through geological materials, but most problems are caused by manufactured chemicals. Fertilizers and pesticides are the major contributors to chemical pollution from agriculture. These chemicals may be applied to soil or foliage over large areas and hence become potential sources of nonpoint pollution; however, fertilizers and pesticides have contributed to a high quality and abundant food supply at reasonable cost for the people of the United States and for export.

Nitrates are a common chemical pollutant of water. Estimates indicate that U.S. cropland received 10.4 million tons of nitrogen in 1987 through nitrogen fertilizers (USDA, 1988b) and a similar amount through animal manures, crop residue, and natural sources. A maximum nitrate concentration greater than 3 mg/L was found in 20 percent of 124 000 wells analyzed over a 25-year period (Madison and Brunett, 1985). Concentrations greater than 3 mg/L usually relate to human activities, such as fertilizer applications and septic systems. The federal drinking water standard of 10 mg/L was exceeded in 6 percent of the samples.

Forty-six pesticides have been detected in ground water and confirmed to come from nonpoint sources (Williams et al., 1988). One or more pesticides have been detected in the ground water of 26 states and attributed to agricultural use. The most commonly detected pesticides are atrazine and aldicarb. Pesticide usage was about 300 000 kg in 1988 (USDA, 1988a). Gianessi and Puffer (1988) reported that the cornbelt states used more than 59 percent of the U.S. total.

The development and implementation of practices and policies to reduce water contamination by agriculture are essential. A better understanding of the fundamental processes affecting the transport and fate of agricultural chemicals must be developed. This knowledge must be used to develop new or improved farming systems that protect, improve, or remediate the quality of water supplies. Engineers have a significant role in the development process.

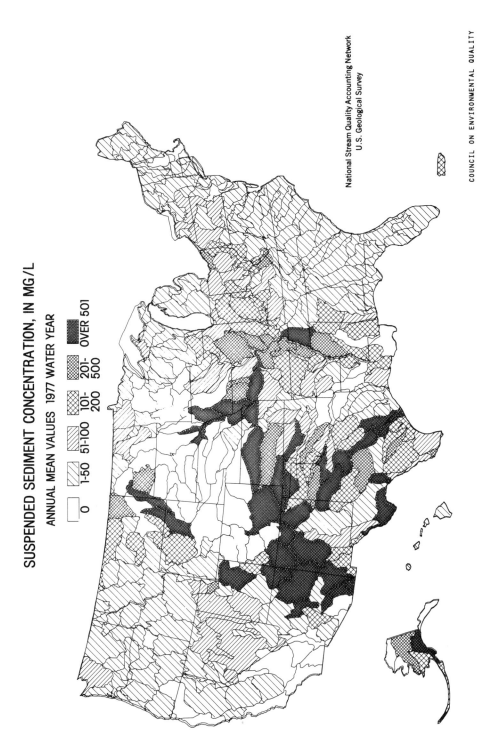

SUSPENDED SEDIMENT CONCENTRATION, IN MG/L

ANNUAL MEAN VALUES 1977 WATER YEAR

0   1-50   51-100   101-200   201-500   OVER 501

National Stream Quality Accounting Network
U.S. Geological Survey

COUNCIL ON ENVIRONMENTAL QUALITY

Fig.17.17   Average annual suspended sediment concentration in mg/L (*Source:* USDA, 1977.)

17.24 Quality of Drainage Water in Humid Areas

In humid areas agricultural drainage water, which originates from precipitation, eventually becomes either runoff from the soil surface or flow from subsurface drains or natural seepage. Some precipitation may flow directly downward where it falls or flows to another area where it moves downward. In either event, water may move to the deeper ground water and thus is a potential source of pollution in well water, which is often used for domestic consumption. In humid areas, pollution in surface runoff potentially can be reduced by dilution with excessive streamflow.

In the eastern states the acidity of precipitation varies from about pH 4 to pH 6. High acidity causes fish kills in some lakes, damage to some trees and forests, and other environmental problems, which are of increasing concern.

In cultivated land in Ohio where extensive pipe drainage systems are installed, annual pipe flow is generally greater than surface runoff from land without subsurface drains. Even though pipe flow was greater, average annual sediment, phosphorus, soluble potassium, atrazine, dicamba, and heptachlor losses were much lower from pipe drains (Schwab et al., 1985). The pH of pipe drain water was 7.1 compared with 6.0 for rainfall, indicating a beneficial effect. The major pollution impact of subsurface drains is the increase in the more soluble materials, especially nitrate nitrogen. An increase of 57 percent was measured compared with losses in surface runoff. Improved surface drainage reduces water movement into the soil and consequently reduces the nitrate losses. Field studies and simulation modeling in North Carolina (Skaggs and Gilliam, 1981) showed a 300 percent difference in nitrate losses among several drainage systems, all of which satisfied drainage design requirements. Nitrate losses for corn were reduced 62 percent by improving surface drainage and by providing controlled drainage, especially during the winter months. Since controlled drainage systems generally require closer drain spacings than conventional systems, the expected increase in nitrates could be reduced by maintaining the outlet water levels as designed. Nitrate losses can also be reduced by using slow-release fertilizers and by timing applications to just meet plant needs. No-tillage cropping systems will greatly reduce sediment loss and, to some extent, loss of nitrates and other fertilizers, but herbicide losses are likely to increase (Schwab et al., 1985). In Louisiana, losses of the herbicides atrazine and metolachlor were 50 percent lower from pipe drain water than from surface runoff (Bengston et al., 1990). The above examples illustrate that one management practice may reduce pollution of one chemical but increase pollution of another.

17.25 Irrigation Effects on Water Quality

In arid and semiarid areas irrigation is necessary for crop production; however, this leads to water quality degradation and salinity problems (Hoffman, 1990). Surface runoff may contain chemicals, fertilizers, or pesticides. In addition, some chemicals are injected into the irrigation water for application to the field and carried in the runoff. Most irrigation waters contain dissolved salts, which remain in the soil after water use by evapotranspiration. These salts must be removed to maintain crop production, and natural or artificial drainage systems are needed to remove excess water and associated salts from the plant root zone. This water is obviously of lower quality than the applied water and is frequently returned to a stream or underground water supply, degrading the quality. Dissolved nitrogen from fertilizer applications is another source of contamination. The nitrogen may be both in surface runoff and combined with salts in the drainage water.

The adverse affects of salinity and nitrogen have long been known. Only recently, however, has the potential impact of trace elements been recognized. These trace elements, including arsenic, boron, cadmium, chromium, lead, molybdenum, and selenium, originate mainly in the geologic materials on site instead of the irrigation water. This impact adds a new dimension to irrigation and drainage management since both salinity and toxic trace elements must be considered when planning the disposal system. Selenium is a prime example.

The contamination of the Kesterson National Wildlife Refuge in the San Joaquin Valley of California by selenium has created awareness of the trace element problem. The environmental issues and impacts are documented by the National Research Council (1989). The wildlife ponds were built in 1971 and received fresh water until 1982. By 1981, the supply was entirely drainage water. Less than two years later, in 1982, reproductive failures and deaths of some aquatic organisms and water fowl were reported. The contamination was caused by increased irrigation development, subsequent installation of subsurface drains, and failure to install an adequate disposal system. It is important to recognize that this event is not isolated and future planning must include measures to alleviate or prevent additional occurrences. As a result of public concern, efforts were initiated to eliminate drainage flows into Kesterson.

17.26 Irrigation Water Quality Criteria

The chemical quality of water determines its suitability for irrigation use [USDA (1954) and Ayers and Westcot (1985)]. The most important characteristics of irrigation water are (1) total concentration of soluble salts, (2) proportion of sodium to other cations, (3) concentration of potentially toxic elements, and (4) bicarbonate concentration as related to the concentration of calcium plus magnesium.

Total soluble salts are commonly indexed by the electrical conductivity of the water expressed in deciSiemens per meter. The relationship between conductivity and parts per million of soluble salts is illustrated in Fig. 17.18. The proportion of sodium to other cations or the sodium hazard of the water is indicated by the sodium-adsorption ratio, or SAR, calculated from

$$SAR = \frac{Na^+}{\sqrt{(Ca^{2+} + Mg^{2+})/2}} \qquad (17.4)$$

where Na^+, Ca^{2+}, and Mg^{2+} represent the concentration in milliequivalents per liter of the respective ions. The suitability of water for irrigation on the basis of conductivity and the sodium-adsorption ratio has been expressed diagrammatically in Fig. 17.19. Table 17.5 shows the suitability for irrigation of the various classes of water.

Boron, though essential to normal growth of plants, is toxic under some conditions in concentrations as low as 0.33 part per million. High concentrations of bicarbonate ions may result in precipitation of calcium and magnesium bicarbonates from the soil solution, increasing the relative proportions of sodium and thus

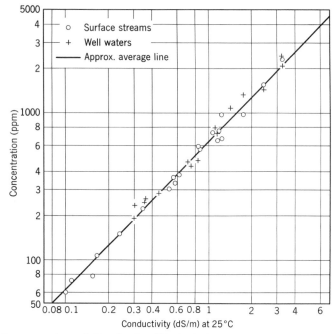

Fig. 17.18   Concentration of dissolved solids in irrigation waters in parts per million as related to conductivity. (*Source:* USDA, 1954.)

the sodium hazard. USDA Handbook No. 60 (USDA, 1954) should be referred to if there are potential problems with toxic elements or bicarbonate ion concentration.

WATER RIGHTS

Much of the confusion regarding water rights stems from the failure to make a distinction among the several types of naturally occurring waters. From the legal standpoint water may be classified as (1) diffused surface water, (2) water in well-defined surface channels, (3) water in well-defined underground aquifers, and (4) underground percolating water. Diffused surface water and underground percolating water, because of their diverse nature, are normally regulated by common or civil law rather than by legislative action. In some western states, however, diffused surface water is treated the same as water in well-defined channels. In most states diffused surface water is considered the property of the landowner, who may use it in any way without regard to its effect on the water supply of other owners. Especially in the eastern states the law of diffused surface water has been concerned with the damage caused by such water and the fixing of responsibility for such damage. Underground percolating water is defined as that subsurface water that flows in small pores or filters through the soil in such a way that its course or direction cannot be easily determined.

Two basic divergent doctrines regarding the right to use water exist, namely, riparian and appropriation. They are recognized either separately or as a combina-

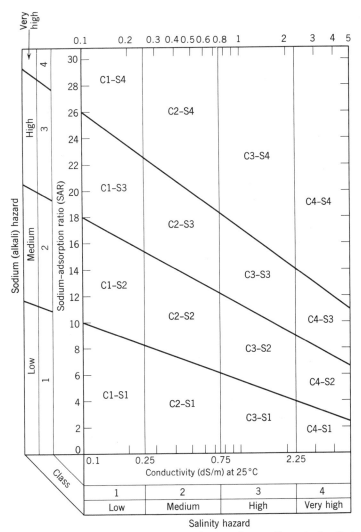

Fig. 17.19 Classification of irrigation waters with regard to sodium and salinity hazards. (*Source:* USDA, 1954.)

tion of both doctrines in different states. A comparison of the salient features of these two doctrines is made in Table 17.6.

17.27 Riparian Doctrine

Riparian doctrine may have originated in early English water law, which was borrowed in part from Roman civil law. The riparian doctrine in its American form recognizes the right of a riparian owner to make reasonable use of the stream's flow, provided the water is used on riparian land. Riparian land is that which is contiguous to a stream or other body of surface water. Land ownership accompanies the right of access to and use of the water, and this right is not lost by nonuse. Reasonable use of water generally implies that the landowner may use all that he

Table 17.5   Suitability of Waters for Irrigation

Class	Salinity or Conductivity	Sodium-Adsorption Ratio[a]
Low 1	*Low-salinity water* (C1) can be used for irrigation with most crops on most soils with little likelihood that soil salinity will develop. Some leaching is required, but this occurs under normal irrigation practices except in soils of extremely low permeability.	*Low-sodium water* (S1) can be used for irrigation on almost all soils with little danger of the development of harmful levels of exchangeable sodium; however, sodium-sensitive crops such as stone-fruit trees and avocados may accumulate injurious concentrations of sodium.
Medium 2	*Medium-salinity water* (C2) can be used if a moderate amount of leaching occurs. Plants with moderate salt tolerance can be grown in most cases without special practices for salinity control.	*Medium-sodium water* (S2) will present an appreciable sodium hazard in fine-textured soils having high cation-exchange capacity, especially under low-leaching conditions, unless gypsum is present in the soil. This water may be used on coarse-textured or organic soils with good permeability.
High 3	*High-salinity water* (C3) cannot be used on soils with restricted drainage. Even with adequate drainage, special management for salinity control may be required and plants with good salt tolerance should be selected.	*High-sodium water* (S3) may produce harmful levels of exchangeable sodium in most soils and will require special soil management— good drainage, high leaching, and organic matter additions. Gypsiferous soils may not develop harmful levels of exchangeable sodium from such waters. Chemical amendments may be required for replacement of exchangeable sodium, except that amendments may not be feasible with waters of very high salinity.
Very high 4	*Very-high-salinity water* (C4) is not suitable for irrigation under ordinary conditions, but may be used occasionally in very special circumstances. The soils must be permeable, drainage must be adequate, irrigation water must be applied in excess to provide considerable leaching, and very salt-tolerant crops should be selected.	*Very-high-sodium water* (S4) is generally unsatisfactory for irrigation purposes except at low and perhaps medium salinity, where the solution of calcium from the soil or use of gypsum or other amendments may make the use of these waters feasible.

[a]Sometimes the irrigation water may dissolve sufficient calcium from calcareous soils to decrease the sodium hazard appreciably, and this should be taken into account in the use of C1-S3 and C1-S4 waters. For calcareous soils with high pH values or for noncalcareous soils, the sodium status of waters in classes C1-S3, C1-S4, and C2-S4 may be improved by the addition of gypsum to the water. Similarly, it may be beneficial to add gypsum to the soil periodically when C2-S3 and C3-S2 waters are used.

Source: USDA (1954).

Table 17.6 Comparison of Water Rights Laws[a]

Characteristic	Riparian	Doctrine of Appropriation
Acquisition of water right	By ownership of riparian land	By permit from state (state ownership)
Quantity of water	Reasonable use	Restricted to that allowed by permit
Types of use allowed	Domestic, livestock, etc., but not precisely defined	Some beneficial use required
Loss of water by nonuse	No	Yes, but continued use not always required
Location where water may be used	On riparian land, but some exceptions	Anywhere, unless specified in permit

[a]Generally applicable only for surface water in well-defined channels and for water in well-defined underground aquifers. Some state laws on ground water deviate from the above.

or she needs for drinking, for household purposes, and for watering livestock. For percolating ground water most states recognize absolute ownership and use of percolating water is not restricted. This principle is known as "common law riparian." In some states the riparian doctrine is modified by applying the principle of "reasonable use." The California Supreme Court under this principle definitely established that no proprietor can absorb all the water of the stream so as to allow none to flow down to a neighbor. In California, the riparian doctrine was further modified by establishing "correlative rights." Under this doctrine the landowner's use of ground water not only must be reasonable in consideration of the similar rights of others, but must be correlated with the uses of others in times of shortage.

In states that do not have statutory laws governing water rights, the riparian doctrine is based on previous court decisions. Many of the eastern states have modified the riparian doctrine by regulating use through the issuance of permits for specified amounts of water. Others restrict use in certain areas of the state.

17.28 Doctrine of Appropriation

This doctrine is normally applied to prior rights. It is based on the priority of development and use; that is, the first to develop and put water to beneficial use has the prior right to continue his or her use. The right of prior appropriation is acquired mainly by filing a claim in accordance with the laws of the state. The water must be put to some beneficial use, but the appropriator has the right to water required to satisfy her or his needs at the given time and place. This principle assumes that it is better to let individuals, prior in time, take all the water rather than to distribute inadequate amounts to several owners. Water rights are not limited to riparian land and may be lost by nonuse and abandonment.

17.29 Water Rights Law by States

Water rights doctrines vary from state to state; however, because of increased demand for water and the need for better utilization, many states have modified their water laws. Generally these laws have given more authority to state agencies to protect the public values associated with water resources by establishing water duties or other stringent criteria for use (for discussion and examples see ASAE, 1986).

REFERENCES

American Society of Agricultural Engineers (ASAE) (1986). *Water Resources Law.* Proceedings of the National Symposium on Water Resources Law. ASAE, St. Joseph, MI.

Ayers, R. S., and D. W. Westcot (1985). *Water Quality for Agriculture.* Irrig. and Drain. Paper 29/1. FAO, Rome, Italy.

Bengston, R. L., L. M. Southwick, G. H. Willis, and C. E. Carter (1990). "The Influence of Subsurface Drainage Practices on Herbicide Losses." *ASAE Trans.* **30**, 415–418.

Bennison, E. W. (1947). *Ground Water, Its Development, Uses and Conservation.* E. E. Johnson, St. Paul, MN.

Bos, M. G., J. A. Replogle, and A. J. Clemmens (1984). *Flow Measuring Flumes for Open Channel Systems.* Wiley, New York.

Bouwer, H. (1978). *Groundwater Hydrology.* McGraw-Hill, New York.

Brater, C. F., and H. W. King (1976). *Handbook of Hydraulics,* 6th ed. McGraw-Hill, New York.

Clemmens, A. J., and J. A. Replogle (1980). *Constructing Simple Measuring Flumes for Irrigation Canals.* Farmers' Bulletin No. 2268. Science and Education Admin., USDA, Washington, DC.

Driscoll, F. G. (1986). *Groundwater and Wells,* E. E. Johnson, St. Paul, MN.

Ferris, J. G. (1959). "Ground Water." In *Hydrology,* C. E. Wisler and E. F. Brater (eds.), Chapter 6, pp. 127–191. Wiley, New York.

Gianessi, L. P., and C. M. Puffer (1988). *Use of Selected Pesticides for Agricultural Crop Production in the United States, 1982–1985.* Quality of the Environ. Div., Resources for the Future, Inc., Washington, DC.

Hansen, V. E., O. W. Israelsen, and G. E. Stringham (1980). *Irrigation Principles and Practices,* 4th ed., Wiley, New York.

Hoffman, G. J. (1990). *Environmental Impacts of Subsurface Drainage.* Proceedings of 4th International Workshop on Land Drainage. Int. Commission on Irrigation and Drainage, Cairo, Egypt.

Hutchins, W. A. (1939). "Water Rights for Irrigation in Humid Areas." *Agr. Eng.* **29**, 431–432, 436.

——(1942). *Selected Problems in the Law of Water Rights in the West,* USDA Misc. Publ. 418. Washington, DC.

Langsford, W. M. (1936). *The Use of an Elbow in a Pipe Line for Determining the Rate of Flow in the Pipe.* University of Illinois Eng. Expt. Sta. Bull. 289.

Lauritzen, C. W., and A. A. Thayer (1966). *Rain Traps for Interception and Storage of Water for Livestock,* USDA Agr. Inf. Bull. Washington, DC.

Linsley, R. K., and J. B. Franzini (1979). *Water-Resources Engineering,* 3rd ed. McGraw-Hill, New York.

Madison, R. J., and J. O. Brunett (1985). *Overview of the Occurrence of Nitrate in Ground Water of the United States,* U.S. Geol. Surv. Water Supply Paper 2275.

Mancl, K., M. Sailus, and L. Wagenet (1991). *Private Drinking Water Supplies:*

Quality, Testing, and Options for Problem Waters. NRAES-47. Northeast Regional Agricultural Engineering Service, Ithaca, NY.

Meinzer, O. E. (1923). *The Occurrence of Ground Water in the United States.* U.S. Geol. Surv. Water Supply Paper 489.

Muskat, M. (1942). *The Effect of Casing Perforations on Well Productivity.* Am. Inst. of Mining Met. Eng. Tech. Publ. 1528.

National Research Council (1989). *Irrigation-Induced Water Quality Problems.* National Academic Press, Washington, DC.

Parshall, R. L. (1950). *Measuring Water in Irrigation Channels with Parshall Flumes and Small Weirs.* USDA Agr. Cir. 843.

Replogle, J. A. (1971). "Critical-Depth Flumes for Determining Flow in Canals and Natural Channels. *ASAE Trans.* **14**, 428–433.

Schwab, G. O., N. R. Fausey, E. D. Desmond, and J. R. Holman (1985). *Tile and Surface Drainage of Clay Soils.* Res. Bull. 1166. Ohio Agr. Res. and Dev. Center, Ohio State University, Wooster, OH.

Schwalen, H. C., and R. J. Shaw (1961). *Water in the Santa Cruz Valley. Arizona.* Arizona Agr. Expt. Sta. Rep. No. 205.

Skaggs, R. W., and J. W. Gilliam (1981). "Effects of Drainage Design and Operation on Nitrate Transport." *ASAE Trans.* **24**, 929–934, 940.

Skogerboe, G. V. (1973). *Selection and Installation of Cutthroat Flumes for Measuring Irrigation and Drainage Water.* Colorado State University Eng. Expt. Sta. Tech. Bull. 120.

Thomas, R. O. (1959). "Legal Aspects of Ground Water Utilization." *J. Irrig. and Drain Division.* ASCE. **85**(IR-4), 41–63.

Thompson, C. B. (1958). "Importance of Phreatophytes in Water Supply." *J. Irrig. and Drain Division.* Paper No. 1502. *ASCE.* **84**(IR-1).

U.S. Department of Agriculture (USDA) (1954). Handbook No. 60. *Diagnosis and Improvement of Saline and Alkali Soils.* GPO, Washington, DC.

———(1977). *Handbook of Agriculture Charts,* Agr. Handbook No. 524. GPO, Washington, DC.

———(1988a). *ARS Strategic Groundwater Plan. 1. Pesticides* (Rep.). USDA–ARS, Washington, DC.

———(1988b). *Agricultural Resources, Inputs and Outlook* (Rep.). AR-9. Economic Research Service, Washington, DC.

U.S. Environmental Protection Agency (EPA) (1990). *Federal Register 40 CFR Parts 141 to 143 National Primary and Secondary Drinking Water Regulations.* Office of Drinking Water, Washington, DC.

Williams, W. P., P. W. Holden, D. W. Parsons, and M. N. Lorber (1988). *Pesticides in Ground Water Data Base 1988 Interim Report.* EPA, Office of Pesticides Programs, Washington, DC.

PROBLEMS

17.1 Determine the stream discharge for the velocities and gage widths of the stream shown in Fig. 17.11 if the depths of the channel at the point of gaging from left to right were changed to 0.5 m (1.64 ft), 2.0 (6.56), 1.5 (4.92), and 0.5 (1.64), respectively. Record and tabulate data as shown in Table 17.3 for the velocities as given. Compute the average stream velocity.

17.2 Determine the discharge of a stream having a cross-sectional area of 18.6 m² (200 ft²) by the float method. Trial runs for surface floats to travel 90 m (300 ft) were 122, 128, 123, 124, and 128 s each.

17.3 Determine the discharge of a stream having a cross-sectional area of 9.0 m² (97 ft²) and a wetted perimeter of 9.1 m (29.8 ft) using the slope–area method. The channel has some weeds and stones with straight banks and is flowing at full stage. The difference in elevation of the water surface at points 100 m (328 ft) apart is 0.085 m (0.28 ft).

17.4 Determine the capacity of a Parshall flume having a throat width W of 0.38 m (1.25 ft) for $H_a = 0.40$ m (1.30 ft) and $H_b = 0.25$ m (0.90 ft). See Parshall (1950) or a hydraulics handbook.

17.5 Determine the discharge rate of a 0.1-m (0.33-ft)-diameter sharp-edged orifice if the head to the center of the submerged orifice is 0.2 m (0.66 ft) and the discharge coefficient is 0.6.

17.6 Compute the flow rate into a gravity well 600 mm (24 in.) in diameter if the depth of the water-bearing stratum is 24 m (80 ft), the drawdown is 9 m (30 ft), the soil hydraulic conductivity K is 75 mm/h (3 iph), and the radius of influence R is 180 m (600 ft).

17.7 Compute the flow rate into a well completely penetrating a 6-m (20-ft)-deep confined aquifer in which the piezometric surface is 15.0 m (50 ft) above the top of the aquifer. Diameter, drawdown, K, and R are the same as in Problem 17.6.

17.8 An irrigation reservoir has a storage capacity of 10 ha-m (80 ac-ft). If the seasonal irrigation requirement for the crop is 600 mm (24 in.) and the seepage and evaporation losses are 60 percent of the stored water, how many hectares (acres) can be irrigated?

17.9 Derive the equation for the flow rate into a gravity well completely penetrating the aquifer.

17.10 Derive the equation for the flow rate into a well completely penetrating a confined horizontal aquifer with a uniform constant depth.

CHAPTER 18

Irrigation Principles

Human dependence on irrigation can be traced to earliest biblical references. Irrigation in very early times was practiced by the Egyptians, the Asians, and Native Americans. For the most part, water supplies were available to these people only during periods of heavy runoff. Current concepts of irrigation have been made possible only by the application of modern power sources to deep-well pumps and by the storage of large quantities of water in reservoirs. Thus, by use of either underground or surface reservoirs it is now possible to bridge over the years and provide consistent water supplies.

Increasing demands for water, limited availability, and concerns about water quality make effective use of water essential. Because irrigation is a major water user, it is very important that irrigation systems be planned, designed, and operated efficiently. This requires a thorough understanding of the relationships among plants, soils, water supply, and system capabilities.

CROP WATER NEEDS

The water requirements and time of maximum demand vary with different crops. Although growing crops are continuously using water, the rate of evapotranspiration depends on the kind of crop, the degree of maturity, and the atmospheric conditions, such as radiation, temperature, wind, and humidity. Where sufficient water is available, the soil water content should be maintained for optimum growth. The rate of growth at different soil water contents varies with different soils and crops. Some crops are able to withstand drought or high water contents better than others. During the early stages of growth the water needs are generally low but increase rapidly during the maximum growing period to the fruiting stage. During the later stages of maturity water use decreases and irrigation is usually discontinued when the crops are ripening.

18.1 Crop Water Requirements

To make maximum use of available water supplies, the irrigator must have a knowledge of the total seasonal water requirements of crops and how water use varies during the growing season. The seasonal requirement is necessary to select crops and areas that match the available water supply. Knowledge of the variation during the season aids in scheduling irrigations. Table 18.1 illustrates seasonal evapotranspiration and water requirements for crops grown near Deming, New Mexico. It should be noted that expected effective rainfall is considered in determining the field irrigation requirement. The duration and length of periods of inadequate precipitation during the growing season in humid and subhumid regions largely determine the economic feasibility of irrigation. In the Northern Hemisphere water deficiency during the months of June, July, and August is more serious than in earlier or later months.

Estimates of evapotranspiration must be known when planning an irrigation system. The methods of Penman, Jensen-Haise, and Blaney-Criddle have been discussed in Chapter 3 (see Hoffman et al., 1990, or Jensen et al., 1990). Evapotranspiration may be determined by field measurements of soil water content as in Fig. 18.1 for three crops grown in the Salt River Valley of Arizona. Wheat, being a fairly short-season crop in this region, has the lowest seasonal use of 655 mm, with the high water requirement occurring during March and April. Alfalfa, a long-season crop, has a seasonal use of 1888 mm, and in this area grows during the entire year with the exception of December and January. Cotton, a tropical crop, has its highest seasonal use during the hottest portion of the summer and a seasonal use of 1046 mm. Similar data are available in other regions.

18.2 Effective Rainfall

Rainfall must be considered when determining the crop water needs that must be supplied by irrigation. Not all rainfall is effective, but only the portion that contributes to evapotranspiration. Effective rainfall estimates should consider local conditions. Rainfall on a wet soil profile is ineffective in meeting evapotranspiration but may contribute to the leaching requirement. Rainfall that produces runoff has reduced effectiveness.

A method commonly used for estimating effective rainfall was developed by the SCS (1970). This method bases effective rainfall on monthly evapotranspiration, monthly rainfall, and the soil water deficit, which also may be the net irrigation depth. The following equations describe the relationships developed by the SCS:

$$P_e = f(D)[1.25\ P_m^{0.824} - 2.93]\ [10^{0.000955ET}] \tag{18.1}$$

$$f(D) = 0.53 + 0.0116D - 8.94 \times 10^{-5}D^2 + 2.32 \times 10^{-7}\ D^3 \tag{18.2}$$

where P_e = estimated effective rainfall for a 75-mm soil water deficit depth in mm,
P_m = mean monthly rainfall in mm,
ET = average monthly evapotranspiration in mm,
$f(D)$ = adjustment factor for soil water deficits or net irrigation depths (equals 1.0 for $D = 75$ mm),
D = soil water deficit or net irrigation depth in mm,

Note that P_e is limited to the lowest of P_m, ET, or P_e from Eqs. 18.1 and 18.2.

Table 18.1 Seasonal Evapotranspiration and Irrigation Requirements for Crops Near Deming, New Mexico[a]

Crop	Length of Growing Season (days)	Evapotranspiration Depth (mm)	Effective Rainfall Depth (mm)	Evapotranspiration Less Rainfall (mm)	Water Application Efficiency (%)	Irrigation Requirement Depth (mm)
Alfalfa	197	915	152	763	70	1090
Beans (dry)	92	335	102	233	65	358
Corn	137	587	135	452	65	695
Cotton	197	668	152	516	65	794
Grain (spring)	112	396	33	363	65	558
Sorghum	137	549	135	414	65	637

[a]Average frost-free period is April 15 to October 29. Irrigation prior to the frost-free period may be necessary for some crops.
Source: Jensen (1973).

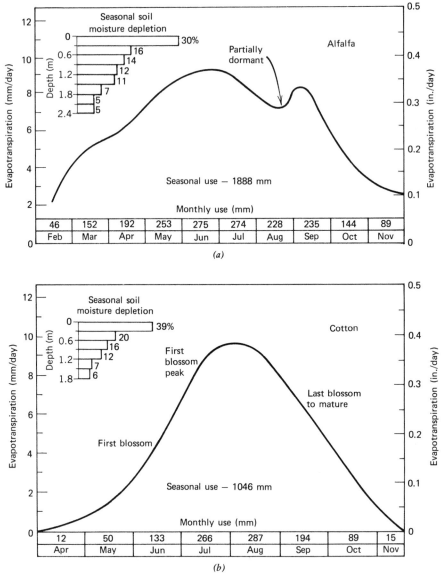

Fig. 18.1 Average evapotranspiration and seasonal water depletion with depth for (*a*) alfalfa, (*b*) cotton, and (*c*) wheat at Mesa and Tempe, Arizona. (Redrawn and adapted from Erie et al., 1982.)

SOILS AND SALINITY

Knowledge of soil characteristics is very important for crop production. The soil is a reservoir for water and chemicals, including plant nutrients, and provides a medium to support the plants. For irrigation, the water-holding capacity and salt content of the soil must be considered.

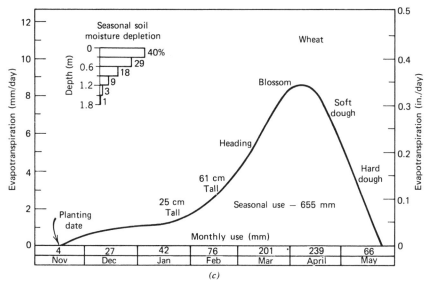

Fig. 18.1   (*Continued*)

18.3   The Soil Water Reservoir

In planning and managing irrigation it is helpful to think of the soil's capacity to store available water as the soil water reservoir. The reservoir is filled periodically by irrigations, then slowly depleted by evapotranspiration. Water application in excess of the reservoir capacity is wasted unless it is used for leaching (Sect. 18.5). Irrigation must be scheduled to prevent the soil water reservoir from becoming so low as to inhibit plant growth.

Irrigation can raise the soil water content to the field capacity. In either sprinkler or surface irrigation the infiltration capacity and the permeability of the soil will determine how fast water can be applied. In sprinkler irrigation the water should be applied at a rate lower than the infiltration capacity. In surface irrigation, the soil surface must be flooded to allow water to enter the soil. In some cases where early spring runoff is available beyond that which may be stored in surface reservoirs, fields are sometimes irrigated with surface runoff channeled through the irrigation ditches. This fills the soil water reservoir and conserves other water supplies for use later in the season.

For irrigation design and management, the water-holding capacity of the soil reservoir must be known. Representative values of soil physical properties, including their water-holding abilities, are listed in Table 18.2 by soil texture. Field capacity (FC) is the water content after a soil is wetted and allowed to drain 1 to 2 days and represents the upper limit of water available to plants. Permanent wilting point (PWP) represents the lower limit of water available to plants, usually defined as 1.5 MPa tension. Neither FC nor PWP can be precisely defined since both vary with soils and plants. The difference between FC and PWP is known as available water (AW) and can be estimated using

$$AW = (FC_v - PWP_v)D_r/100 \qquad (18.3)$$

Table 18.2 Representative Physical Properties of Soils

Soil Texture	Saturated Hydraulic Conductivity, K_s[a] (mm/h)	Total Pore Space (% by vol)	Apparent Specific Gravity (A_s)	Field Capacity, FC_v (% by vol)	Permanent Wilting, PWP_v (% by vol)	Available Water (% by vol)	Available Water (mm/m)
Sandy	50 (25–250)	38 (32–42)	1.65 (1.55–1.80)	15 (10–20)	7 (3–10)	8 (6–10)	80 (70–100)
Sandy loam	25 (12–75)	43 (40–47)	1.50 (1.40–1.60)	21 (15–27)	9 (6–12)	12 (9–15)	120 (90–150)
Loam	12 (8–20)	47 (43–49)	1.40 (1.35–1.50)	31 (25–36)	14 (11–17)	17 (14–20)	170 (140–190)
Clay loam	8 (3–5)	49 (47–51)	1.35 (1.30–1.40)	36 (31–42)	18 (15–20)	18 (16–22)	190 (170–220)
Silty clay	3 (0.25–5)	51 (49–53)	1.30 (1.25–1.35)	40 (35–46)	20 (17–22)	20 (18–23)	210 (180–230)
Clay	5 (1–10)	53 (51–55)	1.25 (1.20–1.30)	44 (39–49)	21 (19–24)	23 (20–25)	230 (200–250)

[a]Saturated hydraulic conductivities vary greatly with soil structure and structural stability, even beyond the normal ranges shown.
Note: Normal ranges are shown in parentheses.
Source: V. E. Hansen, O. A. Israelson, G. Stingham, *Irrigation Principles and Practice*, Copyright © 1980 by John Wiley & Sons, Inc., New York, p. 52. Reprinted by permission of John Wiley & Sons, Inc.

where FC_v and PWP_v = volumetric field capacity and permanent wilting point percentages, respectively,

D_r = depth of the root zone or depth of a layer of soil within the root zone (L),

AW = depth of water available to plants (L).

Since dry weight (gravimetric) water contents are easier to obtain by sampling, it is useful to relate volumetric water content P_v to soil dry weight content P_d in percent by $P_v = P_d A_s$, where A_s is the soil apparent specific gravity. It is important to note that plants can remove only a portion of the available water before growth and yield are affected. This portion is known as readily available water (RAW) and, for most crops, ranges between 40 and 65 percent of AW.

18.4   Salinity

The presence of soluble salts in the root zone can be a serious problem especially in arid regions. In subhumid regions, where irrigation is provided on a supplemental basis, salinity is usually of little concern because rainfall is sufficient to leach out any accumulated salts; however, all water from surface streams and underground sources contains dissolved salts. The salt applied to the soil with irrigation water remains in the soil unless it is flushed out in drainage water or is removed in the harvested crop. Usually the quantity of salt removed by crops is so small that it will not make a significant contribution to salt removal or enter into determinations of leaching requirements.

Salt-affected soils may be classified as saline, sodic, or saline–sodic soils.

Saline soils contain sufficient soluble salt to interfere with the growth of most plants. Sodium salts are in relatively low concentration in comparison with calcium and magnesium salts. Saline soils often are recognized by the presence of white crusts on the soil, by spotty stands, and by stunted and irregular plant growth. Saline soils generally are flocculated, and the permeability is comparable to that of similar nonsaline soils.

The principal effect of salinity is to reduce the availability of water to the plant. In cases of extremely high salinity, there may be curling and yellowing of the leaves, firing in the margins of the leaves, or actual death of the plant. Long before such effects are observed, the general nutrition and growth physiology of the plant will have been altered.

Sodic soils are relatively low in soluble salts, but contain sufficient exchangeable (adsorbed) sodium to interfere with the growth of most plants. Exchangeable sodium is adsorbed on the surfaces of the fine soil particles. It is not leached readily until displaced by other cations, such as calcium or magnesium.

As the proportion of exchangeable sodium increases, soils tend to become dispersed, less permeable to water, and of poorer tilth. High-sodium soils usually are plastic and sticky when wet, and are prone to form clods and crusts on drying. These conditions result in reduced plant growth, poor germination, and, because of inadequate water penetration, poor root aeration and soil crusting.

Saline–sodic soils contain sufficient quantities of both total soluble salt and adsorbed sodium to reduce the yields of most plants. As long as excess soluble salts are present, the physical properties of these soils are similar to those of saline soils. If the excess soluble salts are removed, these soils may assume the properties of sodic soils.

Both sodic and saline–sodic soils may be improved by the replacement of the excessive adsorbed sodium by calcium or magnesium. This usually is done by applying soluble amendments that supply these cations. Acid-forming amendments, such as sulfur or sulfuric acid, may be used on calcareous soils since they react with limestone (calcium carbonate) to form gypsum, a more soluble calcium salt.

Leaching is the only way by which the salts added to the soil by the irrigation water can be removed satisfactorily. Sufficient water must be applied to dissolve the excess salts and carry them away by subsurface drainage.

With the necessity of using additional water beyond the needs of the plant to provide sufficient leaching, it is imperative under irrigation that there be adequate drainage of water passing through the root zone. Natural drainage through the underlying soil may be adequate. In cases where subsurface drainage is inadequate, open or pipe drains must be provided.

Water will rise 0.6 to 1.5 m or more in the soil above the water table by capillarity. The height to which water will rise above a free-water surface depends on soil texture, structure, and other factors. Water reaching the surface evaporates, leaving a salt deposit typical of saline soils.

Some crop plants can tolerate large amounts of salt. Others are more easily injured. The relative salt tolerance of a number of crop plants is shown in Table 18.3. The tolerance of crops listed may vary somewhat, depending on the particu-

Table 18.3 Salt Tolerance Levels for Crops[a]

| Crop | Yield Potential | | | | |
| | 100% | | 90% | | |
	ECe	ECw	ECe	ECw	Max ECe
Field crops					
Barley[b]	8.0	5.3	10.0	6.7	28
Corn	1.7	1.1	2.5	1.7	10
Cotton	7.7	5.1	9.6	6.4	27
Sorghum	4.0	2.7	5.1	3.4	18
Soybeans	5.0	3.3	5.5	3.7	10
Wheat[b]	6.0	4.0	7.4	4.9	20
Vegetable crops					
Beans	1.0	0.7	1.5	1.0	7
Lettuce	1.3	0.9	2.1	1.4	9
Potato, sweet potato	1.6	1.1	2.5	1.7	10
Tomato	2.5	1.7	3.5	2.3	13
Forage crops					
Alfalfa	2.0	1.3	3.4	2.2	16
Bermuda grass	6.9	4.6	8.5	5.7	23
Sudan grass	2.8	1.9	5.1	3.4	26
Fruit crops					
Date palm	4.0	2.7	6.8	4.5	32
Grape	1.5	1.0	2.5	1.7	12
Orange, grapefruit, lemon	1.7	1.1	2.3	1.6	8

[a]All values are in dS/m at 25°C.
[b]During germination and seedling stage, ECe should not exceed 4 or 5 dS/m. Data may not apply to new semidwarf varieties of wheat.
Source: Ayers and Westcot (1985).

lar variety grown, the cultural practices used, and climatic factors. The term *ECe* in the table denotes the electrical conductivity of the saturated extract of the soil in deciSiemens per meter (dS/m) at 25°C. *ECw* is the electrical conductivity of the irrigation water in deciSiemens per meter at 25°C. One deciSiemens per meter represents about 670 ppm of dissolved salts. In general, a yield reduction of 10 percent as a result of salinity is considered acceptable since there are many other factors that might well be limiting in determining the maximum yield of a given crop. Similar data for additional crops are available from Ayers and Westcot (1985) and Tanji (1990).

18.5 Leaching

The traditional concept of leaching involves the ponding of water to achieve more or less uniform salt removal from the entire root zone; however, Ayers and Westcot (1985) have shown that salt accumulation can take place for short periods in the lower root zone without adverse effects. As can be noted in Fig. 18.1, most of the water transpired by the plant is taken from the upper portion of the root zone. This area will be leached to a considerable degree by normal applications of irrigation water and by rainfall that may come at any time of the year. The same amount of water when applied with more frequent irrigations is more effective in removing salts from this critical upper portion of the root zone than from the lower root zone. Thus, high-frequency sprinkler irrigation, microirrigation, or surface irrigation should be effective for salinity control. Other concepts that may be helpful in controlling salinity are the use of soil or water amendments, deep tillage, and irrigation before planting.

Ayers and Westcot (1985) summarized several methods for estimating the leaching requirement for specific crops. One method is

$$LR = ECw/[5(ECe) - ECw] \qquad (18.4)$$

where   LR = leaching requirement expressed as a portion of the infiltrated water,

      ECw = salinity of the irrigation water in dS/m,

       ECe = average soil salinity tolerated by the specific crop from Table 18.3 in dS/m.

LR represents that portion of the infiltrated water that must pass through the soil and percolate below the root zone. Thus, the depth of water infiltrated must equal (1 + LR) times the soil water deficit to maintain a salt balance in the soil appropriate to the crop and the accepted potential yield reduction. With some irrigation applications, sufficient water moves through the root zone to leach the salts and additional water is not needed. It is good practice to monitor the actual salt content of the soil through chemical analyses of soil samples to ensure that neither inadequate nor excessive quantities of water are being applied.

IRRIGATION MANAGEMENT

Irrigation system designers must consider the operation and management requirements for their designs. The performance or efficiency of the system, water delivery requirements, and irrigation scheduling must be understood. A comprehensive review of irrigation management considerations is given by Hoffman et al. (1990).

18.6   Irrigation Efficiencies

Efficiency is an output divided by an input and is usually expressed as a percentage. An efficiency figure is meaningful only when the output and input are clearly defined. There are three basic irrigation efficiency concepts.

(1) Water conveyance efficiency:

$$E_c = 100W_d/W_i \tag{18.5}$$

where W_d = water delivered by a distribution system,
W_i = water introduced into the distribution system.

The water-conveyance efficiency definition can obviously be applied along any reach of a distribution system. For example, a water-conveyance efficiency could be calculated from a pump discharge to a given field or from a major diversion work to a farm turnout.

(2) Water-application efficiency:

$$E_a = 100W_s/W_d \tag{18.6}$$

where W_s = water stored in the soil root zone by irrigation,
W_d = water delivered to the area being irrigated.

This efficiency may be calculated for an individual furrow or border, for an entire field, or for an entire farm or project. When applied to areas larger than a field, it overlaps the definition of conveyance efficiency.

(3) Water-use efficiency:

$$E_u = 100W_u/W_d \tag{18.7}$$

where W_u = water beneficially used,
W_d = water delivered to the area being irrigated.

The concept of beneficial use differs from that of water stored in the root zone in that leaching water would be considered beneficially used though it moved through the soil moisture reservoir. Sometimes water-use efficiency is based on dry plant weight produced by a unit volume of water.

Another useful measurement of the effectiveness of irrigation is the uniformity of water distribution. The uniformity coefficient is

$$UC = 1 - y/d \tag{18.8}$$

where y = average of the absolute values of the deviations in depth of water infiltrated or caught from the average depth of water infiltrated or caught,
d = average depth of water infiltrated or caught.

This coefficient indicates the degree to which water has been applied and penetrated to a uniform depth throughout the field. Note that each value should represent an equal area and that when the deviation from the average depth is zero, the uniformity coefficient is 1.0. UC values above 0.8 are acceptable.

☐ *Example 18.1*

If 42 m³/s (1480 cfs) is pumped into a distribution system and 38 m³/s (1340 cfs) is delivered to a turnout 3 km (1.86 mi) from the pumps, what is the conveyance efficiency of the portion of the distribution system used in conveying this water?

Solution.   Substituting into Eq. 18.5,

$$E_c = 100 \times 38/42 = 90 \text{ percent}$$ ☐

☐ *Example 18.2*

Delivery of 0.5 m³/s (17.6 cfs) to a 30-ha (74-ac) field is continued for 40 h. Tailwater flow is estimated at 0.1 m³/s (3.6 cfs). Soil water measurement after the irrigation indicates that 0.16 m (6.3 in.) of water has been stored in the root zone. Compute the application efficiency.

Solution.   Apply Eq. 18.6.

$$W_d = (0.5) (3600 \text{ s/h}) (40 \text{ h}) = 72\ 000 \text{ m}^3 (58.4 \text{ ac-ft})$$
$$W_s = (0.16) (30 \text{ ha}) (10\ 000 \text{ m/ha}) = 48\ 000 \text{ m}^3 (38.9 \text{ ac-ft})$$
$$E_a = 100 \times 48\ 000/72\ 000 = 67 \text{ percent}$$ ☐

☐ *Example 18.3*

A uniformity check is taken by probing at 30-m (100-ft) stations down one border. The depths of penetration recorded were as follows:

Station (m)	Penetration (m)	Deviation from Mean (m)
0 + 00	0.95	0.10
0 + 30	0.98	0.13
0 + 60	0.98	0.13
0 + 90	0.92	0.07
1 + 20	0.89	0.04
1 + 50	0.89	0.04
1 + 80	0.83	−0.02
2 + 10	0.77	−0.08
2 + 40	0.74	−0.11
2 + 70	0.68	−0.17
3 + 00	0.77	−0.08
3 + 30	0.83	−0.02
3 + 60	0.86	0.01
Sum 11.09	Sum of absolute values	1.0
Mean 0.85		0.08

Compute the uniformity coefficient.

Solution.   Apply Eq. 18.8.

$$UC = 1 - 0.08/0.85 = 0.91$$ □

A second measure is the distribution uniformity:

$$DU = \frac{\text{average low-quarter depth of water infiltrated or caught}}{\text{average depth of water infiltrated or caught}}$$ (18.9)

The average low-quarter depth is the average of the lowest one fourth of all values where each value represents an equal area. DU values above 0.7 are considered acceptable.

□ *Example 18.4*

A solid-set sprinkler system with 15.2 × 12.2-m (50 × 40-ft) head spacing was operated for 4 h. Twenty catch cans were placed under the system on a 3.05 × 3.05-m (10 × 10-ft) spacing and the following depths of water were measured in the cans immediately after irrigation stopped. Assuming evaporation was equal from all cans, determine the distribution uniformity.

Depths (mm)				
50	44	40	42	47
44	42	39	40	44
43	37	30	32	35
48	42	36	40	45

Solution.   Apply Eq. 18.9 using the five lowest depths.

$$\text{Average low-quarter depth} = (37 + 30 + 36 + 32 + 35)/5 = 34$$
$$\text{Average depth} = (50 + 44 + \cdots + 45)/20 = 820/20 = 41$$
$$DU = 34/41 = 0.83$$ □

18.7   Irrigation Requirement

The irrigation requirement (IR) is the total amount of water that must be supplied over a growing season to a crop that is not limited by water, fertilizer, salinity, or diseases.

$$IR = [(ET - P_e)(1 + LR)]/E_a$$ (18.10)

where IR = seasonal irrigation requirement (L),
   ET = seasonal evapotranspiration (L),
   LR = leaching requirement as defined by Eq. 18.4,
   P_e = effective rainfall from Eqs. 18.1 and 18.2 (L),
   E_a = application efficiency (decimal).

Equation 18.10 assumes that the soil water contents at the beginning and end of the season are similar.

□ *Example 18.5*

Corn is being grown and sprinkler irrigated with water having an electrical conductivity of 1.2 dS/m; the average depth of irrigation is 50 mm, and the application efficiency is 70 percent. The monthly ET and rainfall are as follows:

Month	ET (mm)	Rainfall (mm)
May	75	100
June	150	100
July	200	90
August	200	50
September	75	50
Total	700	390

Compute the seasonal irrigation requirement.

Solution.   Assume no yield reduction is preferred and apply Eq. 18.4.

$$LR = 1.2/[5(1.7) - 1.2] = 0.16$$

The effective rainfall from Eqs. 18.1 and 18.2 for May is

$$f(50) = 0.53 + 0.0116(50) - 8.94 \times 10^{-5}(50)^2 + 2.32 \times 10^{-7}(50)^3 = 0.92$$
$$P_e = 0.92 \, [1.25(100)^{0.824} - 2.93][10^{0.000955(75)}] = 57 \text{ mm}$$

May = 57 mm, June = 67, July = 69, August = 41, September = 31, and the total is 265 mm. Now solve Eq. 18.10.

$$IR = [(700 - 265)(1 + 0.16)]/0.70 = 721 \text{ mm}$$

Thus the seasonal irrigation requirement can be taken as 720 mm.   □

18.8 Irrigation Scheduling

Irrigations must be scheduled according to water availability and crop need. If adequate water supplies are available, irrigations are usually provided to obtain optimum or maximum yield; however, overirrigation should be avoided as this can decrease yields by reducing soil aeration and leaching fertilizers while increasing water and energy costs. In addition, overirrigation can contribute to high water tables and water pollution. If water supplies are limited and/or expensive, the irrigation scheduling strategy becomes one of maximizing economic return. In practice, much irrigation water is applied on a routine schedule based on the experience of the farm manager.

Irrigation scheduling requires knowing when to irrigate and how much water to apply. When to irrigate can be determined on the basis of plant or soil indicators or water balance techniques. How much water to apply can be based on soil water measurements or water balance techniques.

Since the objective of irrigation is to provide water for plant growth, plant indicators can directly show the need for water. Growth and appearance are visual indicators of water status. Slow growth of leaves and stems may indicate water stress. Irregular growth patterns may suggest excessive wet and dry cycles. Plant wilting and dark color are also indicators of water stress. Leaf or canopy temperatures, which can be easily measured with infrared thermometers, are good indicators. Increases in leaf temperature relative to air temperatures indicate transpiration is slowing, hence leaves are hotter than those of well-watered plants. Leaf water potentials become lower (more negative) when plants are water stressed; however, these measurements are difficult. Care must be taken when interpreting these signs since diseases or nutrient deficiencies may cause similar reactions.

Soil indicators include feel and appearance, tensiometers, porous blocks, gravimetric sampling, and neutron probes. Feel and appearance require obtaining a soil sample with a probe and then judging the water content by color and consistency when squeezed. With experience this method can indicate both the need for irrigation and the amount to apply. Tensiometers measure soil water tension in the wet range (0 to −80 kPa) which can serve to indicate the need for irrigation. Gravimetric sampling or laboratory measurements are required to convert tension readings to depths to be applied. Gravimetric sampling is a direct method for determining the water content of soil samples and the soil water deficit, but is somewhat labor intensive. Porous blocks buried in the soil change their water content according to that of the surrounding soil. Changes in electrical resistance or thermal conductivity are sensed to indicate the soil water status. These devices must be calibrated for reliable measurements of water content but are simple to read. Neutron probes give direct measurements of water content or water depth, but should be calibrated. In addition, training in radiation safety is required.

Water balance techniques can be used to obtain a record of the estimated water in the root zone. When water depletion reaches a level known as the management-allowed depletion (MAD), irrigation is scheduled. Management-allowed depletion represents the farm manager's decision on how much water can be removed from the root zone and typically equals the water readily available to the plant. Crop and soil characteristics, water availability and costs, irrigation system capabilities, and other factors influence the value selected for MAD. The soil water balance may be determined from soil water measurements. If estimates of evapotranspiration are available, the soil water balance can be calculated from

$$D_i = D_{i-1} + (\text{ET} - P_e)_i \tag{18.11}$$

where D_i and D_{i-1} = the total depth of water removed from the soil root zone at the end of days i and $i-1$, respectively,

 ET = the evapotranspiration calculated from climatic data for day i,

 P_e = the effective rainfall for day i.

Historic data can be used to estimate ET and P_e and a projected irrigation schedule developed. The water balance can be updated from current climatic data also. Irrigations are scheduled when D_i reaches the threshold value defined by MAD.

☐ *Example 18.6*

Determine the date and amount of water to apply to a field having a soil water deficit of 50 mm at the end of the day on June 14. The soil is a loam, the effective root zone depth is 1.0 m, and MAD is 50 percent of the available water.

Solution.   For the ET and P_e values given, apply Eq. 18.11 to calculate D_i. For example $D_{15} = 50 + 7 - 0 = 57$, $D_{16} = 57 + 6 - 5 = 58$, etc.

Date (June)	ET (mm)	P_e (mm)	D_i (mm)
15	7	0	57
16	6	5	58
17	8	0	66
18	7	0	73
19	8	6	75
20	9	0	84
21	8	0	92
22	10	0	102

From Table 18.2, the available water is 170 mm/m, or 170 mm for the 1-m root zone. MAD = $170 \times 0.5 = 85$ mm. Irrigation should begin June 21, with a net application of 84 mm.   ☐

In some cases irrigation districts, government agencies, or irrigation consultants provide irrigation water management services to farm operators. These services can provide the irrigation manager with recommendations for effective management of irrigation activities. Evapotranspiration rates also may be available from government agencies. Computer programs may be used to facilitate data acquisition and handling. Programs are available that collect current climatic data or use historic data to calculate evapotranspiration, compute the soil water balance, and project when to irrigate (Appendix I).

IRRIGATION METHODS

The methods of applying water may be classified as subirrigation, surface irrigation, sprinkler irrigation, and microirrigation.

18.9   Subirrigation

In special situations water may be applied below the soil surface by developing or maintaining a water table that allows water to move up through the root zone by capillary action. This is essentially the same practice as controlled drainage discussed in Chapters 13 and 14. Controlled drainage becomes subirrigation if water must be supplied to maintain the desired water level. Water may be introduced into the soil profile through open ditches, mole drains, or pipe drains. The open ditch method is most widely used. In some river valleys and near lakes, subirrigation is a natural process. Water table maintenance is suitable where the soil in the plant root zone is quite permeable and there is either a continuous impermeable layer or a natural water table below the root zone. Since subirrigation allows no opportunity for leaching and establishes an upward movement of water, salt accumulation is a hazard; thus the salt content of the water should be low.

18.10 Surface Irrigation

By far the most common method of applying irrigation water, especially in arid regions, is by flooding the surface. Surface methods include wild flooding, where the flow of water is essentially uncontrolled, and surface application, where flow is controlled by furrows, corrugations, border dikes, contour dikes, or basins. Except in the case of wild flooding, the land should be carefully prepared before irrigation water is applied. To conserve water, the rate of water application should be carefully controlled and the land properly graded (Chapter 19).

18.11 Sprinkler Irrigation

Lightweight portable pipes with slip joint connections were common for water distribution; however, in view of the high labor cost in moving these systems, such applications are becoming more limited to high-value crops. Mechanical-move systems are now widely accepted. These may be either intermittent or continuous. Solid-set and permanent systems are suitable for intensively cultivated areas growing a high-income crop, such as flowers, fruits, and vegetables. Sprinkler irrigation systems provide reasonably uniform application of water. On coarse-textured soils, water application efficiency may be twice as high as with surface irrigation.

Sprinkler irrigation can be used for temperature control. Irrigation water, especially if supplied from wells, is often considerably warmer than the soil and air near the surface under frost conditions. The heat of fusion released by water freezing on plant parts keeps the temperature from falling below 0°C. Conversely, irrigation may be used for cooling, particularly when germination occurs under high temperatures, or to delay premature blossoming of fruit trees when warm weather occurs before the frost danger has past. The cooling effect of evaporation lowers the temperature of plant parts (Chapter 20).

18.12 Microirrigation

Increasing use is being made of microirrigation (trickle or drip) systems that apply water at very low rates, often to individual plants. Such rates are achieved through the use of specially designed emitters or porous tubes. A typical emitter might apply water at from 2 to 10 L/h and is usually installed on or just below the soil surface. Perforated or porous tubes apply 1 to 5 L/min per 100 m of tube and are often installed 0.1 to 0.3 m below the soil surface. These systems provide an opportunity for efficient use of water because of minimum evaporation losses and because irrigation is limited to the root zone. Because of their high cost, their use is generally limited to high-value crops. Since the distribution pipes are usually at or near the surface, operation of field equipment is difficult. Both sprinkler and microirrigation systems are well adapted to application of agricultural chemicals, such as fertilizers and pesticides, with the irrigation water (Chapter 21).

18.13 Comparison of Irrigation Methods

Table 18.4 compares different types of irrigation systems in relation to various site and situation factors.

Table 18.4 Comparison of Irrigation Systems in Relation to Site and Situation Factors

Site and Situation Factors	Improved Surface Systems		Sprinkler Systems			Microirrigation Systems
	Redesigned Surface Systems	Level Basins	Intermittent Mechanical-Move	Continuous Mechanical-Move	Solid-Set and Permanent	Emitters and Porous Tubes
Infiltration rate	Moderate to low	Moderate	All	Medium to high	All	All
Topography	Moderate slopes	Small slopes	Level to rolling	Level to rolling	Level to rolling	All
Crops	All	All	Generally shorter crops	All but trees and vineyards	All	High value required
Water supply	Large streams	Very large streams	Small streams nearly continuous	Small streams nearly continuous	Small streams	Small streams, continuous and clean
Water quality	All but very high salts	All	Salty water may harm plants	Salty water may harm plants	Salty water may harm plants	All—can potentially use high salt waters
Efficiency	Average 60–70%	Average 80%	Average 70–80%	Average 80%	Average 70–80%	Average 80–90%
Labor requirement	High, training required	Low, some training	Moderate, some training	Low, some training	Low to seasonal high, little training	Low to high, some training
Capital requirement	Low to moderate	Moderate	Moderate	Moderate	High	High
Energy requirement	Low	Low	Moderate to high	Moderate to high	Moderate	Low to moderate
Management skill	Moderate	Moderate	Moderate	Moderate to high	Moderate	High
Machinery operations	Medium to long fields	Short fields	Medium field length, small interference	Some interference circular fields	Some interference	May have considerable interference
Duration of use	Short to long	Long	Short to medium	Short to medium	Long term	Long term, but durability unknown
Weather	All	All	Poor in windy conditions	Better in windy conditions than other sprinklers	Windy conditions reduce performance, good for cooling	All
Chemical application	Fair	Good	Good	Good	Good	Very good

Source: Fangmeier and Biggs (1986).

Efficient surface irrigation requires grading of the land surface to control the flow of water. The extent of grading required depends on the topography. In some soil and topographic situations, the presence of unproductive subsoils may make grading for surface irrigation unfeasible. The utilization of level basins where large streams of water are available generally provides high irrigation efficiencies.

Sprinkler irrigation is particularly adaptable to hilly land where grading for surface irrigation is not feasible. It is appropriate for most circumstances where the infiltration rate exceeds the rate of water application. With sprinkler irrigation, the rate of water application can be easily controlled. Sprinkler irrigation systems usually have a relatively high cost of installation. With mechanical-move systems labor can be substantially reduced. In some cases disease problems have resulted from moistened foliage. Evaporation losses with sprinkler irrigation are not excessively high even in arid regions. A well-designed sprinkler irrigation system can provide a high efficiency of water application.

A well-designed microirrigation system can provide a high efficiency of water application. It is especially well suited to tree fruit and high-value crops. Water must be clean and uncontaminated, usually achieved by a filtration system. Microirrigation lends itself well to automation and has a low labor requirement. Since microirrigation systems usually operate at low pressure, energy requirements are generally lower than with sprinkler systems. Some low-pressure sprinklers operate at pressures comparable to those of microirrigation systems.

REFERENCES

Ayers, R. S., and D. W. Westcot (1985). *Water Quality for Agriculture*. Irrig. and Drain. Paper No. 29, Rev. 1. FAO, Rome, Italy.

Burman, R. D., P. R. Nixon, J. L. Wright, and W. O. Pruitt (1983). "Water Requirements." In *Design and Operation of Farm Irrigation Systems*, M. E. Jensen, (ed.). ASAE, St. Joseph, MI.

Erie, L. J., O. F. French, D. A. Bucks, and K. Harris (1982). *Consumptive Use of Water by Major Crops in the Southwestern United States*. USDA–ARS Cons. Res. Rep. No. 29. GPO, Washington, DC.

Fangmeier, D. D., and E. N. Biggs (1986). *Alternative Irrigation Systems*. Rep. 8555. Cooperative Extension Serv. University of Arizona, Tucson, AZ.

Hansen, V. E., O. W. Israelsen, and G. E. Stringham (1980). *Irrigation Principles and Practices*, 4th ed. Wiley, New York.

Hoffman, G. J., T. A. Howell, and K. H. Solomon (1990). *Management of Farm Irrigation Systems* (Monograph). ASAE, St. Joseph, MI.

Jensen, M. E. (1973). *Consumptive Use of Water and Irrigation Water Requirements*. ASCE, New York.

———(ed.) (1983). *Design and Operation of Farm Irrigation Systems*. Monograph No. 3. ASAE, St. Joseph, MI.

Jensen, M. E., R. D. Burman, and R. G. Allen (eds.) (1990). *Evapotranspiration and Irrigation Water Requirements*. ASCE, New York.

Tanji, K. K. (ed.) (1990). *Agricultural Salinity Assessment and Management*. ASCE, New York.

U.S. Soil Conservation Service (SCS) (1970). *Irrigation Water Requirements*. Tech. Release No. 21. Washington, DC.

PROBLEMS

18.1   Determine the evapotranspiration and irrigation requirement for small grain where the monthly evapotranspiration coefficient $k = 0.65$ in the Blaney-Criddle formula. The average monthly temperatures for the 3-month growing season are 21.6, 23.4, and 24.6°C, respectively. The percentages of daytime hours for the same period are 8.7, 9.3, and 9.5, and the average rainfall is 33, 49, and 54 mm (1.3, 1.9, and 2.1 in.), respectively. Assume that the water application efficiency is 60 percent.

18.2   A 75-mm (3-in.) application of water measured at the pump increased the average water content of the top 0.6 m (2 ft) of soil from 18 to 23 percent (dry weight basis). If the average dry density of the soil is 1200 kg/m³ (75 pcf), what is the water application efficiency?

18.3   A flow of 5 m³/s (177 cfs) is diverted from a river into a canal. Of this amount 4 m³/s (140 cfs) is delivered to farm land. The surface runoff from the irrigated area averages 0.7 m³/s (25 cfs) and the contribution to ground water is 0.4 m³/s (14.1 cfs). What is the water-conveyance efficiency? What is the water-application efficiency?

18.4   Determine the water-application efficiency, distribution uniformity, and uniformity coefficient if a stream of 85 L/s (3 cfs) was delivered to the field for 2 h, runoff averaged 42 L/s (1.5 cfs) for 1 h, and depth of penetration of the water varied linearly from 1.6 m (5.2 ft) at the upper end to 1.0 m (3.3 ft) at the lower end of the field. The root zone depth is 1.6 m (5.2 ft).

18.5   Determine the leaching requirement and depth of water application to an alfalfa field if a 10 percent yield reduction is acceptable, the salinity of the irrigation water is 1.2 dS/m, 15 days elapsed since the last irrigation, and the average evapotranspiration rate of alfalfa is 9 mm/day. Assume no runoff from the irrigation.

18.6   Determine the irrigation requirement for furrow-irrigated cotton using the evapotranspiration data from Fig. 18.1b. Assume the only rainfall is 50 mm during the month of August. The irrigation water has an ECw = 1.4 dS/m, and 100 mm is the normal application depth.

18.7   Determine the date and amount of the next irrigation for cotton if the last irrigation was July 31, evapotranspiration is 7.2 mm/day, effective rainfall is 20 mm on August 5 and 15 mm on August 12, soil is a clay loam 1 m deep, and MAD is 65 percent of the available water.

CHAPTER 19

Surface Irrigation

Surface irrigation is the predominant method of irrigation in the United States and in most other countries with large irrigation areas. A 1989 U.S. irrigation survey ("1989 Irrigation Survey," 1990) reported that 58 percent of the irrigation was accomplished with surface methods. In the western states, where this percentage is higher, the major water supply is surface runoff, usually stored in reservoirs. Since this water must be conveyed for considerable distances over rough terrain, conveyance canals and control structures are key parts of most irrigation systems in arid regions. The hydraulic principles involved in the design of control structures are presented in Chapter 9 and the design of canals in Chapter 13. Ground water also provides an important source of water for surface irrigation (Chapter 17).

DISTRIBUTION OF WATER ON THE FARM

The farm water supply is normally delivered either from surface storage by conveyance ditches or from irrigation wells. Sometimes surface storage and underground supplies are combined to provide an adequate water supply at the farm.

19.1 Surface Ditches

A system of open ditches often distributes the water from the source on the farm to the field as shown in Fig. 19.1. These ditch systems should also be carefully designed as discussed in Chapter 13 so as to provide adequate head (elevation) and capacity to supply water at all areas to be irrigated. The amount of land that can be irrigated is often limited by the quantity of water available as well as by the design and location of the ditch system. Where irrigation is used as an occasional supplement to rainfall, these ditches may be temporary. To minimize water losses there is an increasing tendency to line ditches with impermeable materials. This practice is particularly applicable in the more arid regions, where irrigation water supplies are limited and crop needs are dependent largely on irrigation water.

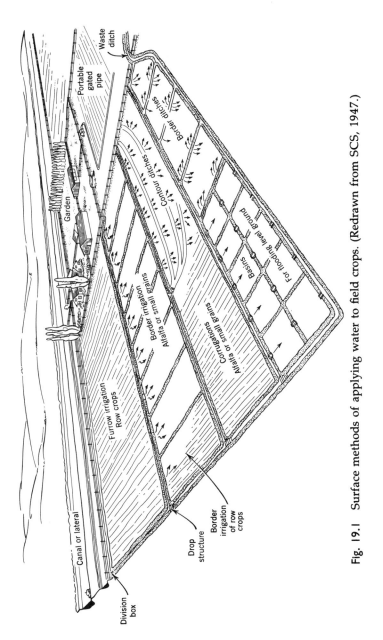

Fig. 19.1 Surface methods of applying water to field crops. (Redrawn from SCS, 1947.)

19.2 Devices to Control Water Flow

Control structures, such as illustrated in Fig. 19.2, are essential in open ditch systems to (1) divide the flow into two or more ditches, (2) lower the water elevation without erosion, and (3) raise the water level in the ditch so that it will have adequate head for removal. Various devices are used to divert water from the irrigation ditch and to control its flow to the appropriate basin, furrow, or border. Valves may be installed in the side or bottom of the ditch during construction. Other devices shown in Fig. 19.3 are spiles, gate takeouts for border irrigation, and siphon tubes. These siphons, usually aluminum or plastic, carry the water from the ditch to the surface of the field and have the advantage of metering the quantity of water applied. Figure 19.4 gives the rate of flow that can be expected from siphons of various diameters and head differences between the inlet and outlet.

19.3 Underground Pipe

Since open channel distribution systems provide continuing problems of maintenance, constitute an obstruction to farming operations, and provide a water surface subject to evaporation losses, underground pipe distribution systems shown in Fig. 19.5a are becoming increasingly popular. In these systems water flows from the distribution pipe upward through riser pipes and irrigation valves to appropriate basins, borders, or furrows. In Fig. 19.5b a multiple-outlet riser controls the distribution of water to several furrows. Alfalfa valves regulate the flow to a header ditch from which water can be distributed to the field.

19.4 Portable Pipe

Surface irrigation with portable pipe or large-diameter plastic tubing may be advantageous particularly in circumstances where water is applied infrequently. Lightweight gated pipe (Fig. 19.5c) provides a convenient and portable method of applying water in furrow irrigation. Portable flumes are sometimes used, particularly with high-value crops, where they may be removed from the fields during certain field operations.

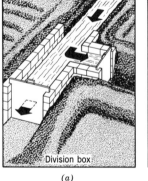  

Division box

Drop

Canvas, plastic, or butyl check

(a) (b) (c)

Fig. 19.2 Devices to control water flow in irrigation ditches. (Adapted from SCS and USBR, 1959.)

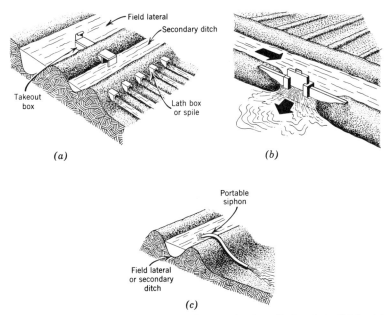

Fig. 19.3   Devices for distribution of water from irrigation ditches into fields. (*a*) Spile or lath box. (*b*) Border takeout. (*c*) Siphons. (Adapted from SCS and USBR, 1959.)

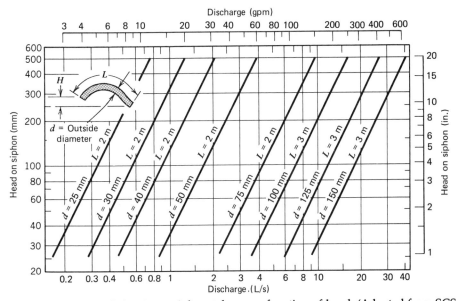

Fig. 19.4   Discharge of aluminum siphon tubes as a function of head. (Adapted from SCS, 1984.)

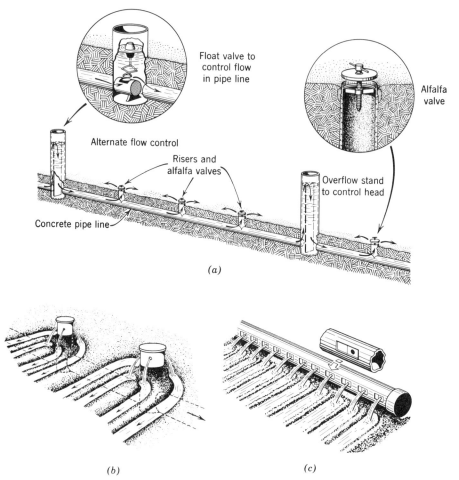

Fig. 19.5 Methods of distribution of water from (*a*) low-pressure underground pipe, (*b*) multiple-outlet risers, and (*c*) portable gated pipe. (Adapted from SCS and USBR, 1959.)

APPLICATION OF WATER

The various surface methods of applying water to field crops are illustrated in Fig. 19.1.

19.5 Flooding

Ordinary flooding is the application of irrigation water from field ditches that may be nearly on the contour or up and down the slope. After the water leaves the ditches, no attempt is made to control the flow by means of levees or other methods of restricting water movement. For this reason ordinary flooding is frequently referred to as "wild flooding." Although the initial cost for land preparation is low, labor requirements are usually high and the efficiency of water application is generally low. Ordinary flooding is most suitable for close-growing crops, particularly where slopes are steep. Contour ditches are usually spaced 15 to

45 m apart, depending on the slope, texture and depth of the soil, size of the stream, and crop to be grown. This method may be used on rolling land where borders, basins, and furrows are not feasible and adequate water supply is not a problem.

19.6 Graded Borders

The graded border method of flooding consists of dividing the field into a series of strips separated by low ridges. Normally, the direction of the strip is in the direction of greatest slope, but in some cases, the borders are placed nearly on the contour. Slopes less than 0.5 percent are best suited for border irrigation; however, slopes up to 4 percent may be used if erosion can be controlled. The strips usually vary from 10 to 20 m in width and are 100 to 400 m in length. Ridges between borders should be sufficiently high to prevent overtopping during irrigation. To prevent water from concentrating on either side of the border, the land should be level perpendicular to the flow. Where row crops are grown in the border strip, furrows confine the flow and eliminate this difficulty.

19.7 Level Borders

The layout of level borders is similar to that described for graded borders, except that the surface is leveled within the area to the irrigated. These areas may be long and narrow or they may be nearly square (often called basins). Modern laser leveling techniques make possible the preparation of smooth level surfaces required for this method of irrigation. Where relatively large rates of flow are available, the field can be quickly covered resulting in high application irrigation efficiencies. Level borders also lend themselves well to preplant irrigation using stream flow diverted during periods of high runoff. In orchard irrigation small leveled basins may include as few as one tree.

19.8 Furrows

Although in the flooding methods water covers the entire surface, irrigation by furrows submerges only from one fifth to one half the surface, resulting in less evaporation and less puddling of the soil, and permitting cultivation sooner after irrigation. Furrows vary in size and are up and down the slope or on the contour. Small, shallow furrows, called corrugations, are particularly suitable for relatively irregular topography and close-growing crops, such as meadow and small grains. Furrows 80 to 200 mm deep are especially suited to row crops since the furrow can be constructed with normal tillage. Contour furrow irrigation may be practiced on slopes up to 12 percent, depending on the crop, the erodibility of the soil, and the size of the irrigation stream.

Furrows may have significant runoff if a constant inflow rate is maintained throughout the application interval. To reduce runoff, the inflow stream can be *cutback* (reduced) after water reaches the end of the field and the application efficiency can be greatly improved. This procedure increases the labor requirement because cutback must be made during the irrigation. Unless the supply flow can be decreased, additional labor is required to use the flow remaining from the cutback stream. An alternative to cutback is installation of a runoff reuse system.

In cases where water advance is too slow, *surge irrigation* may be advantageous. Here intermittent rather than continuous streams are delivered to furrows. Between surges most or all of the water infiltrates. The next surge advances faster across the wetted portion because the infiltration rate is lower and roughness may be reduced. Fast advance improves uniformity, and, with proper design and management, runoff and deep percolation can be reduced, which increases application efficiencies. Special surge valves are available to facilitate surge irrigation. These valves alternate the flow between two sets of furrows to form the surges. The water is delivered to the furrows through gated pipe connected to the surge valves.

DESIGN AND EVALUATION

Recognition and understanding of the variables involved in the hydraulics of surface irrigation are essential to effective design. Hansen et al. (1980) have listed, with reference to Fig. 19.6, the pertinent variables as (1) size of stream, (2) rate of advance, (3) length of run and time involved, (4) depth of flow, (5) intake rate, (6) slope of land surface, (7) surface roughness, (8) erosion hazard, (9) shape of flow channel, (10) depth of water to be applied, and (11) fluid characteristics.

Since the hydraulics of surface irrigation are complex, empirical procedures are often employed in design of surface irrigation systems. In designing a system it is recommended that local practices be explored and locally available data be considered. For example, extension services, experiment stations, and the SCS have prepared "irrigation guides" suggesting design procedures for many states. Selection of design values, such as border width, depends on judgment and experience in managing water under specific soil, slope, and crop conditions.

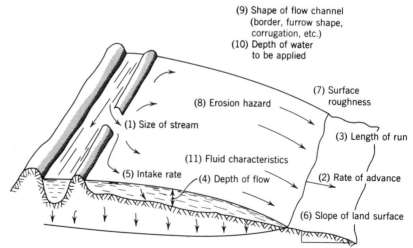

Fig. 19.6 Schematic view of flow in surface irrigation indicating the variables involved. (*Source:* Hansen et al., 1980.)

19.9 Graded Border Design

Probably the most commonly used reference in the United States for the design of graded and level border irrigation systems is Chapter 4 of the *National Engineering Handbook on Irrigation* published by the SCS (1974) (also see Hart et al., 1983). This publication presents the derivation of equations relating the infiltration, roughness, irrigation efficiency, length of run, time of application, and stream size.

The length of run and slope are generally defined by the field situation. The field efficiency will depend on management practices of the irrigator and also on the preparation of the field for irrigation. On gently sloping, well-leveled, uniformly graded fields an efficiency of 60 to 75 percent is usually feasible. SCS (1974) gives suggested design efficiencies for various situations.

Manning's equation is used in design with an n value of 0.04 for a smooth, bare soil surface. The value 0.10 is usually accepted for small grain and similar crops if the rows run lengthwise with the border strip. A value of 0.15 is suggested for alfalfa and broadcast small grain; dense sod and small grain crops drilled across the border strip have an n of about 0.25.

The brief design procedure summarized below for graded borders with free outflow is based on SCS (1974) procedures, which also are reported by Hart et al. (1983), James (1988), and Cuenca (1989). The design assumes nearly equal infiltration opportunity times over the border length, and the entire border will infiltrate the desired net application depth.

Infiltration is probably the most important design variable for surface irrigation and the most difficult to determine accurately. Soils are classified into intake families as shown in Fig. 19.7 for borders (SCS, 1974). The equation for these families is

$$Z = aT^b + 7 \tag{19.1}$$

where Z = infiltrated volume per unit area in mm,
               T = time water is on the soil surface in min,
   a and b = coefficients.

Values of a and b are found in Table 19.1 for furrows. Though not identical to the families for borders, they are assumed to be interchangeable within the normal accuracy of infiltration measurements. The intake family can be determined by conducting field infiltration tests and plotting the results in Fig. 19.7 to select the closest family, or Eq. 19.1 can be fitted to the data directly. For a desired net depth of application Z_n, Eq. 19.1 can be solved for the net intake opportunity time letting $T = T_n$.

The inflow rate is determined for a unit width of border, m³/s per m or m²/s, from

$$q = \frac{0.00167(Z_n)(L)}{(T_n - T_L)E} \tag{19.2}$$

where q = unit inflow in m³/s per m of width or m²/s,
   Z_n = desired net application depth in mm,
    L = border length in m,
   T_n = time required to infiltrate Z_n in min,

Table 19.1 SCS Furrow Intake Families and Advance Coefficients

Intake Family	a	b	f	g
0.05	0.5334	0.618	7.16	1.088×10^{-4}
0.10	0.6198	0.661	7.25	1.251×10^{-4}
0.15	0.7110	0.683	7.34	1.414×10^{-4}
0.20	0.7772	0.699	7.43	1.578×10^{-4}
0.25	0.8534	0.711	7.52	1.741×10^{-4}
0.30	0.9246	0.720	7.61	1.904×10^{-4}
0.35	0.9957	0.729	7.70	2.067×10^{-4}
0.40	1.064	0.736	7.79	2.230×10^{-4}
0.45	1.130	0.742	7.88	2.393×10^{-4}
0.50	1.196	0.748	7.97	2.556×10^{-4}
0.60	1.321	0.757	8.15	2.883×10^{-4}
0.70	1.443	0.766	8.33	3.209×10^{-4}
0.80	1.560	0.773	8.50	3.535×10^{-4}
0.90	1.674	0.779	8.68	3.862×10^{-4}
1.00	1.786	0.785	8.86	4.188×10^{-4}
1.50	2.284	0.799	9.76	5.819×10^{-4}
2.00	2.753	0.808	10.65	7.451×10^{-4}

Source: Hart et al. (1983).

T_L = lag time that water remains on the head end of the border after inflow stops in min,

E = application efficiency (ratio of Z_n to gross application depth) in percent.

Lag time is a function of several variables including q. Therefore, a trial-and-error solution is required for slopes less than 0.4 percent. Lag time can be obtained from

$$T_L = \frac{n^{1.2}q^{0.2}}{120\left[S + \dfrac{0.0094nq^{0.175}}{(T_n)^{0.88} S^{0.5}}\right]^{1.6}}$$

(19.3)

where n = Manning roughness coefficient,
S = the border slope in m/m.

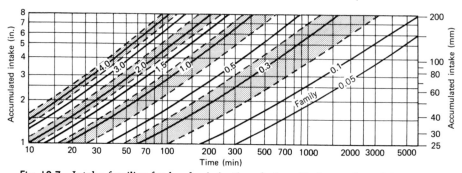

Fig. 19.7 Intake families for border irrigation design. (Redrawn from SCS, 1974.)

For slopes greater than 0.4 percent, T_L is small and may be assumed equal to zero. Inflow time is determined from

$$T_a = T_n - T_L \tag{19.4}$$

where T_a is the time a stream q is delivered to the border in min.

Some limits are necessary to prevent erosion, ensure minimal flow depths, and avoid flow depths that exceed border ridge heights. The maximum nonerosive stream for non-sod-forming crops, such as small grains and alfalfa, is

$$q_{max} = (1.765 \times 10^{-4}) S^{-0.75} \tag{19.5}$$

where q_{max} is the maximum inflow rate in m²/s, and S is the border slope in m/m. For crops that form dense sod, Eq. 19.5 can be increased by a factor of 2.

Flow depths at the head of the border should be 25 percent less than the border ridge heights. The flow depth for high gradient borders with slopes greater than 0.004 m/m is given by

$$d_n = 1000 q^{0.6} n^{0.6} S^{-0.3} \tag{19.6}$$

where d_n is the normal flow depth in mm. For low gradient borders with slopes less than 0.004 m/m, the flow may not reach normal depth but may be computed from

$$d = 2454 T_L^{0.1875} q^{0.5625} n^{0.375} \tag{19.7}$$

where d is the flow depth at the border inlet in mm.

Minimum flow depths are needed to adequately spread water over the entire soil surface. Thus a minimum inflow is required, which can be computed from

$$q_{min} = (5.95 \times 10^{-6}) L S^{0.5} / n \tag{19.8}$$

where q_{min} is the minimum inflow rate in m²/s.

Border lengths are limited by stream size, intake rate, and slope. When erosion limits stream size, q_{max} can be substituted in Eq. 19.2 for q and then solved for the maximum length L_{max}. Further guidelines for length, efficiencies, slopes, and intake rates can be obtained from Booher (1974), Hart et al. (1983), James (1988), Cuenca (1989), and SCS (1974).

□ Example 19.1

Determine the required time of application, stream size per unit border width, depth of flow, maximum stream size, and minimum stream size to begin design of a border system. $L = 200$ m (655 ft), $S = 0.002$ m/m, $Z_n = 75$ mm (3 in.), intake family = 0.5, $n = 0.15$, border ridge height = 150 mm (6 in.), and application efficiency = 65 percent.

Solution.
(1) Find the coefficients $a = 1.196$ and $b = 0.748$ from Table 19.1. Solve Eq. 19.1 or read from Fig. 19.7.

$$T_n = \left(\frac{75 - 7.0}{1.196}\right)^{1/0.748} = 222 \text{ min}$$

(2) Solve Eq. 19.2 assuming $T_L = 0$.

$$q = \frac{0.00167(75)(200)}{(222 - 0)(65)} = 0.00174 \text{ m}^2/\text{s} \ (0.0186 \text{ ft}^2/\text{s})$$

(3) Solve Eq. 19.3 for T_L since $S < 0.004$.

$$T_L = \frac{(0.15)^{1.2} \ (0.00174)^{0.2}}{120 \left[0.002 + \dfrac{(0.0094)(0.15)(0.00174)^{0.175}}{(222)^{0.88} \ (0.002)^{0.5}} \right]^{1.6}} = 5 \text{ min}$$

(4) Solve Eq. 19.2 for new q with $T_L = 5$ min.

$$q = \frac{(0.00167)(75)(200)}{(222 - 5) \ (65)} = 0.00178 \text{ m}^2/\text{s} \ (0.019 \text{ ft}^2/\text{s})$$

This differs from the previously computed value by less than 5 percent so continued iteration is not necessary.

(5) Determine the time of application from Eq. 19.4.

$$T_a = (222 - 5) = 217 \text{ min}$$

(6) Depth of flow is obtained from Eq. 19.7.

$$d = 2454 \ (5)^{0.1875} \ (0.00178)^{0.5625} \ (0.15)^{0.375} = 46 \text{ mm} \ (1.8 \text{ in.})$$

Since $d < (0.75)(150 \text{ mm})$, the flow depth is satisfactory.

(7) Find maximum flow rate from Eq. 19.5.

$$q_{max} = 1.765 \times 10^{-4}/(0.002)^{0.75} = 0.0187 \text{ m}^2/\text{s} \ (0.20 \text{ ft}^2/\text{s})$$

Thus, q is less than q_{max} and the flow rate is satisfactory.

(8) Obtain the minimum flow rate from Eq. 19.8.

$$q_{min} = (5.95 \times 10^{-6})(200)(0.002)^{0.5}/0.15 = 0.00035 \text{ m}^2/\text{s} \ (0.0037 \text{ ft}^2/\text{s})$$

The flow rate q is greater than the minimum allowed. Thus the design variables have been determined and all limits satisfied with $q = 0.00178$ m³/s per m (0.019 ft³/s per ft) and application time of 217 min. $\square$

19.10 Furrow Irrigation Design

This section summarizes the design equations for graded furrows developed by the SCS (1984) and as presented by Hart et al. (1983). It is assumed that (1) the entire furrow will infiltrate the desired net application depth, (2) recession lag time can be neglected because the slope exceeds 0.05 percent, and (3) outflow is not restricted.

The length of furrows depends on infiltration capacity, slope, and size of the stream. Excessively long furrows result in deep percolation losses and erosion in the upper ends. Suggested lengths of furrows for various slopes and soil textures are given in Table 19.2.

Furrow slopes of 2 percent or less are recommended to avoid excessive erosion. In humid regions slopes must be smaller, and maximum slopes can be estimated from

$$S_{max} < 67/(P_{30})^{1.3} \tag{19.9}$$

where P_{30} is the 30-min rainfall in mm for a 2-year return period (Chap. 2) and S_{max} is the maximum allowable furrow slope in percent. The maximum allowable stream size to avoid excessive erosion can be obtained from

$$Q_{max} < 0.60/S \tag{19.10}$$

where Q_{max} is the maximum furrow inflow in L/s, and S is the furrow slope in percent. Inflows into furrows with small slopes are often limited by capacity and inflow rates are lower than computed by Eq. 19.10. More extensive design recommendations may be obtained from SCS (1984), Booher (1974), or Cuenca (1989).

With furrow irrigation only a portion of the soil surface is wetted and the quantity of water infiltrated is influenced by furrow shape and flow depth. The furrow wetted perimeter is used to define the wetted area; however, since the intake has a horizontal component, adjustments are required and an empirical relationship for wetted perimeter (called the adjusted wetted perimeter) based on typical furrow shapes is

$$p = 0.265 \, [Qn/S^{0.5}]^{0.425} + 0.227 \tag{19.11}$$

where p = the adjusted wetted perimeter in m,
Q = the furrow inflow rate in L/s,
n = the Manning roughness coefficient,
S = furrow slope in m/m.

Table 19.2 Suggested Maximum Lengths of Cultivated Furrows

Furrow Slope (%)	Average Depth of Water Applied (mm)											
	75	150	225	300	50	100	150	200	50	75	100	125
	Clays				Loams				Sands			
	Length (m)											
0.05	300	400	400	400	120	270	400	400	60	90	150	190
0.1	340	440	470	500	180	340	440	470	90	120	190	220
0.2	370	470	530	620	220	370	470	530	120	190	250	300
0.3	400	500	620	800	280	400	500	600	150	220	280	400
0.5	400	500	560	750	280	370	470	530	120	190	250	300
1.0	280	400	500	600	250	300	370	470	90	150	220	250
1.5	250	340	430	500	220	280	340	400	80	120	190	220
2.0	220	270	340	400	180	250	300	340	60	90	150	190

Source: Booher (1974).

The value of p cannot exceed the furrow spacing W.

The time for water to advance along the furrow, based on a regression analysis of field measurements, is

$$T_T = Xe^B/f = X(2.718)^B/f \qquad (19.12)$$

where $T_T =$ the advance time to point X in min,
$X =$ distance along the furrow from the upper end in m,
$B = gX/(Q\ S^{0.5})$,
$S =$ slope in m/m,
$f, g =$ furrow advance coefficients from Table 19.1.

The infiltration opportunity time along the furrow is described by

$$T_o = T_a - T_T + T_r \qquad (19.13)$$

where $T_o =$ intake opportunity time at point X in min,
$T_a =$ time of application, which is constant, in min,
$T_T =$ advance time to point X in min,
$T_r =$ recession time in min (assumed zero for sloping furrows with free outflow).

Combining Eqs. 19.12 and 19.13 assuming $T_r = 0$ yields

$$T_o = T_a - (X/f)\ (2.718)^B \qquad (19.14)$$

Since the lower end of the field is to be adequately irrigated, the application time equals

$$T_a = T_T + T_n \qquad (19.15)$$

where T_n is the time to infiltrate the desired net application depth in min. The average intake opportunity time is obtained by integrating Eq. 19.14 from 0 to X and dividing by X.

$$T_{0-x} = T_a - \left[\frac{0.0929\ [(B-1)(2.718)^B + 1]}{fX\ (0.305B/X)^2} \right] \qquad (19.16)$$

where T_{0-x} is the average intake opportunity time over the length X in min. Setting X equal to L, the field length, in Eq. 19.16 yields the average intake opportunity time for the entire furrow T_{0-L}. The gross application depth is

$$Z_g = 60QT_a/WL \qquad (19.17)$$

where Z_g is the gross application depth in mm and W is the furrow spacing in m. The quantity of water infiltrated is obtained by multiplying Z from Eq. 19.1 by the wetted perimeter from Eq. 19.11. This is expressed as an average equivalent depth over the furrow spacing:

$$Z_a = [a(T_{0-L})^b + 7]\ p/W \qquad (19.18)$$

where Z_a is the average equivalent depth infiltrated in mm. The time to infiltrate the desired net depth T_n is obtained by rearranging Eq. 19.18 and substituting Z_n, the desired net application depth for the field, for Z_a and T_n for T_{0-L}.

$$T_n = \left[\left(\frac{Z_n W}{p} - 7\right)/a\right]^{1/b} \tag{19.19}$$

where Z_n is in mm. Surface runoff is the difference between Z_g and Z_a or

$$RO = Z_g - Z_a \tag{19.20}$$

where RO is the average equivalent runoff depth in mm. Deep percolation, which is the infiltrated depth that exceeds the desired net application depth, is computed from

$$DP = Z_a - Z_n \tag{19.21}$$

where DP is the average infiltrated depth in excess of the desired net application depth in mm. The application efficiency is

$$E = 100 Z_n / Z_g \tag{19.22}$$

The procedure for design of graded furrows is an iterative process where Q (furrow inflow rate) is assumed and application efficiency is obtained. Then Q is adjusted until an acceptable application efficiency is obtained. This process may be expedited by a computer program or by developing a series of design charts for the previous equations.

□ *Example 19.2*
___

Perform the preliminary design of a graded furrow system by determining the inflow time, deep percolation, and application efficiency. The field is in an arid region, length = 300 m (984 ft), $S = 0.003$ m/m, $n = 0.04$, intake family from field measurements = 0.4, furrow spacing $W = 0.8$ m (2.6 ft), and desired net application depth $Z_n = 75$ mm (3 in.).

Solution.
(1) Assume the inflow is 1 L/s and from Table 19.1 find $a = 1.064$, $b = 0.736$, $f = 7.79$, and $g = 2.23 \times 10^{-4}$. Check Q_{max} from Eq. 19.10.

$$Q_{max} = 0.6/0.3 = 2.0 \text{ L/s (32 gpm)} > 1.0 \text{ L/s (16 gpm)}$$

(2) Calculate the adjusted wetted perimeter from Eq. 19.11.

$$p = 0.265 [(1)(0.04)/(0.003)^{0.5}]^{0.425} + 0.227 = 0.46 \text{ m (1.51 ft)}$$

(3) Determine the advance time to the end of the furrow from Eq. 19.12 (with $X = 300$ m).

$$B = \frac{(2.23 \times 10^{-4})(300)}{(1)(0.003)^{0.5}} = 1.221$$

$$T_T = 300(2.718)^{1.221}/7.79 = 130 \text{ min}$$

(4) Determine time to infiltrate the desired net depth from Eq. 19.19.

$$T_n = \left[\frac{(75)(0.8)/(0.46) - 7}{1.064}\right]^{1/0.736} = 638 \text{ min}$$

(5) Calculate the time of application from Eq. 19.15.

$$T_a = 130 + 638 = 768 \text{ min}$$

(6) Determine gross application depth from Eq. 19.17.

$$Z_g = 60(1)(768)/(0.8)(300) = 192 \text{ mm} \ (7.56 \text{ in.})$$

(7) Calculate the average intake opportunity time for the entire field from Eq. 19.16.

$$T_{0-L} = 768 - \frac{0.0929\ [(1.221 - 1)(2.718)^{1.221} + 1]}{(7.79)(300)[0.305(1.221)/300]^2} = 722 \text{ min}$$

(8) Determine the average infiltrated depth from Eq. 19.18.

$$Z_a = [(1.064)(722)^{0.736} + 7]\,\frac{0.46}{0.8} = 82 \text{ mm} \ (3.22 \text{ in.})$$

(9) The surface runoff is obtained from Eq. 19.20.

$$RO = 192 - 82 = 110 \text{ mm} \ (4.33 \text{ in.})$$

Deep percolation is calculated from Eq. 19.21.

$$DP = 82 - 75 = 7 \text{ mm} \ (0.28 \text{ in.})$$

Finally, the application efficiency is calculated from Eq. 19.22.

$$E = 100\,\frac{75}{192} = 39 \text{ percent}$$

Analysis of these results indicates the application efficiency is low because runoff is excessive but deep percolation is small. Runoff can be decreased by reducing Q; however, this may increase deep percolation. Therefore, a Q is needed that minimizes the sum of runoff and deep percolation. In this example, Q should be decreased and the process repeated. An alternative solution would be to design and install a system to collect and reuse the runoff.   □

19.11 Computer Models

Computer programs can be used to perform the previous calculations for both the border and furrow irrigation designs. An alternative is to solve the flow equations describing the flow of water over the soil surface, which are based on the principles of conservation of mass and Newton's second law. The resulting two partial differential equations, known as the Saint-Venant equations, are (Henderson, 1966; Strelkoff, 1969)

$$\frac{\partial Q}{\partial x} + \frac{\partial A}{\partial t} + I_x = 0 \qquad (19.23)$$

$$\frac{1}{g}\frac{\partial V}{\partial t} + \frac{V}{g}\frac{\partial V}{\partial x} + \frac{\partial y}{\partial x} = S_0 - S_f + \frac{I_x V}{2gA} \qquad (19.24)$$

where x = distance (L),
t = time (T),
Q = flow rate (L^3/T),
A = cross-sectional area of flow (L^2),
I_x = volume rate of infiltration per unit length of channel (L^2/T),
$V = Q/A$, the average velocity (L/T),
y = flow depth (L),
S_0 = channel bottom slope (L/L),
S_f = channel friction slope (L/L),
g = ratio of weight to mass (L/T^2).

Models are available that solve these equations (Strelkoff, 1990) (Appendix I). Models simulate an irrigation; however, users can conduct various simulations to compare, develop, or evaluate designs.

19.12 Evaluation of Existing Systems

Although the design of new irrigation systems constitutes an important engineering activity, it should be noted that even greater opportunities are often available in the evaluation and improvement of existing systems. Many of these older systems were designed before much was known about intake rates and water-holding capacities of soils, and when water supplies were not as limited as they are now becoming. Fields or portions of fields that do not receive enough water have limited production potential. Excessive irrigation not only wastes water, but leaches water-soluble chemicals and nutrients, and may cause drainage problems.

Evaluation of existing systems can be approached in a number of ways. A simple method of determining underirrigation is by use of the soil auger or tube sampler. Observation of the *opportunity time* for infiltration in various parts of the field may be helpful. More sophisticated methods involve application of such equipment as portable flumes, meters, and infiltrometers in carefully conducted diagnostic procedures, such as described by Merriam and Keller (1978).

REFERENCES

Booher, L. J. (1974). *Surface Irrigation*. Publ. No. 95. FAO, Rome, Italy.

Cuenca, R. H. (1989). *Irrigation System Design: An Engineering Approach*. Prentice-Hall, Englewood Cliffs, NJ.

Hansen, V. E., O. W. Israelsen, and G. E. Stringham (1980). *Irrigation Principles and Practices*, 4th ed. Wiley, New York.

Hart, W. E., H. G. Collins, G. Woodard, and A. S. Humphreys (1983). "Design and Operation of Gravity or Surface Systems." In *Design and Operation of Farm Irrigation Systems*, M. E. Jensen (ed.). Monograph No. 3. ASAE, St. Joseph, MI.

Henderson, F. M. (1966). *Open Channel Flow*. Macmillan, New York.

"1989 Irrigation Survey." (1990). *Irrig. J.* **40**(1), 34.

James, L. G. (1988). *Principles of Farm Irrigation System Design*. Wiley, New York.

Merriam, J. L., and J. Keller (1978). *Farm Irrigation System Evaluation: A Guide for Management*. Agr. and Irrig. Eng. Dept., Utah State University, Logan, UT.

Strelkoff, T. (1969). "One Dimensional Equations of Open Channel Flow." *Proc. ASCE J. Hydr. Div.* **95** (HY-3), 861–876.

———(1990). *SRFR: A Computer Program for Simulating Flow in Surface Irrigation—Furrows-Basins-Borders*. WCL Report No. 17. U.S. Water Cons. Lab., Phoenix, AZ.

U.S. Soil Conservation Service (SCS) (1947). *First Aid for the Irrigator*. USDA Misc. Publ. 624.

———(1957). "Instructions and Criteria for Preparation of Irrigation Guides." In *Engineering Handbook for Western States and Territories*, Sect. 15, Part I. Eng. and Watershed Planning Unit, Portland, OR.

———(1974). "Border Irrigation." In *National Engineering Handbook*, Sect. 15, Chap. 4. Washington, DC.

———(1984). "Furrow Irrigation." In *National Engineering Handbook*, 2nd ed., Sect. 15, Chap. 5. Washington, DC.

U.S. Soil Conservation Service (SCS) and U.S. Bureau of Reclamation (USBR) (1959). *Irrigation on Western Farms*. Agr. Inf. Bull. 199. Washington, DC.

PROBLEMS

19.1 Determine the time of application and stream size for a 10-m (33-ft)-wide border and the depth of flow for a 140-m (460-ft) border length with a slope of 0.2 percent. The field application efficiency is 70 percent, the depth of application is 50 mm (2 in.), the intake family is 1.0, the roughness coefficient is 0.04, and the border ridge height is 100 mm (4 in.).

19.2 Determine the time of application, stream size, and depth of flow if the slope in Problem 19.1 is decreased to 0.05 percent.

19.3 What stream size, time of application, and minimum ridge height are needed to irrigate a border 250 m (820 ft) long if intake family is 0.5, roughness is 0.15, slope is 0.1 percent, application efficiency is 60 percent, and net depth of application is 75 mm (3 in.)?

19.4 Determine the discharge of a 25-mm (1-in.)- and a 75-mm (3-in.)-diameter siphon tube for a head of 200 mm (7.9 in.) from the graph. Compare the relative flow using the theoretical flow rates, assuming the same friction loss in both tubes and the Manning equation applies.

19.5   If the flow into the border described in Problem 19.1 is discharged by 50-mm (2-in.)-diameter, 2-m (6.3-ft)-long siphon tubes under a head of 150 mm (5.9 in.), how many siphon tubes are required to irrigate the 10-m (33-ft)-wide border?

19.6   Repeat Example 19.2 with an inflow of 0.6 L/s (9.5 gpm).

19.7   Determine the inflow time, deep percolation, runoff, and application efficiency for furrows 250 m (820 ft) long; slope is 0.005 m/m, intake family is 0.6, row width is 1 m (40 in.), desired net depth of application is 100 mm (4 in.), roughness coefficient is 0.04, and inflow rate is 0.5 L/s (7.9 gpm).

CHAPTER 20

Sprinkler Irrigation

Sprinkler irrigation is a versatile means of applying water to any crop, soil, and topographic condition. It is popular because surface ditches and prior land preparation are not necessary and because pipes are easily transported and provide no obstruction to farm operations when irrigation is not needed. Sprinkling is suitable for sandy soils or any other soil and topographic condition where surface irrigation may be inefficient or expensive, or where erosion may be particularly hazardous. Low rates and amounts of water may be applied, such as are required for seed germination, frost protection, delay of fruit budding, and cooling of crops in hot weather. Fertilizers and soil amendments may be dissolved in the water and applied through the irrigation system. The major concerns of sprinkler systems are investment costs and labor requirements (Chapter 18).

20.1 Sprinkler Systems

In general, systems are described according to the method of moving the lateral lines, on which are attached various types of sprinklers. These systems are identified and compared in Table 20.1 and Fig. 20.1. During normal operation, laterals may be moved by hand or mechanically. The sprinkler system may cover only a small part of a field at a time or be a solid-set system in which sprinklers are placed over the entire field. With the solid-set system, all or some of the sprinklers may be operated at the same time. Most rotating sprinklers cover a full circle, but some may be set to operate for any portion of a circle (part circle). Perforated pipes are suitable for distributing water to small acreages of high-value crops where a rectangular pattern is desired.

As shown in Table 20.1, hand-move laterals have the lowest investment cost but the highest labor requirement. With giant sprinklers, the spacing can be increased, thereby reducing labor, but higher pressures are required, which increase the pumping cost. These sprinklers are pulled or transported from one location to another or moved continuously. Hand-move laterals with standard

Table 20.1   Comparison of Sprinkler Irrigation Equipment

Type of System	Relative Investment Cost[a]	Relative Labor Cost	Practical Hours of Operation per Day
Hand-move laterals (standard sprinklers)	0.4	5.0	16
Hand-move laterals (giant sprinklers)	0.5	4.0	12–16
End-pull laterals (tractor tow)	0.5	1.4	16
Boom-type sprinklers (trailer-mounted)	0.6	3.7	12–16
Side-roll laterals (powered-wheel move)	0.7	1.7	18–20
Self-propelled (center-pivot)	1.0	1.0	24
Solid set	3.0–5.0	1.0	24

[a]Based on a 65-ha field, 63 L/s from pump, and 80 percent application efficiency.
Source: Berge and Groskopp (1964).

sprinklers are most suitable for low-growing crops, and are impractical in tall corn because of adverse conditions for moving the pipe.

With the end-pull system shown in Fig. 20.1*a*, the lateral is moved to the next position by pulling the line with a tractor to the opposite side of the main. In moving the lateral, it is pulled at a slight diagonal so as to move the entire length down the field, one half the lateral spacing with each direction pulled. This system is suitable for low-growing crops and in places where adequate moving space is available. It is also practical for tree crops. Labor is greatly reduced (Table 20.1) compared with hand-move systems.

The rotating boom-type system (Fig. 20.1*b*) operates with one trailer unit per lateral at spacings up to 110 m. The trailer and boom unit is moved to the next position along the lateral with a tractor or winch. The lateral line is added or picked up as the trailer is moved progressively through the field away from or toward the main line. The boom-type unit will cover about the same area as a giant sprinkler, but the pressure required is not as high.

The side-roll lateral system (Fig. 20.1*c*) uses the irrigation pipe as the axle of large-diameter wheels that are spaced about 9 m apart. The lateral is moved to the next position by a small gasoline engine mounted at the midpoint of the line or by hand with a lever and ratchet. The side-roll lateral is limited to crops that will not interfere with the movement of the pipe. Unless a long flexible pipe is attached to the main, the lateral must be disconnected for each move. Labor requirements are about the same as for the end-pull system, but much less than for the hand-move systems. Sprinklers that stay in a vertical position regardless of the position of the wheels (Fig. 20.1*c*) eliminate the necessity of having the sprinkler on top as required with the fixed-position sprinklers.

The self-propelled center-pivot system (Fig. 20.1*d*) consists of a radial pipe line supported at a height of 2 to 4 m at intervals of 25 to 75 m. The radial line rotates slowly around a central pivot by either water pressure, electric motors, or oil hydraulic motors. The towers are supported by wheels or skids and are kept in

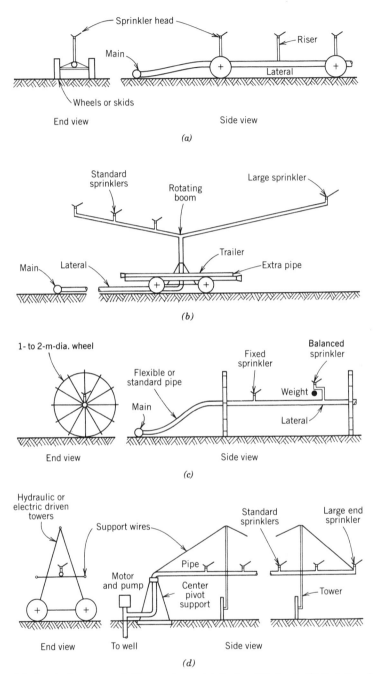

Fig. 20.1 Mechanical-move sprinkler systems. (*a*) End-pull lateral. (*b*) Rotating boom-type unit. (*c*) Side-roll lateral (hand- or mechanical-move). (*d*) Self-propelled (center-pivot) lateral.

alignment by switches or wires. The nozzles may increase in size and increase or decrease in spacing from the pivot to the end of the line, at which a large sprinkler may be placed to obtain the maximum diameter of coverage. The nozzles are selected to provide a uniform depth of application, varying from 13 to 100 mm per revolution. The depth of application is determined by the speed of rotation. The system is best suited to sandy soils, but it will operate in heavier soils if the depth of application is greatly reduced. A common-size system designed for 65 ha is 392 m in length. Such a system will irrigate about 55 of the 65 ha; the remaining 10 ha in the corners is not covered. The major advantage of the center-pivot system is the saving of labor; the major disadvantage is investment cost.

Linear-move laterals use hardware similar to that of center-pivot systems but move in a straight line across large rectangular fields. Water is supplied along the entire length of the field by a ditch or pipeline. If a ditch is used, an engine with pump and generator is mounted on a carriage unit beside the ditch. This unit pumps water from the ditch, generates electrical energy for the drive motors, and guides the system across the field.

Solid-set systems may be operated by changing the flow from one lateral to the next or by sequencing sprinklers (one operating on each line) along the lateral lines. For frost protection, all sprinklers are operated at the same time. Operation can be made automatic with timing devices and solenoid valve controls. As with the self-propelled system, the only labor required is for setup and maintenance. Because of extremely high investment costs (Table 20.1), solid-set systems are practical only for high-value crops.

20.2 Components of Sprinkler Systems

The major components of a hand-move portable sprinkler system are shown in Fig. 20.2. Most components are basically the same as required for the end-pull, rotating boom, and side-roll systems shown in Fig. 20.1. With the rotating boom or

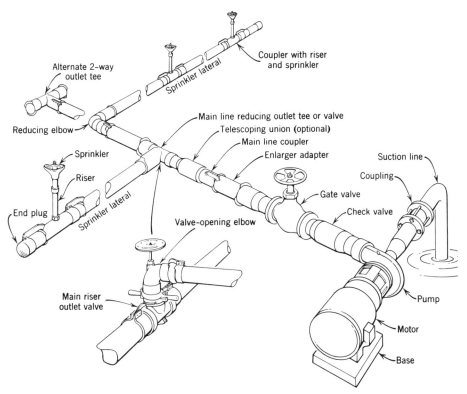

Fig. 20.2 Components of a portable sprinkler irrigation system.

giant sprinkler systems, the lateral spacing and the distance between sprinkler setups along the lateral are much greater than for standard sprinklers. Latches, seals, and other details of these components vary considerably, depending on the manufacturer.

A pump is required to overcome the elevation difference between the water source and the sprinkler nozzle, to counteract friction losses, and to provide adequate pressure at the nozzle for good water distribution. As discussed in Chapter 16, the type of pump will vary with the discharge, pressure, and the vertical distance to the source of water. The pump should have adequate capacity to meet future needs and to allow for wear. For most farm irrigation systems capacity generally varies from 6 to 60 L/s.

The second component of the system, the main line, may be either movable or permanent. Movable mains generally have a lower first cost and can be more easily adapted to a variety of conditions; permanent mains offer savings in labor and reduced obstruction to field operations. Water is taken from the main either through a valve placed at each point of junction with a lateral or in some cases through either an L- or a T-fitting that has been supplied in place of one of the couplings on the main.

The laterals are usually 6- or 9-m lengths of aluminum or other lightweight type pipe connected with couplers. In some cases a coupler is permanently attached to the pipe (Fig. 20.3a).

For rotating sprinklers, the sprinkler heads most often used have two nozzles, one to apply water at a considerable distance from the sprinkler and the other to cover the area near the sprinkler center (Fig. 20.3b). Of the devices to rotate the sprinkler, the most usual, also shown in Fig. 20.3b, taps the sprinkler head with a small hammer activated by the force of the water striking against a small vane connected to it.

A number of sprinkler heads are available for special purposes. Some provide a low-angle jet for use in orchards. Some work at heads as low as 35 kPa, and others operate only in a part circle. Giant sprinkler units discharge 20 to 30 L/s at pressures of 550 to 700 kPa and spray the water up to 50 m. In general, these work at higher pressures than the smaller units and result in greater pumping costs. Because these large units cover a much greater area, a smaller number of moves is required.

Low-pressure sprays (Fig. 20.3c) operate at 70 to 210 kPa. They were developed primarily for use on center-pivot and linear-move systems to reduce the pressure requirement. Droplet sizes are small but application rates tend to be high unless small trusses or spray booms are attached nearly perpendicular to the lateral. Spray heads are mounted on the booms both ahead and behind the lateral to increase the wetted area and decrease the application rate.

20.3 Evaporation Losses

In sprinkler application of irrigation water, evaporation occurs from the spray and from the wet foliage surfaces. Frost (1963) developed relationships between spray losses and atmospheric parameters. He has compared the evapotranspiration rates of various crops under nonsprinkling conditions to the evapotranspiration rates of the same foliage as it is being wetted by sprinkler water. It was found that the evaporation from wet foliage was essentially the same as for dry foliage. In those

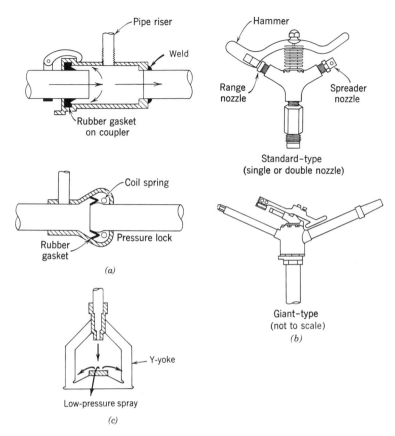

Fig. 20.3   Examples of (*a*) quick-connecting couplers, (*b*) sprinkler heads, and (*c*) low-pressure spray head.

situations where wet-leaf loss was less than dry-leaf loss, energy available for evaporation of water was not reaching the wet leaves at the rate at which it reached the dry leaves. The energy difference was absorbed by the spray.

20.4   Design for Equipment Requirements

Not only should the sprinkler system be properly designed hydraulically and economical in cost, but selection of the system should consider the availability of labor for moving the sprinklers and the pipe. The design of a sprinkler irrigation system involves the maximum rate of application, the irrigation period, and the depth of application. Application at rates in excess of the soil infiltration capacity results in runoff with accompanying poor distribution of water, loss of water, and soil erosion. Maximum water application rates for various soil conditions are given in Table 20.2. These values may serve as a guide where reliable local recommendations are not available.

Applications at rates well below the maximum have been found beneficial. Rates of one half the infiltration rate of the soil combined with nozzle pressures

Table 20.2   Suggested Maximum Water Application Rates for Sprinklers for Average Soil, Slope, and Cultural Conditions

Soil Texture and Profile Conditions	Maximum Water Application Rate for Slope and Cultural Conditions (mm/h)			
	0% Slope		*10% Slope*	
	With Cover	Bare	With Cover	Bare
Light sandy loams uniform in texture to 2 m	32	20	19	11
Light sandy loams over more compact subsoils	25	15	13	8
Silt loams uniform in texture to 2 m	16	10	10	6
Silt loams over more compact subsoil	10	6	5	3
Heavy-textured clays or clay loams	5	3	3	2

Source: Adapted from SCS (1983).

that provide a fine spray have resulted in improved maintenance of soil structure and minimization of soil compaction.

The depth of application and the irrigation period are closely related. Irrigation period is the time required to cover an area with one application of water. The depth of application will depend on the available water-holding capacity of the soil.

Under humid conditions rains may bring the entire field up to a given water level. As the plants use this water, the water level for the entire field decreases. Irrigation must be started soon enough to enable the field to be covered before plants in the last portion to be irrigated suffer from water deficiency.

One recommended system is to commence irrigation when the water level of the field reaches 55 percent of the available water capacity. The net depth of application under this plan is equal to 45 percent of the available water capacity. The irrigation period is set so that the entire irrigated area will be covered before the finishing end of the field reaches a water level below 10 percent of the available water. Typical water-holding capacities are given in Table 18.2. Average root depths and peak rate of water use by crops are given in Table 20.3.

Table 20.3   Root Depth and Peak Rate of Water Use

Crop	Root Depth (m)	Peak Rate of Water Use (mm/day)		
		Cool	Climate Moderate	Hot
Alfalfa	0.9–1.2	5	6	9
Beans	0.3–0.9	3	4	6
Corn	0.6–1.2	5	6	8
Pasture	0.5–0.8	5	6	8
Potatoes	0.3–0.6	4	5	6
Strawberries	0.3–0.5	3	4	6

☐ *Example 20.1*

A sprinkler irrigation system is to be designed to irrigate 16 ha (39.5 ac) of pasture on a silt loam soil near Cleveland, Ohio. The field is flat. Determine the limiting rate of application, the irrigation period, the net depth of water per application, the depth of water pumped per application, and the required system capacity in hectares per day.

Solution. From Table 20.2 the limiting rate of application is 16 mm/h (0.6 iph). The measured available water-holding capacity of the soil is 150 mm/m (1.8 in./ft) and the depth of the root zone from Table 20.3 is about 0.6 m (2 ft). The total available water capacity is thus (0.6 × 150) = 90 mm (3.6 in.). The net depth of application is 0.45 × 90 = 40.5 mm (1.6 in.). Assuming a water application efficiency of 70 percent, the depth of water pumped per application is 40.5/0.70 = 58 mm (2.3 in.). From Table 20.3 the peak rate of use by the crop is 5 mm/day (0.2 ipd). The irrigation period is 8.1 (40.5/5) or rounded to 8 days. To cover this field in 8 days the system must be able to pump and discharge 58 mm on 2.0 (16/8) ha/day. This information is then used as a guide in the selection of equipment. ☐

INTERMITTENT-MOVE SYSTEMS

20.5 General Rules for Sprinkler System Design and Layout

In the absence of experience the following rules may be helpful for design and layout of hand-move or mechanical-move systems other than center-pivot and linear-move. Commercial companies and government agency personnel may also be of great assistance.

In general, sprinkler spacings along the lateral and lateral spacing along the main should be as wide as possible to reduce labor costs. Since greater spacings require higher pressures and thus higher pumping costs, these wide spacings are more easily justified where the power costs are low. Greater spacings also require a higher application rate in which case the infiltration rate may be the limiting factor. The following general rules for layout should be kept in mind: (1) Mains should be laid up and downhill. (2) Laterals should be laid across slope or nearly on the contour. (3) For multiple lateral operation, lateral pipe sizes should be limited to not more than two diameters. (4) If possible, water supply nearest the center of the area should be chosen. (5) For balanced design, lateral operation as shown in Fig. 20.4*b* should be provided. (6) Layout should facilitate and minimize lateral movement during the season. (7) Differences in the number of sprinklers operating for the various setups should be held to a minimum. (8) Booster pumps should be considered where small portions of the field would require a high pressure at the pump. (9) Layout should be modified to apply different rates and amounts of water where soils are greatly different in the design area.

20.6 Layouts for Hand-Move, Side-Roll, and End-Pull Systems

The number of possible arrangements for the mains, laterals, and sprinklers is practically unlimited. The arrangement selected should allow a minimal invest-

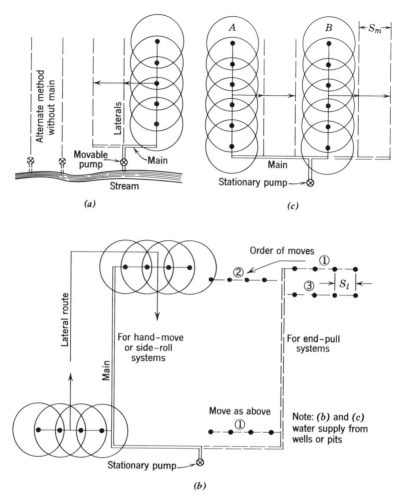

Fig. 20.4 Field layouts of main and laterals for hand-move, side-roll, and end-pull sprinkler systems. (*a*) Fully portable. (*b*) Portable or permanent (buried) main and portable laterals. (*c*) Portable main and laterals.

ment in irrigation pipe, have a low labor requirement, and provide for an application of water over the total area in the required period of time. The most suitable layout can be determined only after a careful study of the conditions to be encountered. The choice will depend to a large extent on the types and capacities of the sprinklers and the pressure required. For medium-pressure systems, the laterals are moved about 18 m at each setting and the sprinklers are spaced about 12 m along each lateral.

Typical layouts for sprinkler irrigation systems are shown in Fig. 20.4. The layout in Fig. 20.4*a* is suitable where the water supply can be obtained from a stream or canal alongside the field to be irrigated. This arrangement either eliminates the main line or requires a relatively short main, depending on the number of moves for the pump. Less pipe is required for this method than for any of the others. The layout illustrated in Figs. 20.4*b* and 20.4*c* is suitable where the water supply is from a well or pond. In Fig. 20.4*b* the two laterals are started at opposite ends of the field and are moved in opposite directions. Since the farther half of the main supplies a maximum of one lateral at a time, the diameter of this section can be reduced. This arrangement is well suited to day and night operation when the required amount of water can be added in about 6 or 8 h. The system shown in Fig. 20.4*c* can be designed for continuous operation. While line A is in operation, the operator moves line B. When the required amount of water has been applied, line B is turned on and then line A is moved. With this procedure the capacity of the pump needs to be adequate to supply only one lateral.

20.7 Capacity of a Sprinkler System

The minimum capacity of a sprinkler system depends on the area to be irrigated, the depth of water application at each irrigation, the efficiency of application, and the application time. The actual capacity of the system is the sum of the discharges of the maximum number of sprinklers operating at one time.

20.8 Sprinkler Capacity

When the rate of application and the spacing of the sprinklers have been determined, the required sprinkler capacity can be computed with the formula

$$q = S_l S_m r \tag{20.1}$$

where q = discharge of each sprinkler (L^3/T),
S_l = sprinkler spacing along the lateral (L),
S_m = sprinkler spacing between lines or along the main (L),
r = rate of application (L/T).

For example, a spacing of 12 × 18 m and an application rate of 10 mm/h require a sprinkler with a capacity of 0.6 L/s. The theoretical discharge of a nozzle may be computed from the orifice flow equation (Chapter 9):

$$q = aC\sqrt{2gh} \tag{20.2}$$

For simplification of calculations, this equation is

$$q = 0.00111 C d_n^2 P^{1/2} \tag{20.3}$$

where q = nozzle discharge in L/s,
C = coefficient of discharge,
d_n = diameter of the nozzle orifice in mm,
P = pressure at the nozzle in kPa.

The coefficient of discharge for well-designed, small nozzles varies from about 0.95 to 0.98. Some nozzles have coefficients as low as 0.80. Normally, the larger the nozzle, the lower is the coefficient. Where the sprinkler has two nozzles, the total discharge is the combined capacity of both.

20.9   Distribution Pattern of Sprinklers

A typical distribution pattern showing the effect of wind for a single sprinkler is illustrated in Fig. 20.5. Since one sprinkler does not apply water uniformly over the area, sprinkler patterns are overlapped to provide more uniform coverage. The distribution pattern shown in Fig. 20.6 illustrates how the overlapping patterns combine to give a relatively uniform distribution between sprinklers. Although Fig. 20.6 shows relatively uniform distribution over the area, wind will skew the pattern so as to give less uniform distribution.

The factors that influence the distribution pattern of sprinklers are nozzle pressure, wind velocity, and speed of rotation. Too low a pressure will result in a "ring-shaped" distribution and a reduction in the area covered; high pressures produce smaller drops with high application rates near the sprinkler. Wind will cause a variable diameter of coverage and also somewhat higher rates near the sprinkler. A high speed of rotation of the sprinkler greatly reduces the area covered and causes excessive wear of the sprinkler. The variation in speed of rotation is not due to the wind, but to changes in frictional resistance attributed to lack of precision in manufacture and to wear. Uniformity of application of a sprinkler

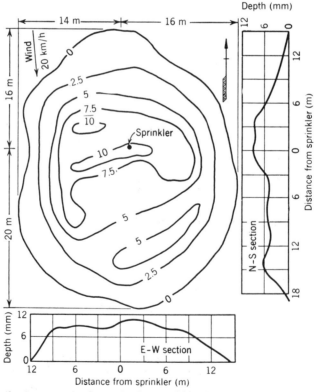

Fig. 20.5   Distribution pattern for a single sprinkler showing the effect of wind. (Nozzle pressure 207 kPa and discharge 1.23 L/s.) (Redrawn from Christiansen, 1948.)

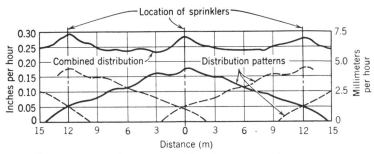

Fig. 20.6 Distribution pattern from a sprinkler showing overlapping to give a relatively uniform combined distribution for no wind. (Redrawn from Gray, 1961.)

system can be expressed by the uniformity coefficient given in Chapter 18. In applying the coefficient to evaluate a sprinkler system, the depths can be the depth of the water applied or the depth of soil water penetration. Depth of water applied evaluates the sprinkler system alone. Depth of penetration combines the effects of the sprinkler system, the topography, and the soil conditions. Observations at many points are made in the field with cans uniformly spaced in the sprinkler area bounded by four adjacent regularly spaced sprinklers. An estimate of the uniformity coefficient can also be determined graphically if the distribution pattern, as shown in Fig. 20.5, is known. An absolutely uniform application would give a uniformity coefficient of 1.0. Uniformity coefficients of 0.85 or more are acceptable. Distribution in wind may be improved by moving the sprinkler lateral only one half the normal spacing, by irrigating at night when the wind is low, or by using only the range nozzle on two-nozzle sprinklers.

20.10 Sprinkler Selection and Spacing

The actual selection of the sprinkler is based largely on design information furnished by manufacturers of the equipment. The choice depends primarily on the diameter of coverage required, pressure available, and capacity of the sprinkler. The data given in Table 20.4 may serve as a guide in selecting the pressure and spacing desired.

Sprinkler discharge and diameter of coverage for a given sprinkler head design are given in Table 20.5. Each manufacturer will recommend a combination of nozzle sizes and pressures to give the best breakup of the stream and distribution pattern for uniform application. Single-nozzle sprinklers are generally recommended only for small nozzles with limited diameter of coverage or for part-circle sprinklers. Because the distribution and coverage depend on the angle of the stream from the horizontal and the rate of rotation, sprinklers should be selected from manufacturer's tables.

The maximum sprinkler spacings for satisfactory uniformity of application are given in Table 20.6 for a distribution pattern that is triangular in cross section or for one similar to that shown in Fig. 20.6. These values may be applied in design when uniformity coefficients are not available.

Table 20.4 Classification of Sprinklers and Their Adaptability

Type of Sprinkler	Very Low Pressure (34–105 kPa)	Low Pressure (100–210 kPa)	Medium Pressure (210–415 kPa)
General characteristics	Special thrust springs or reaction-type arms	Usually single-nozzle oscillating or long-arm dual-nozzle design	Either single- or dual-nozzle design
Range of wetted diameters	6–15 m	18–24 m	23–37 m
Recommended application rate	10 mm/h	5 mm/h	5 mm/h
Jet characteristics (assuming proper pressure–nozzle size relations)	Water drops are large because of low pressure	Water drops are fairly well broken	Water drops are well broken over entire wetted diameter
Water distribution pattern (assuming proper spacing and pressure–nozzle size relation)	Fair	Fair to good at upper limits of pressure range	Very good
Adaptations and limitations	Small acreages confined to soils with intake rates exceeding 13 mm/h and to good ground cover on medium-to-coarse textured soils	Primarily for undertree sprinkling in orchards; can be used for field crops and vegetables	For all field crops and most irrigable soils; well adapted to overtree sprinkling in orchards and groves and to tobacco shades

Source: SCS (1983).

20.11 Size of Laterals and Mains

Laterals and mains should provide the required rate of flow with a reasonable head loss. For laterals the sections at the distant end of the line have less water to carry and may therefore be smaller; however, many authorities advise against "tapering" of pipe diameters in laterals, as it then becomes necessary to keep the

Table 20.4 Continued

High Pressure (350–690 kPa)	Very High Pressure (550–830 kPa)	Undertree Low Angle (70–350 kPa)	Perforated Pipe (30–140 kPa)
Either single- or dual-nozzle design	One large nozzle with smaller supplemental nozzles to fill in pattern gaps; small nozzle rotates the sprinkler	Designed to keep stream trajectories below fruit and foliage by lowering the nozzle angle	Portable irrigation pipe with lines of small perforations in upper third of pipe perimeter
34–70 m	60–120 m	12–27 m	Rectangular strips 3–15 m wide
13 mm/h	16 mm/h	8 mm/h	13 mm/h
Water drops are well broken over entire wetted diameter	Water drops are extremely well broken	Water drops are extremely well broken	Water drops are large because of low pressure
Good *except* where wind velocities exceed 6 km/h	Acceptable in calm air; severely distorted by wind	Fairly good; diamond pattern recommended where laterals are spaced more than one tree interspace	Good; pattern is rectangular
Same as for intermediate-pressure sprinklers except where wind is excessive	Adaptable to close-growing crops that provide a good ground cover; for rapid coverage and for odd-shaped areas; limited to soils with high intake rates	For all orchards or citrus groves; in orchards where wind will distort overtree sprinkler patterns; in orchards where available pressure is not sufficient for operation of overtree sprinklers	For low-growing crops only; unsuitable for tall crops; limited to soils with relatively high intake rates; best adapted to small acreages of high-value crops; low operating pressure permits use of gravity or municipal supply

various pipe sizes in the same relative position. The system may also be less adaptable to other fields and situations.

Pair et al. (1983) recommend that the total pressure variation in the laterals, when practicable, should not be more than ±10 percent of the design pressure. If the lateral runs up or downhill, allowance for this difference in elevation should be made in determining the variation in head. If the lateral runs uphill, less pressure will be available at the nozzle; if it runs downhill, there will be a tendency to balance the loss of head resulting from friction.

Table 20.5   Manufacturer's Sprinkler Characteristics

Nozzle Pressure [kPa (psi)]	Nozzle Diameter [mm (in.)]					
	3.97 × 3.18 (5/32 × 1/8)		4.76 × 3.97 (3/16 × 5/32)		6.35 × 3.97 (1/4 × 5/32)	
	Dia.[a]	L/s	Dia.	L/s	Dia.	L/s
207 (30)	25	0.37	26	0.52	28	0.76
276 (40)	27	0.43	28	0.61	31	0.90
345 (50)	28	0.47	30	0.68	34	1.00
414 (60)	30	0.52	31	0.74	36	1.10

[a]Diameter of coverage in m.
Note: 1 L/s = 15.85 gpm and 1 m = 3.28 ft. Pressures to left and below dashed line recommended for best breakup of stream.

Scobey's (1930) equation for friction or head loss in pipes may be expressed as

$$H_f = \frac{K_s L Q^{1.9}}{D^{4.9}}(4.10 \times 10^6) \qquad (20.4)$$

where H_f = total friction loss in line in m,
K_s = Scobey's coefficient of retardation,
L = length of pipe in m,
Q = total discharge in L/s,
D = inside diameter of pipe in mm.

Although this equation was developed for main line or full flow, it may be adapted to lateral pipe with uniformly spaced sprinkler outlets by multiplying the friction loss H_f by a factor F to obtain the actual loss. Suggested values of F are given in Table 20.7. If the head loss for the main line is 70 kPa, the loss for the same length lateral with eight sprinklers is only 29 kPa (0.41 × 70).

Recommended values of K_s for design purposes are 0.32 for new Transite pipe, 0.40 for steel pipe or portable aluminum pipe and couplers, and 0.42 for portable galvanized steel pipe and couplers.

Table 20.6   Maximum Spacings for Low- or Medium-Pressure Sprinklers[a]

Wind Velocity (km/h)	Lateral Spacing in Percent of the Diameter of Coverage	
	S_l, Along the Lateral	S_m, Along the Main
0	50	65
≤6	45	60
7–12	40	50
≥13	30	40

[a]For laterals normal to the wind direction only.

Table 20.7 Correction Factor F for Friction Losses in Aluminum Pipes with Multiple Outlets[a]

Number of Sprinklers	Correction Factor, F			
	First Sprinkler One Sprinkler Interval from Main		First Sprinkler One-Half Sprinkler Interval from Main	
	$m = 1.9$[a]	$m = 1.75$[b]	$m = 1.9$[a]	$m = 1.75$[b]
1	1.00	1.00	1.00	1.00
2	0.63	0.65	0.51	0.53
4	0.48	0.50	0.41	0.43
6	0.43	0.45	0.38	0.40
8	0.41	0.43	0.37	0.39
12	0.39	0.42	0.36	0.38
16	0.38	0.40	0.36	0.37
20	0.37	0.39	0.35	0.37
30	0.36	0.36	0.35	0.37

[a]Exponent $m = 1.9$ for Q in Eq. 20.4.
[b]Exponent $m = 1.75$ for Q in Eq. 21.5.

The diameter of the main should be adequate to supply the laterals in each of their positions. The rate of flow required for each lateral may be determined by the total capacity of the sprinklers on the lateral. The position of the laterals that gives the highest friction loss in the main should be used for design purposes. The friction loss in the main may be computed by Eq. 20.4. Allowable friction loss in the main varies with the cost of power and the price differential between different diameters of pipe. For small systems with few irrigations per season an approximate maximum friction loss is 4 kPa per 10-m length of pipe. The most economical size should be determined by balancing the increase in pumping costs against the amortized cost difference of the pipe. A direct economic solution was proposed by Keller (1965) and summarized by SCS (1983).

The design capacity for sprinklers on a lateral with uniform spacing should be based on the average operating pressure. Where the friction loss in the laterals is within 20 percent of the average pressure, the average head for design in a sprinkler line can be expressed approximately by (Fig. 20.7)

$$H_a = H_o + 0.25H_f + 0.4H_e \tag{20.5}$$

where H_a = average pressure at the nozzle (L),
H_o = nozzle pressure at the farthest end of the line (L),
H_f = friction head loss in the lateral (L),
H_e = maximum difference in elevation between the junction with the main and the farthest sprinkler on the lateral (L).

The pressure or head H_n required at the junction of the lateral and the main is

$$H_n = H_o + H_f + H_e + H_{rp} \tag{20.6}$$

where H_{rp} = riser height (L).

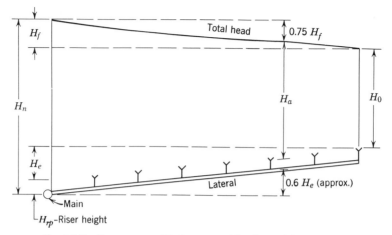

Fig. 20.7   Pressure profile in a sprinkler lateral laid uphill.

By solving for H_o in Eq. 20.5 and substituting in Eq. 20.6,

$$H_n = H_a + 0.75H_f \pm 0.6H_e + H_{rp} \tag{20.7}$$

The constant 0.6 varies from 0.5 to 1.0 for 100 sprinklers to 1 sprinkler, respectively, and is a correction for the percentage of the height at the point where the average pressure exists. Likewise the constant 0.75 for the friction term represents the percentage of the friction loss to this point. If the lateral is located downhill from the main, the elevation term is negative. The relationships in Eqs. 20.5 and 20.7 are shown in Fig. 20.7.

20.12   Pump and Power Units

In selecting a suitable pump (Chapter 16), it is necessary to determine the maximum total head against which the pump is working. This head may be determined by

$$H_t = H_n + H_m + H_j + H_s \tag{20.8}$$

where H_t = total design head against which the pump is working.

H_n = maximum head required at the main to operate the sprinklers on the lateral at the required average pressure, including the riser height (Eq. 20.6),

H_m = maximum friction loss in the main, the suction line, and NPSH of the pump,

H_j = elevation difference between the pump and the junction of the lateral and the main,

H_s = elevation difference between the pump and the water supply after drawdown, all in m or kPa.

The amount of water that will be required is the sum of the capacity of all sprinklers to be operated at the same time. When the total head and rate of pumping are known, the pump may be selected from rating curves or tables furnished by the manufacturer as discussed in Chapter 16.

The following example illustrates the design of a simple sprinkler irrigation system.

☐ Example 20.2

Design a side-roll sprinkler irrigation system to irrigate a square 16-ha (39.5-ac) field for a maximum rate of application of 15 mm/h (0.6 iph), for applying 58 mm (2.3 in.) of water in 8 days or 2.0 ha (4.9 ac) per day as given in Example 20.1. Assume a wind velocity of 6 km/h (3.7 mph), $H_a = 276$ kPa (40 psi), $H_j = 1.0$ m (3.3 ft), $H_e = 0.6$ m (2.0 ft), $H_s = 5.0$ m (16.5 ft), $H_{rp} = 0.8$ m (2.6 ft), NPSH of pump = 2.0 m (6.6 ft), $S_l = 12$ m (40 ft), $S_m = 18$ m (60 ft), allowable variation of pressure on the lateral is 20 percent of the average pressure, and well location is at the center of the field.

Solution. Location and length of laterals and mains and sprinkler layout are shown in Fig. 20.8. Allowing 12 m from first sprinkler to the main line and 8 m

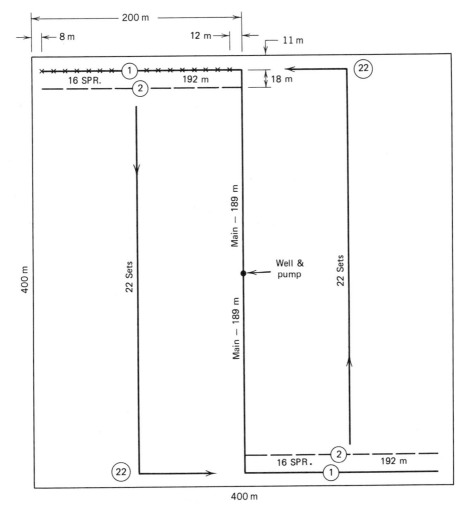

Fig. 20.8 Side-roll sprinkler layout of a 16-ha (39.5-ac) field described in Example 20.2.

from the last sprinkler to the field boundary, 16 sprinklers are required on each lateral (15 spacings).

(1) Calculate the number of lateral settings per day.

$$\frac{2.0 \text{ ha} \times 10\ 000 \text{ m}^2/\text{ha}}{16 \times 12 \text{ m} \times 18 \text{ m}} = 5.8 \text{ or round to } 6$$

Two laterals with three moves per day will be adequate or six laterals if moved once a day. Select two laterals to reduce equipment cost.

(2) Substitute in Eq. 20.1, and determine the discharge per sprinkler.

$$q = \frac{12 \text{ m} \times 18 \text{ m} \times 15 \text{ mm/h} \times 1000 \text{ L/m}^3}{1000 \text{ mm/m} \times 3600 \text{ s/h}}$$

$$= 0.90 \text{ L/s } (14.3 \text{ gpm})$$

(3) Discharge per lateral = $16 \times 0.9 = 14.4$ L/s (228 gpm). Pump or system capacity $q_s = 2 \times 14.4 = 28.8$ L/s.

(4) From Eq. 20.3, calculate the theoretical discharge of a sprinkler with 6.35- and 3.97-mm (¼- and 5/32-in.)-diameter nozzles with $C = 0.95$.

$$q = 0.00111 \times 0.95 \times 276^{1/2} (6.35^2 + 3.97^2)$$

$$= 0.98 \text{ L/s } (15.6 \text{ gpm})$$

Since this is greater than the allowable discharge determined in step 2, this sprinkler is not acceptable. The alternative procedure is to select the same nozzle sizes at 276 kPa from manufacturer's values, such as shown in Table 20.5, for which $q = 0.90$ L/s (14.2 gpm).

(5) From Table 20.6 the diameter of coverage required for a 6 km/h wind for sprinkler spacing along the lateral is 27 m (12 m/0.45), and along the main, 30 m (18 m/0.60). Diameter of coverage from Table 20.5 is 31 m, which is adequate; however, with this diameter of coverage, water will carry beyond the field boundaries and the design layout may need to be modified.

(6) Determine the daily operating time.

$$\frac{58 \text{ mm (application depth)}}{15 \text{ mm/h (application rate)}} = 3.9 \text{ h/set}$$

Assuming it takes 1 h to move each lateral, calculate the daily operating time.

$$(3.9 \text{ h/set} \times 3 \text{ sets/day}) + (3 \text{ moves} \times 2 \text{ h/move}) = 17.7 \text{ h/day}$$

Thus, the system can easily be operated at three sets per lateral per day.

(7) Determine the diameters of the laterals and main. Total allowable variation of pressure in the lateral = $0.20 \times 276 = 55.2$ kPa (8.0 psi). Allowable variation due to friction only

$$= (55.2/9.8) - H_e = 5.6 - 0.6 = 5.0 \text{ m } (7.1 \text{ psi}).$$

Compute the friction loss of a 101.6-mm (4-in.)-outside-diameter lateral (16 gage with 1.30-mm wall thickness) from Eq. 20.4, using $K_s = 0.40$. Read $F = 0.38$ from Table 20.7 for 16 sprinklers. Compute friction loss for other sizes for the lateral and main as follows:

Tubing Outside Diameter [mm (in.)]	Friction Loss [m (ft)]	
	192-m Lateral $H_f \times F$	189-m Main
76.2 (3)	13.5 (44)	35.0 (115)
101.6 (4)	3.2[a] (10)	8.2 (27)
127.0 (5)	1.0 (3)	2.7[a] (9)

[a]Select 101.6-mm (4-in.) lateral and 127.0-mm (5-in.) main.

(8) Determine the pressure or head required at the junction of the lateral and the main for the most remote lateral location and at the highest elevation by substituting in Eq. 20.7 (276 kPa = 28.2 m).

$$H_n = 28.2 + 0.75(3.2) + 0.6(0.6) + 0.8$$

$$= 31.8 \text{ m (104 ft)}$$

(9) Determine the pump capacity and total head by substituting in Eq. 20.8.

$$H_t = 31.8 + 2.0 + 2.7 + 1.0 + 5.0 = 42.5 \text{ m (139 ft)}$$

Select the pump from manufacturer's characteristic curves to deliver 28.8 L/s (456 gpm) at a total head of 42.5 m (60 psi) with an efficiency as high as possible. Allow some factor of safety for possible operation of the pump under other conditions and for loss of efficiency with age.

(10) Determine the size of the power unit. Assume a pump efficiency of 70 percent and substitute in Eq. 16.3.

$$kW = 0.0098 \times 28.8 \text{ L/s} \times 42.5 \text{ m}/0.70$$

$$= 17.12 \text{ kW (22.9 hp)}$$

Select power unit capable of continuously furnishing 17.12 kW, such as a 20-kW electric motor. Assuming a water-cooled internal combustion engine will continuously deliver 70 percent of its rated output, a 25 (17.12/0.7)-kW (33-hp) engine is required.   □

CENTER-PIVOT SYSTEMS

20.13 Layout for Center-Pivot Systems

With a center-pivot system, sprinklers are spaced along the moving lateral, each covering an area equal to $2\pi r S_l$ in one revolution, where r is the distance from the pivot point and S_l is the sum of half the distances from r to the two adjacent

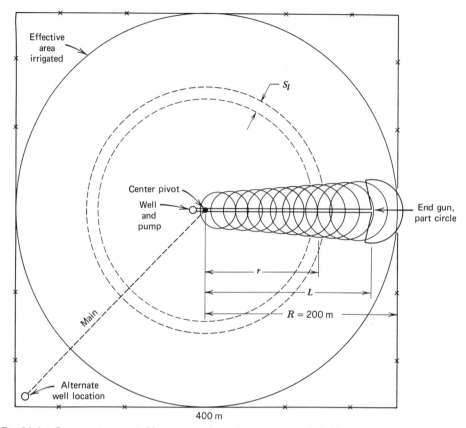

Fig. 20.9   Center-pivot sprinkler system for a 16-ha (39.5-ac) field. Alternate well location is for convenience of supplying four such fields.

sprinklers as shown in Fig. 20.9. To obtain uniform application of water the sprinkler spacing may be constant with a variable sprinkler discharge or with a combination of spacings and discharges. A large part-circle sprinkler (end gun) is usually placed at the end of the lateral to extend the diameter of coverage as far as possible. The length of the lateral line and the size of the end gun are adjusted so that the effective circular area to be irrigated is adequately covered. Note in Fig. 20.9 that the outer area of the sprinkler pattern of the end gun is outside this area. Some manufacturers provide a fold-back extension arm that attaches to the end of the sprinkler lateral (replaces the end gun) to cover most of the corner areas of a square field. An end gun is then located at the end of the extension arm. These systems will also cover odd-shaped fields. Some center-pivot laterals can be pulled to another field by rotating the tower wheels 90 degrees, but during normal sprinkling the lateral rotates about the center pivot as before.

20.14   Capacity of a Center-Pivot System

For 100 percent efficiency of application, the capacity of a center-pivot system is given by

$$q_s = d_s a/t = d_s \pi R^2/t \qquad (20.9)$$

where q_s = capacity of the system (L³/T),
$\qquad d_s$ = depth of application per revolution (L),
$\qquad a$ = area to be irrigated (L²),
$\qquad R$ = radius of a circular area as in a center-pivot system (L),
$\qquad t$ = total time of operation per revolution (T).

For a center-pivot system the area covered by one revolution of the lateral by an infinitesimally small sprinkler is $2\pi r dr$, where r is the distance from the sprinkler to the pivot point. The sprinkler capacity for this infinitesimal area is

$$dq = [d_s(2\pi r)dr]/t \qquad (20.10)$$

As given by Chu and Moe (1972), the integration of Eq. 20.10 from the limits r to R gives the flow rate in the lateral at any point r as

$$q_r = (\pi R^2 d_s/t)\,(1 - r^2/R^2) \qquad (20.11)$$

At the outer end of the lateral at length L, the discharge required for the end gun can be obtained by substituting in Eq. 20.11, $r = L$. The discharge of the sprinklers along the lateral is thus the difference between Eq. 20.9 and Eq. 20.11. Discharge from Eq. 20.11 is theoretically correct for an infinite number of sprinklers applying the average water depth uniformly along the lateral. In practice the distribution will not be as ideal because of the limited number of sprinklers. Chu and Moe (1972) found that the theoretical distribution was close to experimental data.

20.15 Size of Center-Pivot Laterals

For center-pivot systems Chu and Moe (1972) found $F = 0.54$. This F is higher than those in Table 20.7 because the flow in a center-pivot lateral is more than in a regular lateral due to the end gun. The friction loss of the lateral is also adjusted for the size of the end gun by taking the length of the lateral equal to the radius of the irrigated area (Example 20.3). The size of the lateral is determined by pipe cost, energy cost, structural strength of the pipe, and limits on allowable friction loss adjusted for elevation differences along the line. The selected pipe diameter should be the closest nominal size available commercially.

For center-pivot systems with an end gun Chu and Moe (1972) derived the pressure distribution along the lateral, from which

$$H_r = H_f\,[1 - 1.875(x - 2x^3/3 + x^5/5)] + H_L \qquad (20.12)$$

where H_r = pressure at a distance r from the pivot (L),
$\qquad x = r/R$, distance from the pivot divided by the radius of the field to be irrigated,
$\qquad H_L$ = pressure at the end of the lateral that is also the nozzle pressure for the end gun (L),
$\qquad H_f$ = friction loss in the center-pivot lateral (L).

This equation neglects the velocity head and assumes the pressure at R is equal to that at distance L and Scobey's coefficient $m = 2.0$. The theoretical equation is a good approximation of field measurements. From Eq. 20.12 nozzle pressure can be computed for a selected sprinkler location along the lateral. With knowledge of the pressure, the nozzle size can be selected, and the area covered can be determined.

20.16 Center-Pivot Application Rates

Since the area covered by a sprinkler is largest at the outer end of a center-pivot lateral, the sprinkler discharge is largest. As a result, the application rate also is highest at the end of the lateral. If an elliptical water application pattern is assumed, the average application rate is (SCS, 1983)

$$I = \frac{7200rQ}{R^2\, w} \tag{20.13}$$

where I = average application rate at radius r in mm/h,
r = radius from pivot in m,
R = radius irrigated by the center pivot in m,
Q = system capacity in L/s,
w = wetted diameter of sprinkler at radius r in m.

The maximum application rate I_m is given by

$$I_m = 4I/\pi \tag{20.14}$$

These application rates should be compared with the infiltration characteristics of the soil to determine if runoff will occur and, thus, the suitability of the center-pivot system.

☐ Example 20.3

Design a center-pivot irrigation system to irrigate the circular area of the square 16-ha (39.5-ac) pasture at a maximum rate of application of 15 mm/h (0.6 iph) with a net depth of 40.5 mm (1.6 in.) of water as given in Example 20.1. Assuming an application efficiency of 70 percent, the total depth to be applied is 40.5/ 0.70 = 58 mm (2.3 in.). The field is to be irrigated in 2 days with one revolution per day. With a portable unit other fields can be irrigated with the same equipment.

Solution. The maximum area irrigated is a 400-m (1312-ft)-diameter circular area or 125 663 m² (31.0 ac).
    The system capacity from Eq. 20.9 using 70 percent application efficiency is

$$q_s = \frac{58\text{ mm} \times 125\ 663\text{ m}^2 \times 1000\text{ L/m}^3}{2\text{ rev.} \times 24\text{ h/rev} \times 3600\text{ s/h} \times 1000\text{ mm/m}}$$

$$= 42.2\text{ L/s (669 gpm)}$$

Assuming that the end gun covers an area of about 15 m (49 ft) beyond the end of the lateral (obtain from manufacturer), $L = 200 - 15 = 185$ m (607 ft) and the discharge of the end gun from Eqs. 20.9 and 20.11 for $r = L$ is

$$q_r = 42.2\ [1 - (185/200)^2] = 6.1\text{ L/s (97 gpm)}$$

Select an end gun from the manufacturer's catalog to deliver 6.1 L/s at a pressure of 42 m (60 psi).

Friction loss in 152-mm (6.0-in.)-outside-diameter steel tubing (12 gage, 2.77 mm wall) from Eq. 20.4, with $L = R$ and $F = 0.54$, is

$$H_f = \frac{0.4 \times 200 \, (42.2)^{1.9} \times 4.10 \times 10^6 \times 0.54}{(152 - 5.54)^{4.9}}$$

$$= 5.3 \text{ m } (17.4 \text{ ft})$$

Total head at the center pivot including 4.0-m sprinkler height and $H_e = 0.6$ m is

$$H_t = 42 + 5.3 + 4.0 + 0.6 = 51.9 \text{ m } (73.7 \text{ psi})$$

Nozzle size may be determined at any point along the lateral from the pressure head and the required sprinkler capacity. For example, at a distance of 100 m (328 ft) from the center pivot for $x = 0.5$, the nozzle pressure from Eq. 20.12 is

$$H_r = 5.3 \, [1 - 1.875 \, (0.5 - 2(0.5)^3/3 + (0.5)^5/5)] + 42$$

$$= 43.1 \text{ m } (141 \text{ ft})$$

The required capacity for a 10-m (32.8-ft) spacing of the sprinkler at $r = 100$ m from Eq. 20.9 is

$$q = \frac{58 \times 2 \times 3.14 \times 100 \times 10 \times 1000}{2 \times 24 \times 3600 \times 1000} = 2.11 \text{ L/s } (33.4 \text{ gpm})$$

Center-pivot systems are usually sold as a complete system from each manufacturer, and thus the example is intended to show only some of the more important design parameters. □

SPECIAL SPRINKLER APPLICATIONS

20.17 Sprinkler Systems for Environmental Control

Sprinkling has been successful for protecting small plants from wind damage, soil from blowing, and plants from frosts or freezing, and for reducing high air and soil temperatures. Since the entire area usually needs protection at the same time, solid-set systems are required. The rate of application should be as low as possible or just enough to achieve the desired control. Small pipes and low-volume sprinklers may be desired to reduce costs, but normal sprinkler systems may be modified for dual use. Because water is applied without regard to irrigation requirements, natural drainage should be adequate or a good drainage system should first be provided.

Especially in organic or sandy soils where onions, carrots, lettuce, and other small seed crops are grown, the soil dries out quickly and the seed may be blown away or covered too deeply for germination. When such plants are small they are also easily damaged by wind-blown soil particles. Protection for such conditions can be provided with a sprinkler system that will apply low rates up to 2.5 mm/h. Operation at night when winds are usually at a minimum will provide more uniform coverage.

Low-growing plants can be protected from freezing injury, which is likely in either the early spring or the late fall. Sprinkling has been most successful against radiation frosts. Water must be applied continuously at about 2.5 mm/h until the plant is free of ice. Sprinkling should be started before the temperature reaches 0°C at the plant level. Strawberries have been protected from temperatures as low as −6°C. Tomatoes, peppers, cranberries, apples, cherries, and citrus have been successfully protected. Tall plants, such as trees, may suffer limb breakage when ice accumulates, but in some areas low-level undertree sprinkling has provided some control. Rates of application may be reduced by increasing the normal sprinkler spacing. A slightly higher pressure may be desirable to increase the diameter of coverage and to give better breakup of the water droplets.

Sprinkling during the day to reduce plant stress has been successful with many plants, such as lettuce, potatoes, green beans, small fruits, tomatoes, cucumbers, and muskmelons. This practice is sometimes called "misting" or "air conditioning" irrigation. Maximum stress in the plant usually occurs at high temperatures, at low humidity, with rapid air movement, on bright cloudless days, and/or with rapidly growing crops on dry soils. Under these conditions crops at a critical state of growth as during emergence, flowering, or fruit enlargement may benefit greatly from low application of water during the midday. At 27°C water loss was reduced 80 percent with an increase in humidity from 50 to 90 percent. Measured temperature reductions in the plant canopy of about 11°C were attained by misting in an atmosphere of 38 percent relative humidity. Green bean yields were increased 52 percent by midday misting during the bloom and pod development period. Potatoes and corn respond to sprinkling, especially when temperatures exceed 30°C. The tasseling period is a critical time for corn. Small quantities of water applied frequently to strawberries increased the quality and the yield by as much as 55 percent. For low-growing crops the same sprinkler system can be adapted for frost protection. Misting in a greenhouse or under a lath house to reduce transpiration of nursery plants for propagation increases plant growth and root development.

During periods of high incoming radiation, soil temperatures may be 20°C greater than the ambient air temperature. Seedlings emerging through soil temperatures as high as 50°C frequently die as a result of high transpiration. Small applications of water at this critical period often ensure emergence and good stands. Another benefit from sprinkling at this stage is enhancement of the effectiveness of herbicides applied to control weeds.

20.18 Sprinkler Systems for Fertilizer, Chemical, or Waste Applications

Fertilizers, soil amendments, and pesticides may be injected into the sprinkler line as a convenient means of applying these materials to the soil or crop. This method primarily reduces labor costs and in some cases may improve the effectiveness and timeliness of application. Liquid manure and sewage wastes are applied with sprinklers for disposing of unwanted material. A large amount of specialized commercial equipment is on the market. Cannery wastes are usually sprinkled on wooded areas or on land in permanent grass. Good subsurface drainage is required.

Liquid and dry fertilizers have been successfully applied with sprinklers. Dry material must first be dissolved in a supply tank. The liquid may be injected on the suction side of the pump, forced under pressure into the discharge line, or injected into the discharge line by a differential pressure device, such as a venturi section. The material is applied for a short time during the irrigation set. Sprinkling should be continued for at least 30 min after the material has been applied to rinse the supply tank, pipes, and crop. For center-pivot or other moving systems the material must be applied continuously or until the field has been completely covered.

For waste disposal systems the solids should be well mixed and small enough so as not to plug the nozzles, which necessitates an effective nonplugging screen on the suction side of the pump. Liquid must be stored in a lagoon or other holding pond. Equipment should be resistant to corrosion from chemicals that may be present in the water. The system should be designed to apply water during subfreezing weather, or sufficient storage should be provided during the nonoperation period. Such systems are usually solid-set and are operated for long periods. Application rates should be lower than the infiltration rate so that the water does not run off the surface causing stream pollution. Excessive applications should be avoided to prevent ground water contamination. Federal and state guidelines must be consulted regarding the application of chemicals and wastes to crops and soils.

20.19   Sprinkler Irrigation Systems for Turf

Sprinkler systems are generally used to irrigate turf, such as golf courses, parks, athletic fields, and lawns. These systems are usually permanent installations and the design criteria are similar to those for agricultural crops. There are several important differences: (1) Triangular sprinkler layouts may be used. (2) Heads are installed flush with the soil surface so they do not interfere with use and maintenance. (3) Sprinklers usually have a "pop-up" feature where the nozzles rise up to 50 mm above the base, so spray is not affected by the grass. (4) Irregular-shaped areas make designs more difficult. (5) Spray is not allowed beyond the boundaries onto roads, walkways, or adjacent property, yet the entire area must be irrigated. To conserve fresh water supplies, treated waste effluent waters are being applied to turf. Installations must meet standards and codes for safety and prevent contamination of the water supply. For example, backflow prevention devices are required so that water cannot flow back into the supply system.

REFERENCES

Berge, I. O., and M. D. Groskopp (1964). *Irrigation Equipment in Wisconsin.* Univ. of Wisconsin Spec. Cir. 90.

Christiansen, J. E. (1948). *Irrigation by Sprinkling.* California Agr. Expt. Sta. Bull. 670.

Chu, S. T., and D. L. Moe (1972). "Hydraulics of a Center Pivot System." *ASAE Trans.* **15**, 894–896.

Dillon, R. C., E. A. Hiler, and G. Vittetoe (1972). "Center-Pivot Sprinkler Design Based on Intake Characteristics." *ASAE Trans.* **15**, 996–1001.

Frost, K. R. (1963). "Factors Affecting Evapotranspiration Losses During Sprin-kling." *ASAE Trans.* **6**, 282–283, 287.

Gray, A. S. (1961). *Sprinkler Irrigation Handbook,* 7th ed. Rain Bird Sprinkler Mfg. Corp., Glendora, CA.

Hagan, R. M., H. R. Haise, and T. W. Edminster (eds.) (1967). *Irrigation of Agricul-tural Lands.* Monograph No. 11. American Society of Agronomy, Madison, WI.

Jensen, M. E. (ed.) (1983). *Design and Operation of Farm Irrigation Systems.* Mono-graph No. 3. ASAE, St. Joseph, MI.

Jensen, M. C., and A. M. Frantini (1957). "Adjusted 'F' Factors for Sprinkler Lateral Design." *Agr. Eng.* **38**, 247.

Keller, J. (1965). "Selection of Economical Pipe Sizes for Sprinkler Irrigation Systems." *ASAE Trans.* **8**, 186–190.

Pair, C. H., W. H. Hinz, K. R. Frost, R. E. Sneed, and T. J. Schiltz (eds.) (1983). *Irrigation,* 5th ed. Sprinkler Irrigation Assoc., Silver Springs, MD.

Scobey, F. C. (1930). *The Flow of Water in Riveted Steel and Analogous Pipes.* USDA Tech. Bull. 150.

U.S. Soil Conservation Service (SCS) (1983). "Sprinkler Irrigation." In *National Engineering Handbook,* Sec. 15: "Irrigation," Chap. 11. Washington, DC.

PROBLEMS

20.1 Determine the required capacity of a sprinkler system to apply water at a rate of 13 mm/h (0.5 iph). Two 186-m (620-ft) sprinkler lines with 16 sprinklers each at 12-m (40-ft) spacing on the line and 18-m (60-ft) spacing between lines are required.

20.2 Allowing 1 h for moving each 186-m (620-ft) sprinkler line described in Problem 20.1, how many hours would be required to apply a 50-mm (2-in.) application of water to a square 16-ha (39.5-ac) field? How many 10-h days are required?

20.3 Determine the discharge rate for one sprinkler operating at 280 kPa (40.6 psi) and having two nozzles, 4.0 and 2.8 mm (5/32 and 7/64 in.) in diameter, with a discharge coefficient of 0.96.

20.4 Compute the depth rate of application for a 1.0-L/s (15.8-gpm) sprinkler head if the sprinkler spacing is 18×24 m (60×80 ft)?

20.5 Compute the total friction loss for a sprinkler system having a 102-mm (4-in.)-diameter main 250 m (820 ft) long and one 76-mm (3-in.) lateral 120 m (394 ft) long. The pump delivers 8.0 L/s (127 gpm) with 12 sprinklers on the lateral. Sixteen-gage aluminum pipe (1.3-mm or 0.051-in. wall thickness) and standard couplers are used.

20.6 Design a sprinkler irrigation system for a square 16-ha (39.5-ac) field to irrigate the entire field within a 14-day period. Not more than 16 h per day is available for moving the pipe and sprinkling. Depth of application is to be 60 mm (2.4 in.) at each setting at a rate not to exceed 9.0 mm/h (0.35 iph). A well 25 m (82 ft) deep located in the center of the field will provide the following drawdown–discharge characteristics obtained from a well test:

10 m (33 ft) — 12 L/s (190 gpm)
15 m (50 ft) — 15 L/s (238 gpm)
20 m (65 ft) — 19 L/s (300 gpm)

Design for an average pressure of 276 kPa (40 psi) at the nozzle. Highest point in the field is 2 m (6.5 ft) above the well site, and 1-m (3-ft) risers are needed on the sprinklers. Assuming a pump efficiency of 60 percent and assuming that the engine will furnish 70 percent of its rated output for continuous operation, determine the rated output for a water-cooled internal combustion engine.

20.7 Determine the total pumping head for a sprinkler irrigation system on level land. The average operating pressure at the nozzle is 276 kPa (40 psi); the friction loss in the main is 6 m (20 ft), and in the lateral, 3.7 m (12 ft); the drawdown of the well is 4.3 m (14 ft) at the required discharge of 32 L/s (507 gpm); the riser height is 1.5 m (5 ft); and friction loss in all valves is 3.0 m (10 ft). Determine the power requirements for the pump if it operates at 65 percent efficiency.

20.8 Derive the equation for sprinkler nozzle discharge, $q = 0.00111Cd_n^2P^{1/2}$. (*Hint:* Pressure is proportional to velocity head.)

20.9 Assuming a triangular depth-distribution pattern for a rectangular placement of sprinklers and a diameter of coverage of 30 m (100 ft), compute the uniformity coefficient for a 12 × 18-m (40 × 60-ft) sprinkler spacing using 6-m (20-ft) grids for depth measurements. Assume also that the depth of application at the sprinkler is 25 mm (1 in.) and zero at 15 m (50 ft).

20.10 A 12.1-ha (30-ac) field is to be irrigated at a maximum rate of 10 mm/h (0.4 iph) with a sprinkler system. The root zone is 1 m (3.3 ft) deep and the available water capacity of the soil is 200 mm/m (2.4 in./ft) of depth. The water application efficiency is 70 percent, and the soil is to be irrigated when 45 percent of the available water capacity is depleted. The peak rate of water use is 5.0 mm/day (0.2 ipd). Determine the net depth of application per irrigation, depth of water to be pumped, days to cover the field, and area to be irrigated per day.

20.11 If the pressures at opposite ends of a sprinkler lateral are 290 and 262 kPa (42 and 38 psi), what would be the discharge of the distal sprinkler provided the sprinkler at the 290-kPa (42-psi) end discharged 0.8 L/s (12.7 gpm)?

20.12 Twenty sample cans are uniformly spaced in the area covered by four sprinklers. The following depths were caught in the cans: 26, 23, 21, 27, 24, 22, 22, 26, 21, 20, 19, 23, 20, 18, 17, 20, 26, 22, 15, and 18 mm. Determine the uniformity coefficient and the distribution uniformity.

20.13 A center-pivot system is to irrigate the circular area of a square 64-ha (158-ac) level field. Peak water use is 6 mm/day, application efficiency is 75 percent, and maximum operation time is 22 h/day. Assume 161.6-mm (6.36-in.) lateral inside diameter (steel pipe), 3-m lateral height, and end gun with 30-m (100-ft) wetted diameter at 35 m (50 psi) pressure. Determine total discharge, pressure at the pivot, and discharge of the end gun. Also find the discharge and average application rate of a sprinkler at 370-m (1214-ft) radius assuming a 28-m (92-ft) wetted diameter and a 6-m (19.7-ft) spacing between sprinklers.

CHAPTER 21

Microirrigation

Microirrigation is a method for delivering slow, frequent applications of water to the soil using a low-pressure distribution system and special flow-control outlets. Microirrigation is also referred to as drip, subsurface, bubbler, or trickle irrigation, and all have similar design and management criteria.

These systems deliver water to individual plants or rows of plants. The outlets are generally placed at short intervals along small tubing, and unlike surface or sprinkler irrigation, only the soil near the plant is watered. The outlets include emitters, orifices, bubblers, and sprays or microsprinklers with flows ranging from 2 to over 200 L/h.

According to Karmeli and Keller (1975), microirrigation research began in Germany about 1860. In the 1940s it was introduced in England especially for watering and fertilizing plants in greenhouses. With the increased availability of plastic pipe and the development of emitters in Israel in the 1950s, it has since become an important method of irrigation in Australia, Europe, Israel, Japan, Mexico, South Africa, and the United States (California, Hawaii, and Florida). According to the "1989 Irrigation Survey" in *Irrigation Journal* (1990) California had 190 000 ha, and Florida had 160 000 ha; the U.S. total was over 540 000 ha.

Microirrigation has been accepted mostly in the more arid regions for watering high-value crops, such as fruit and nut trees, grapes and other vine crops, sugar cane, pineapples, strawberries, flowers, and vegetables. Although successfully used on cotton, sorghum, and sweet corn, microirrigation is not as well adapted to field crops.

21.1 Advantages and Disadvantages of Microirrigation

With microirrigation only the root zone of the plant is supplied with water, and with proper system management deep percolation losses are minimal. Soil evaporation may be lower because only a portion of the surface area is wet. Like solid-set sprinkler systems, labor requirements are lower and the systems can be readily automated. Reduced percolation and evaporation losses result in a greater economy of water use. Weeds are more easily controlled, especially for the soil area that is not irrigated. Bacteria, fungi, and other pests and diseases that depend on a moist environment are reduced as the above-ground plant parts normally are completely dry. Because soil is kept at a high water level and the water does not contact the plant leaves, use of more saline water may be possible with less stress and damage to the plant, such as leaf burn. Field edge losses and spray evaporation, such as occur with sprinklers, are reduced with these systems. Low rates of water application at lower pressures are possible so as to eliminate runoff. With some crops, yields and quality are increased probably as a result of maintenance of a high temporal soil water level adequate to meet transpiration demands. Crop yield experiments have shown wide differences varying from little or no difference to 50 percent increase compared with other methods of irrigation. Crop quality may also be improved. Some fertilizers and pesticides may be injected into the system and applied in small quantities, as needed, with the water. With good system design and management, this practice can minimize chemical applications and reduce chemical movement to the ground water supply.

The major disadvantages of microirrigation are high cost and clogging of system components, especially emitters, by particulate, biological, and chemical matter. Emitters are not well suited to certain crops and special problems may be caused by salinity. Salt tends to accumulate along the fringes of the wetted surface strip (Fig. 21.1). Since these systems normally wet only part of the potential soil-root volume, plant roots may be restricted to the soil volume near each emitter as shown in Fig. 21.1. The dry soil area between emitter lateral lines may result in dust formation from tillage operations and subsequent wind erosion. Compared with surface irrigation systems, more highly skilled labor is required to operate and

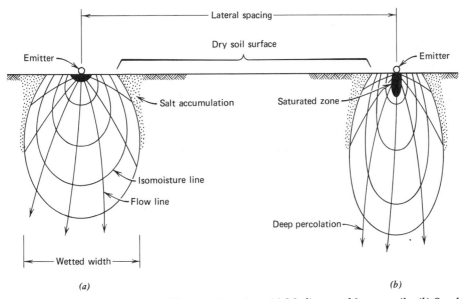

Fig. 21.1   Soil wetting pattern with microirrigation. (*a*) Medium and heavy soils. (*b*) Sandy soils. (Adapted from Karmeli and Keller, 1975, and FAO, 1973.)

maintain the filtration equipment and other specialized components. Rodents may also damage tubing or other plastic components.

21.2 Layout and Components of Microirrigation Systems

System layouts are similar to sprinkler systems (Chapter 20). As with sprinkler systems many arrangements are possible. The one given in Fig. 21.2 shows split-line operation for the upper left quadrant of the 16-ha orchard described in Fig. 20.9. The well is located in the center of the larger field. The layout would be similar for the other quadrants. Tree rows are parallel to the laterals. Sections 1, 2, 3, and 4 of the 4-ha quadrant in Fig. 21.2 could be operated independently of each other or in any combination since each section has its own control valve.

As shown in Figs. 21.2 and 21.3, the primary components of a microirrigation system are a pump, a filter, an injector, a main and submain, a manifold, and a lateral line to which the emitters are attached. The manifold is a line to which the laterals are connected. Pressure regulators, pressure gages, a water meter, flushing valves, time clocks, and automating control devices are other desirable components. The manifold, submain, and main may be laid on the surface or buried underground. The manifold is usually flexible pipe if laid on the surface or rigid pipe if buried. The main lines may be any type of pipe, such as polyethylene (PE), polyvinyl chloride (PVC), galvanized steel, or aluminum. The lateral lines that have emitters are usually flexible PVC or PE tubing. They generally range from 10

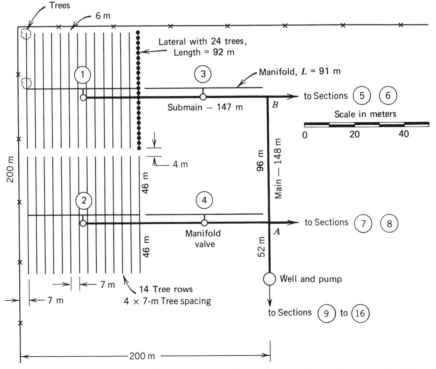

Fig. 21.2 Microirrigation system layout for a 4-ha quadrant of a 16-ha orchard with a well in the center of a square field.

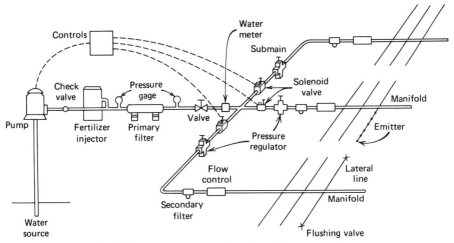

Fig. 21.3   Components of a microirrigation system.

to 32 mm in diameter and have emitters spaced at short intervals appropriate for the crop to be grown.

An efficient filter is the most important component of the microirrigation system because of emitter clogging. Most water should be cleaner than drinking water. Microirrigation systems generally require screen, gravel, or graded sand filters. Recommendations of the emitter manufacturer should be followed in selecting the filtration system. In the absence of such recommendations the net opening diameter of the filter should be smaller than one tenth to one fourth of the emitter opening diameter. For clean ground water an 80- to 200-mesh filter may be adequate. This filter will remove soil, sand, and debris, but should not be used with high-algae water. For high silt and algae water a sand filter backed up with a screen filter may be required. A sand separator ahead of the filter may be necessary if the water contains considerable sand. In-line strainers with replaceable screens and cleanout plugs may be adequate with small amounts of sand. Secondary filters may be installed at the inlet to each manifold. These are recommended as a safety precaution should accidents during cleaning or filter damage allow particles or unfiltered water to pass into the system. Filters must be cleaned and serviced regularly. Pressure loss through the filter should be monitored as an indication for maintenance.

Lateral lines may be located along the row of trees, with several emitters required for each tree as shown in Fig. 21.4. Many laterals have multiple emitters, such as the "spaghetti" tubing or "pigtail" lines shown in Fig. 21.4c. One or two laterals per row (Fig. 21.4a or 21.4b) may be provided, depending on the size of the trees. With small trees a single line is adequate.

Many types and designs of emitters are commercially available, some of which are shown in Fig. 21.5. The emitter controls the flow from the lateral. The pressure is greatly decreased by the emitter; this loss is accomplished by small openings, long passageways, vortex chambers, manual adjustment, or other mechanical devices. Some emitters may be pressure-regulated by changing the length or cross section of passageways or size of orifice. These emitters (Fig. 21.5c) give nearly a constant discharge over a wide range of pressures. Some are self-cleaning and

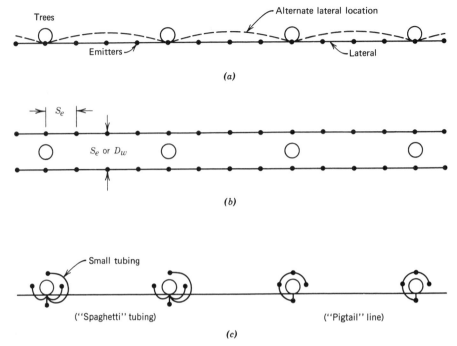

Fig. 21.4 Lateral and emitter locations for an orchard. (*a*) Single lateral for each row of trees. (*b*) Two laterals for each tree row. (*c*) Multiple-exit emitters.

flush automatically. Porous pipe or tubing may have many small openings as shown in Figs. 21.5*e* to 21.5*g*. The actual size is much smaller than indicated in the drawing. Some holes are barely visible to the naked eye. The double-tube lateral shown in Fig. 21.5*g* has more openings in the outer channel than in the main flow channel. Such tubes have thin walls and are low in cost. In Hawaii they are often discarded after the crop is harvested and replaced with new lines. Most emitters are placed on the soil surface, but they may be buried at shallow depths for protection.

21.3 Emitter Discharge

In an orifice-type emitter (Fig. 21.5*d*) the flow is fully turbulent and the discharge can be determined from the sprinkler nozzle equation (Eq. 20.3). The discharge of any emitter may be expressed by the power-curve equation (Karmeli and Keller, 1975)

$$q = Kh^x \tag{21.1}$$

where q = emitter discharge (L³/T),
K = constant for each emitter,
H = pressure head (L or F/L²),
x = emitter discharge exponent.

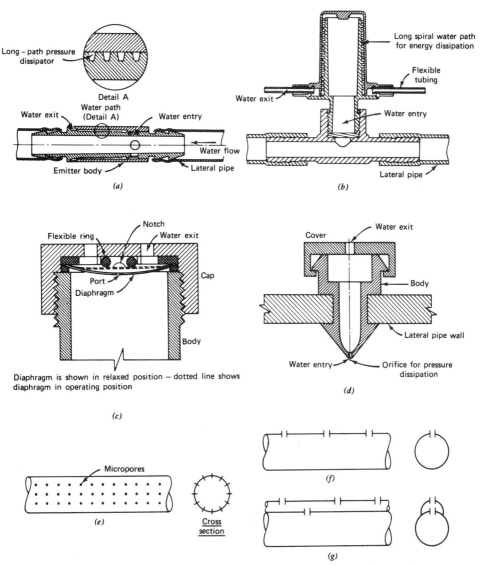

Fig. 21.5 Types of microirrigation laterals and emitters. (*a*) In-line long-path single-exit emitter. (*b*) In-line long-path multiple-exit emitter. (*c*) Self-flushing emitter. (*d*) On-line orifice-type emitter. (*e*) Porous tubing. (*f*) Single-tube lateral. (*g*) Double-tube lateral. (*a* to *d* redrawn from Karmeli and Keller, 1975.)

The exponent x can be determined by measuring the slope of the log–log plot of head versus discharge. With x known, K can be determined from Eq. 21.1. In fully turbulent flow $x = 0.5$, and in a laminar flow regime $x = 1.0$. In a fully pressure-compensating emitter K is a constant for a wide range of pressure and $x = 0$. Because of the large number of emitters available, it may be more convenient to determine discharge directly from manufacturer's curves. Some are shown in Fig. 21.6 from which K and x were computed using Eq. 21.1. An average $x = 0.63$ was

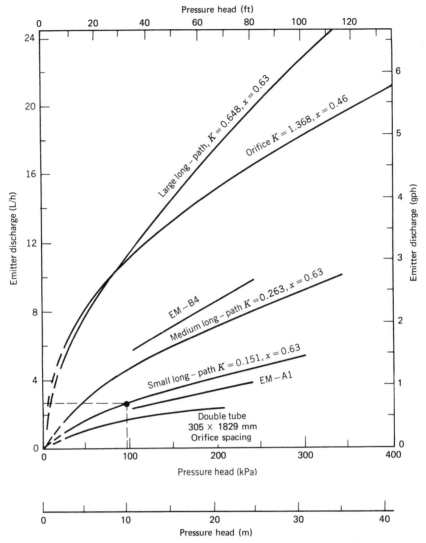

Fig. 21.6 Discharge of various emitters versus pressure head. Values of K and x are for pressure in kPa and flow in L/h. (Adapted from Karmeli and Keller, 1975; Walker, 1979; and Davis, 1976.)

used to compute K for the small, medium, and large long-path emitters using h in kPa. These were rated at 2, 4, and 8 L/h, respectively. Double-tube laterals are typically rated as discharge per length, for example, 3 L/min per 100 m at 70 kPa pressure. Manufacturer's data should be obtained for the actual discharge-versus-pressure rating. The EM-A1 and EM-B4 emitters have straight lines with $x = 1.0$, indicating laminar flow. Emitter discharge usually varies from about 1 to 30 L/h and pressures range from 15 to 280 kPa. Average diameters of openings for emitters range from 0.0025 to 0.25 mm. Emitters made from thermoplastic material may vary in discharge depending on the temperature. Thus, discharge curves should be corrected for temperature.

21.4 Water Distribution from Emitters

Microirrigation was developed to provide more efficient application of water. An ideal system should provide a uniform discharge from each emitter. Application efficiency depends on the variation of emitter discharge, pressure variation along the lateral, and seepage below the root zone or other losses, such as soil evaporation. Emitter discharge variability is greater than that for sprinkler nozzles because of smaller openings and lower design pressures. Such variability may result from the design of the emitter, materials, and care in manufacture. Solomon (1979) found that the statistical coefficient of variation may range from 0.02 to 0.4. The coefficients of variation (C_v = standard deviation/mean) should be available for emitters and provided by the manufacturer. ASAE (1989) guidelines for classification of emitter uniformity are shown in Table 21.1.

Microirrigation systems must deliver the water required to each plant with minimum losses to obtain high efficiencies. This is achieved by having a high uniformity of water delivery by each section of the system having a separate control valve. Thus, Sections 1, 2, 3, and 4 in Fig. 21.2 should each be designed for a high uniformity. The uniformity varies with pressure, emitter variation, and numbers of emitters per plant. This is defined by the emission uniformity.

$$EU = 100[1 - 1.27C_v/(n^{0.5})]q_{min}/q_{avg} \qquad (21.2)$$

where EU = emission uniformity,
   C_v = manufacturer's coefficient of variation,
   n = number of emitters per plant for trees and shrubs or 1 for line sources,
   q_{min} = minimum emitter discharge rate for the minimum pressure in the section,
   q_{avg} = average or design emitter discharge rate for the section.

Recommended design values for EU are shown in Table 21.2.

☐ Example 21.1

Determine the emission uniformity of a system section that uses an emitter with $K = 0.3$, $x = 0.57$, $C_v = 0.06$, and two emitters per plant with average pressure of 100 kPa and minimum pressure of 90 kPa.

Table 21.1 Recommended Classification of Manufacturer's Coefficient of Variation, C_v

Emitter Type	C_v Range	Classification
Point source	<0.05	Excellent
	0.05–0.07	Average
	0.07–0.11	Marginal
	0.11–0.15	Poor
	>0.15	Unacceptable
Line source	<0.10	Good
	0.10–0.20	Average
	>0.20	Marginal to unacceptable

Source: ASAE (1989).

Table 21.2 Recommended Ranges of Design Emission Uniformity, (EU)

Emitter Type	Spacing (m)	Topography	Slope (%)	EU Range (%)
Point source	>4	Uniform	<2	90–95
		Steep or undulating	>2	85–90
Point source	<4	Uniform	<2	85–90
		Steep or undulating	>2	80–90
Line source	All	Uniform	<2	80–90
		Steep or undulating	>2	70–85

Source: ASAE (1989).

Solution. Substitute Eq. 21.1 into Eq. 21.2

$$EU = 100 \left[1 - \frac{1.27 \, (0.06)}{2^{0.5}} \right] \frac{0.3 \, (90)^{0.57}}{0.3 \, (100)^{0.57}} = 89 \text{ percent}$$

□

21.5 Microirrigation System Design

A major difference between microirrigation and other systems is that not all the area will be irrigated, especially for widely spaced plants. A minimum of 30 percent of the area must be irrigated, with the possible exception where significant rainfall occurs during the irrigation season. For mature trees 75 percent of the area may need to be irrigated, whereas nearly 100 percent of the area is irrigated for closely spaced plants in arid regions.

Emitter spacing and numbers required depends on the wetting pattern and plant spacing. Field tests should be conducted at several representative locations to obtain data on horizontal and vertical water movement. If field measurements are not available, estimates may be obtained from Table 21.3 for the maximum horizontal wetted diameter D_w from a single outlet. For a line source, the outlet spacing S_e should be less than or equal to $0.8D_w$ to overlap the wetting patterns of adjacent emitters along a lateral. For double laterals, a spacing of D_w between laterals will adequately wet the area; however, if outlets are individually spaced, the spacing can be D_w in both directions. The spacing between laterals and between individual outlets should be reduced to the outlet spacing along the lateral S_e if the water is saline. These closer spacings reduce dry areas between emitters where salts might accumulate. The number of emitters required per plant, n, can be obtained from

$$n = \frac{p \times \text{area/tree}}{\text{effective area wetted by one emitter}} \tag{21.3}$$

where p is the percentage of the total area shaded by the crop, area/tree is based on the tree spacing, and effective area wetted by one emitter depends on the wetted diameter, emitter layout, and water quality.

Table 21.3   Estimated Maximum Diameter of the Wetted Circle Formed by a Single Emission Outlet Discharging 4 L/h on Various Soils

Soil or Root Depth and Soil Texture	Homogeneous Soil (m)	Varying Layers	
		Generally Low Density (m)	Generally Medium Density (m)
Depth 0.75 m			
Coarse	0.45	0.75	1.05
Medium	0.90	1.2	1.5
Fine	1.05	1.5	1.8
Depth 1.5 m			
Coarse	0.75	1.4	1.8
Medium	1.2	2.1	2.7
Fine	1.5	2.0	2.4

Source: Adapted from SCS (1984).

☐ Example 21.2

For the layout in Fig. 21.2 with mature trees determine the number of emission devices needed per tree if 35 percent of the area is to be irrigated, salt content of the irrigation water is low, soil is medium textured and layered with medium density, and emitters will be installed on a "pigtail" as in Fig. 21.4c.

Solution.   Assume the effective rooting depth is 0.75 m; then from Table 21.3, $D_w = 1.5$ m. The effective wetted area is assumed to be 0.8(1.5 m) $\times$ 1.5 m, which is substituted into Eq. 21.3.

$$n = \frac{0.35 \times 4 \times 7}{0.8(1.5) \times (1.5)} = 5.4 \text{ or round to 6 emitters per tree}$$

Note that if the irrigation water is saline, the effective wetted area is reduced to 0.8(1.5) $\times$ 0.8(1.5), which provides overlap of the wetting patterns both along and between laterals and reduces salt accumulations within the wetted area.   ☐

The evapotranspiration (ET) of crops under microirrigation is not well defined because if the area is not entirely shaded, ET is somewhat less than under conventional systems that irrigate the entire area. Karmeli and Keller (1975) suggested the following water-use rate for microirrigation design:

$$ET_t = ET \times p/85 \tag{21.4}$$

where ET_t = peak evapotranspiration rate for crops under microirrigation,
      p = percentage of the total area shaded by the crop,
     ET = peak conventional ET rate for the crop.

For example, if a mature orchard shades 70 percent of the area and the peak conventional ET is 7 mm/day, the microirrigation design rate is 5.8 mm/day (7 $\times$ 70/85).

The diameter of the laterals and manifolds should be selected to satisfy Eq. 21.2

and the appropriate EU from Table 21.2. The maximum difference in pressure usually occurs between the control point at the inlet and the pressure at the emitter farthest from the inlet. The inlet is usually at the manifold where the pressure is regulated. In Fig. 21.2 the maximum difference in head loss to the farthest emitter is that for one half the lateral length plus one half the manifold length. Where the manifold is connected to the end of each lateral and the submain is connected to the end of a manifold, the head loss would be computed for their entire length.

For minimum cost, Karmeli and Keller (1975) recommended that on a level area, 55 percent of the allowable head loss should be allocated to the lateral and 45 percent to the manifold. As in sprinkler laterals, allowable head loss should be adjusted for elevation differences along the lateral and along the manifold, unless pressure-compensating devices are used.

The friction loss for mains and submains can be computed from the Hazen–Williams or Darcy–Weisbach equation. From Watters and Keller (1978) the Darcy–Weisbach equation for smooth pipes in microirrigation systems when combined with the Blasius equation for the friction factor is

$$H_f = KLQ^{1.75}D^{-4.75} \tag{21.5}$$

where H_f = friction loss in m,
$\quad\quad K$ = a constant = 7.89×10^5 for SI units for water at 20°C,
$\quad\quad L$ = pipe length in m,
$\quad\quad Q$ = total pipe flow in L/s,
$\quad\quad D$ = inside pipe diameter in mm.

Equation 21.5 applies for continuous sections of plastic pipe. For in-line emitters (Figs. 21.5a and 21.5b), on-line emitters (Fig. 21.5d), and other connectors the head loss should be increased. Such losses may be expressed as equivalent length of lateral pipe. This increase in length L_e can be estimated as follows (Karmeli and Keller, 1975):

(1) L_e = 1.0 to 3.0 m for each in-line emitter (Fig. 21.5a),
(2) L_e = 0.1 to 0.6 m for each on-line emitter (emitter attached by insert through pipe wall, Fig. 21.5d),
(3) L_e = 0.3 to 1.0 m for a solvent-welded Tee connector.

The design process requires determination of the peak ET, water required per plant, emitter selection, design operating pressure, minimum pressure required to maintain the desired EU, pressure variation allowed in a section, and pipe sizes. The allowable head variation in a section Δh can be estimated from (SCS, 1984)

$$\Delta h = 2.5(h_{avg} - h_{min}) \tag{21.6}$$

where h_{avg} = design operating pressure or head,
$\quad\quad h_{min}$ = minimum pressure or head in a section.

For sloping fields, Δh must include both friction losses and elevation differences.

□ *Example 21.3*

Design a microirrigation system for the mature orchard layout shown in Fig. 21.2. Assume the field is level, maximum time for irrigation is 22 h/day, 35 percent of the area is to be irrigated, and peak evapotranspiration rate corrected for shaded area is 5.8 mm/day (0.23 ipd). Sections 1 and 2 are irrigated together every 2 days; Sections 3 and 4 are irrigated on the alternate days. In addition, each quarter of the field has similar sections irrigated every other day.

Solution. Determine the emitter discharge, the required discharge for each section, and the total pumping head to irrigate the field.

(1) From Table 21.2 the minimum EU is 90 percent. The volume of water required per tree is

$$\text{Volume} = \frac{5.8 \text{ mm/day} \times 4 \text{ m} \times 7 \text{ m} \times 1000 \text{ L/m}^3}{1000 \text{ mm/m} \times 0.90} = 180 \text{ L/day (48 gpd)}$$

Since each tree is irrigated every other day, 360 L must be delivered at each irrigation to each tree.

(2) From Example 21.2, six emitters are used per tree and each emitter must deliver 60 L (16 gal) in a maximum of 22 h or $60/22 = 2.733$ L/h (0.72 gph).

(3) From Fig. 21.6 select the small long-path emitter with $K = 0.151$ and $x = 0.63$. Substituting into Eq. 21.1, $h = (2.73/0.151)^{1/0.63} = 99$ kPa or, rounded, 100 kPa (10.2 m or 33.4 ft). This is the average operating pressure.

(4) The discharge in each of the lines is as follows:

Line	Number of Trees	Number of Emitters	Required Discharge [L/s (gpm)]
Half-lateral	12	72	0.055 (0.876)
Half-manifold	168	1008	0.77 (12.2)
Submain, B to Section 1	336	2016	1.54 (24.4)
Main, A to B	672	4032	3.08 (48.8)
Main, pump to A	1344	8064	6.16 (97.6)

(5) Determine the minimum discharge of the emitter in the farthest lateral with $C_v = 0.05$ by rearranging Eq. 21.2.

$$q_{min} = 90(2.73)/\{100[1 - 1.27(0.05)/6^{0.5}]\} = 2.52 \text{ L/h (0.67 gph)}$$

The minimum head in the last lateral in Section 1 is obtained from Eq. 21.1.

$$h = (2.52/0.151)^{1.0/0.63} = 87 \text{ kPa (8.87 m or 29.1 ft)}$$

(6) The pressure variation allowed in Section 1 is determined from Eq. 21.6.

$$\Delta h = 2.5 (100 - 87) = 32 \text{ kPa (3.2 m or 10.8 ft)}$$

The maximum allowable inlet pressure to a section is $87 + 32 = 119$ kPa.

(7) Assume a barbed Tee is used to connect the "pigtail" to the lateral which has a friction loss equivalent to 2 m of pipe. Since there are 12 trees per half-lateral, the equivalent lateral length is $46 + 12 \times 2 = 70$ m (230 ft). The friction loss allowed in the half-lateral is $0.55 \times 32 = 17.6$ kPa (1.8 m or 5.9 ft) and $0.45 \times 32 = 14.4$ kPa (1.5 m or 4.9 ft) in the half-manifold.

(8) Compute the friction loss in each of the lines from Eq. 21.5 by selecting a diameter to keep the loss within the allowable limits previously specified. Assume PE pipe for the lateral and PVC for the manifold and mains.

Line	Q (L/s)	Pipe i.d.[a] [mm (in.)]	$L + L_e$ (m)	F[b]	H_f (m)
Half-lateral	0.055	15.8 (0.622)	46 + 24	0.38	0.30
Half-manifold	0.77	26.6 (1.049)	45.5 + 3.5	0.40	1.66
Submain, B to Section 1	1.54	52.5 (2.067)	147	1	1.7
Main, A to B	3.08	52.5 (2.067)	96	1	3.7
Main, pump to A	6.16	62.7 (2.469)	50	1	2.8

[a]Actual diameters of the next largest commercial pipe size were used.
[b]From Table 20.7 for multiple-outlet lines and m = 1.75.

(9) Determine the pressure head at the inlet to the manifold using Eq. 20.7 and values from Steps 3 and 8 above.

$$H = 10.2 + 0.75(0.30 + 1.66) + 0 + 0 = 11.8 \text{ m (38.7 ft)}$$

(10) The total head at the outlet from the pumping station is

$$11.8 + 1.7 + 3.7 + 2.8 = 20.0 \text{ m (65.6 ft).}$$

To irrigate the entire field, the pump must deliver 12.3 L/s (2×6.16). After allowance for pump wear, losses through filters, pressure regulators, valves, and other devices, the pump must net 20 m of head. Computer programs are available to aid the design of microirrigation systems. (Appendix I).

21.5 Microirrigation of Landscape Plants

Landscape plantings, which are often arranged in small groups or as individual plants, are well suited to microirrigation. Water can be supplied to meet the needs of each plant. Simple systems may be connected to a valve on a house and consist of a backflow preventer, a small filter, a pressure regulator, and a lateral with emitters. Complex systems must be designed for large areas such as parks and golf courses, around large structures, or along highways. In designing these systems, care must be taken to have plants with similar water requirements on the same control valve. Plants in greenhouses and nurseries also are easily irrigated with microirrigation systems since water and nutrients can be delivered to individual plants. Backflow prevention devices are required to prevent contamination of the water supply.

REFERENCES

American Society of Agricultural Engineers (ASAE) (1989). *Design, Installation, and Performance of Trickle Irrigation Systems*. EP405.1. ASAE, St. Joseph, MI.

Benami, A., and A. Ofen (1983). *Irrigation Engineering*. Irrig. Eng. Sci. Pub., Haifa, Israel.

Davis, D. D. (1976). *Rain Bird Irrigation Systems Design Handbook*. Rain Bird Sprinkler Mfr. Corp., Glendora, CA.

Howell, T. A., and E. A. Hiler (1974). "Trickle Irrigation Lateral Design." *ASAE Trans.* **17**, 902–908.

"1989 Irrigation Survey." (1990). *Irrig. J.* **40**(1), 23–34.

Karmeli, D., and J. Keller (1975). *Trickle Irrigation Design*. Rain Bird Sprinkler Mfr. Corp. Glendora, CA.

Nakayama, F. S., and D. A. Bucks (eds.) (1986). *Trickle Irrigation for Crop Production*. Elsevier, Amsterdam.

Solmon, K. (1979). "Manufacturing Variation of Trickle Emitters." *ASAE Trans.* **22**, 1034–1038, 1043.

United Nations' Food and Agriculture Organization (FAO) (1973). *Trickle Irrigation*. FAO, Rome, Italy.

U.S. Soil Conservation Service (SCS) (1984). "Trickle Irrigation." In *National Engineering Handbook*, Sect. 15: "Irrigation," Chap. 7. Washington, DC.

Walker, W. R. (1979). *Sprinkler and Trickle Irrigation*, 3rd ed. (mimeo.). Dept. of Agr. and Chem. Eng., Colorado State University, Fort Collins, CO.

Watters, G. Z., and J. Keller (1978). *Trickle Irrigation Tubing Hydraulics*. Paper 78-2015. ASAE, St. Joseph, MI.

Wu, I. P., and D. D. Fangmeier (1974). *Hydraulic Design of Twin-Chamber Trickle Irrigation Laterals*. Arizona Agr. Expt. Sta. Tech. Bull. 216.

Wu, I. P., and H. M. Gitlin (1974). *Design of Drip Irrigation Lines*. Hawaiian Agr. Expt. Sta. Tech. Bull. 96.

PROBLEMS

21.1 Determine the emission uniformity for large long-path emitters if the manufacturer's coefficient of variation is 0.1. There are four emitters per plant, average operating pressure is 80 kPa, and minimum pressure is 70 kPa.

21.2 If the conventional peak ET is 7.6 mm/day (0.3 in./day) and 75 percent of the area is shaded by trees in an orchard, determine the design ET rate, volume of water required per tree per day, and application rate in L/h per tree for a microirrigation system. Assume EU = 0.92, a tree spacing of 3 × 6 m (10 × 20 ft), 20 h/day operation, and irrigation interval of 2 days.

21.3 For the orchard in Problem 21.2, determine the number of emitters required per tree. Assume a coarse-textured layered soil of low density, a 1.5-m root zone, and 40 percent of the area is to be irrigated.

21.4 From the data in Example 21.3, determine the maximum allowable lateral length if the maximum allowable loss is 1.8 m (5.9 ft) and the inside diameter is 15.8 mm (0.622 in.).

21.5 From the data in Example 21.3, determine the friction loss in the 15.8-mm-diameter half-lateral if 72 in-line emitters are uniformly spaced along the line as in Fig. 21.4a.

21.6 From the data in Example 21.3, redesign the delivery system for half-laterals 92 m long. Assume the manifolds remain 91 m long.

APPENDIX A

Runoff Determination

Table A.1 Time of Concentration for Small Watersheds

Maximum Length of Flow [m (ft)]	Watershed Gradient (%)					
	0.05	0.1	0.5	1.0	2.0	5.0
	Time of Concentration (min)[a]					
152 (500)	18	13	7	6	4	3
305 (1 000)	30	23	11	9	7	5
610 (2 000)	51	39	20	16	12	9
1220 (4 000)	86	66	33	27	21	15
1830 (6 000)	119	91	46	37	29	20
2440 (8 000)	149	114	57	47	36	25
3050 (10 000)	175	134	67	55	42	30
6100 (20 000)	306	234	117	97	74	52

[a]Computed from Eq. 4.2.

Table A.2 Runoff Coefficients for Urban Areas

Type of Drainage Area	Runoff Coefficient, C
Business	
Downtown areas	0.70–0.95
Neighborhood areas	0.50–0.70
Residential	
Single-family areas	0.30–0.50
Multiunits, detached	0.40–0.60
Multiunits, attached	0.60–0.75
Suburban	0.25–0.40
Apartment dwelling areas	0.50–0.70

Table A.2   Runoff Coefficients for Urban Areas
(*Continued*)

Type of Drainage Area	Runoff Coefficient, C
Industrial	
Light areas	0.50–0.80
Heavy areas	0.60–0.90
Parks, cemeteries	0.10–0.25
Playgrounds	0.20–0.35
Railroad yard areas	0.20–0.40
Unimproved areas	0.10–0.30
Streets	0.70–0.95
Brick	
Drives and walks	0.75–0.85
Roofs	0.75–0.95

Source: Chow, V. T. (1962). *Hydrologic Determination of Waterway Areas for the Design of Drainage Structures in Small Drainage Basins.* Illinois Eng. Expt. Sta. Bull. 462.

APPENDIX B

Manning Velocity Formula

Table B.1 Roughness Coefficient *n* for Manning Formula

Line No.	Type and Description of Conduits	n Value[a]		
		Minimum	*Design*	*Maximum*
	Channels, Lined			
1	Asphaltic concrete, machine placed		0.014	
2	Asphalt, exposed prefabricated		0.015	
3	Concrete	0.012	0.015	0.018
4	Concrete, rubble	0.017		0.030
5	Metal, smooth (flumes)	0.011		0.015
6	Metal, corrugated	0.021	0.024	0.026
7	Plastic	0.012		0.014
8	Shotcrete	0.016		0.017
9	Wood, planed (flumes)	0.010	0.012	0.015
10	Wood, unplaned (flumes)	0.011	0.013	0.015
	Channels, Earth			
11	Earth bottom, rubble sides	0.028	0.032	0.035
	Drainage ditches, large, no vegetation			
12	(*a*) <0.8 m, hydraulic radius	0.040		0.045
13	(*b*) 0.8–1.2 m, hydraulic radius	0.035		0.040
14	(*c*) 1.2–1.5 m, hydraulic radius	0.030		0.035
15	(*d*) >1.5 m, hydraulic radius	0.025		0.030
16	Small drainage ditches	0.035	0.040	0.040
17	Stony bed, weeds on bank	0.025	0.035	0.040
18	Straight and uniform	0.017	0.0225	0.025
19	Winding, sluggish	0.0225	0.025	0.030

continued

Table B.1 Roughness Coefficient *n* for Manning Formula (*Continued*)

Line No.	Type and Description of Conduits	Minimum	Design	Maximum
	Channels, Vegetated			
	(grassed waterways) (Chapter 7)			
	Dense, uniform stands of green vegetation more than 250 mm long			
20	(*a*) Bermuda grass	0.04		0.20
21	(*b*) Kudzu	0.07		0.23
22	(*c*) Lespedeza, common	0.047		0.095
	Dense, uniform stands of green vegetation cut to a length less than 60 mm			
23	(*a*) Bermuda grass, short	0.034		0.11
24	(*b*) Kudzu	0.045		0.16
25	(*c*) Lespedeza	0.023		0.05
26	Sorghum, 1-m rows	0.04		0.15
27	Wheat, mature poor	0.08		0.15
	Natural Streams			
28	(*a*) Clean, straight bank, full stage, no rifts or deep pools	0.025		0.033
29	(*b*) Same as (*a*) but some weeds and stones	0.030		0.040
30	(*c*) Winding, some pools and shoals, clean	0.035		0.050
31	(*d*) Same as (*c*), lower stages, more ineffective slopes and sections	0.040		0.055
32	(*e*) Same as (*c*), some weeds and stones	0.033		0.045
33	(*f*) Same as (*d*), stony sections	0.045		0.060
34	(*g*) Sluggish river reaches, rather weedy or with very deep pools	0.050		0.080
35	(*h*) Very weedy reaches	0.075		0.150
	Pipe			
36	Cast iron, coated or uncoated	0.011	0.013	0.015
37	Clay or concrete drain tile (76–760 mm dia.)	0.011	0.013	0.020
38	Concrete or clay vitrified sewer pipe	0.01	0.014	0.017
39	CPT, Corrugated plastic tubing, 76–203 mm dia.		0.015	
40	CPT, 254–305 mm dia.		0.017	
41	CPT, >305 mm dia.		0.02	
42	CPT, smooth wall inside		0.009	
43	Metal, corrugated, ring	0.021	0.025	0.026
44	Metal, corrugated, helical	0.013	0.015	
45	Steel, riveted and spiral	0.013	0.016	0.017
46	Wood stave	0.010	0.013	
47	Wrought iron, black	0.012		0.015
48	Wrought iron, galvanized	0.013	0.016	0.017

[a]Selected from numerous references.

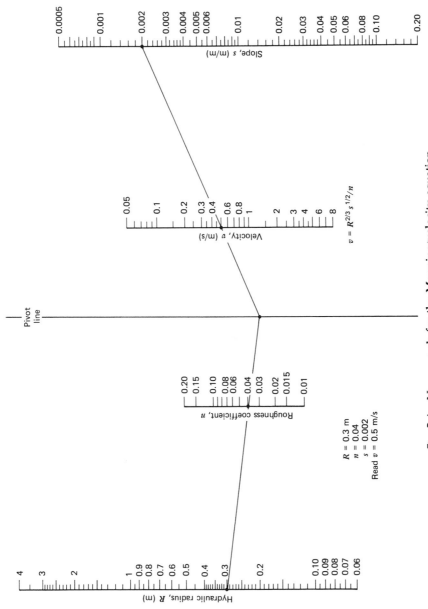

Fig. B.1 Nomograph for the Manning velocity equation.

Pipe and Conduit Flow

Table C.1   Friction Loss Coefficients for Circular or Square Pipe at Bends

$\dfrac{R}{D} = \dfrac{Bend\ Radius\ to\ Pipe\ Center\ Line}{Pipe\ Diameter}$	Bend Coefficient, K_b	
	45° Bend	90° Bend
0.5	0.7	1.0
1	0.4	0.6
2	0.3	0.4
5	0.2	0.3

Source: U.S. Soil Conservation Service (1951). *Engineering Handbook*, Section 5: "Hydraulics." SCS, Washington, DC.

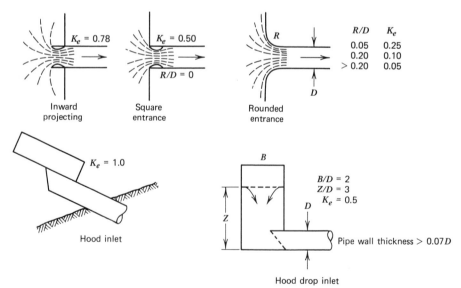

Fig. C.1   Entrance loss coefficients for pipe conduits. (*Source:* U.S. Soil Conservation Service (SCS) (1951). *National Engineering Handbook Hydraulics*, Sect. 5. Washington, DC; F. T. Mavis (1943). *The Hydraulics of Culverts*. Pennsylvania Eng. Expt. Sta. Bull. 56; Blaisdell, F. W. and C. A. Donnelly (1956). "Hood Inlet for Closed Conduit Spillways." Agr. Eng. 37, 670–672, and K. Yalamanchili and F. W. Blaisdell (1975). *Hydraulics of Closed Conduit Spillways, The Hood Inlet*. ARS-NC-23. USDA, Washington, DC. See Chapter 9.)

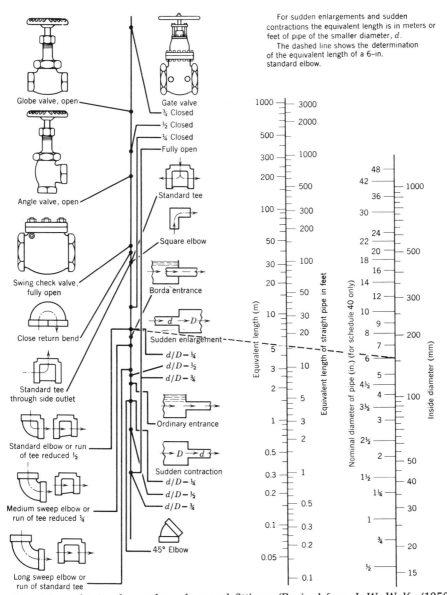

For sudden enlargements and sudden contractions the equivalent length is in meters or feet of pipe of the smaller diameter, d.

The dashed line shows the determination of the equivalent length of a 6-in. standard elbow.

Globe valve, open

Angle valve, open

Swing check valve, fully open

Close return bend

Standard tee through side outlet

Standard elbow or run of tee reduced $\frac{1}{2}$

Medium sweep elbow or run of tee reduced $\frac{1}{4}$

Long sweep elbow or run of standard tee

Gate valve
$\frac{3}{4}$ Closed
$\frac{1}{2}$ Closed
$\frac{1}{4}$ Closed
Fully open

Standard tee

Square elbow

Borda entrance

$d \rightarrow D$
Sudden enlargement
$d/D - \frac{1}{4}$
$d/D - \frac{1}{2}$
$d/D - \frac{3}{4}$

Ordinary entrance

$D \rightarrow d$
Sudden contraction
$d/D - \frac{1}{4}$
$d/D - \frac{1}{2}$
$d/D - \frac{3}{4}$

45° Elbow

Equivalent length (m)

Equivalent length of straight pipe in feet

Nominal diameter of pipe (in.) (for schedule 40 only)

Inside diameter (mm)

Fig. C.2 Minor friction losses for valves and fittings. (Revised from J. W. Wolfe (1950). *Friction Losses for Pipe and Fittings*. Oregon Agr. Expt. Sta. Bull. 181.)

Table C.2 Head Loss Coefficients for Circular Pipe Flowing Full (SI units)

$$K_c = \frac{1\ 244\ 522n^2}{d^{4/3}}, \qquad \text{where } d = \text{Diameter (mm)}$$

Pipe Inside Diameter [mm (in.)]	Flow Area (mm²)	Manning Coefficient of Roughness, n				
		0.010	0.013	0.016	0.020	0.025
13 (0.5)	133	4.071	6.881	10.423	16.286	25.447
25 (1)	491	1.702	2.877	4.358	6.810	10.641
51 (2)	2 043	0.658	1.112	1.685	2.632	4.113
76 (3)	4 536	0.387	0.653	0.990	1.546	2.416
102 (4)	8 171	0.261	0.441	0.669	1.045	1.632
127 (5)	12 668	0.195	0.329	0.499	0.780	1.218
152 (6)	18 146	0.153	0.259	0.393	0.614	0.959
203 (8)	32 365	0.104	0.176	0.267	0.417	0.652
254 (10)	50 671	0.0774	0.131	0.198	0.309	0.484
305 (12)	73 062	0.0606	0.102	0.155	0.242	0.379
381 (15)	114 009	0.0451	0.0761	0.115	0.180	0.282
457 (18)	164 030	0.0354	0.0598	0.0905	0.141	0.221
533 (21)	223 123	0.0288	0.0487	0.0737	0.115	0.180
610 (24)	292 247	0.0241	0.0407	0.0616	0.0962	0.150
762 (30)	456 037	0.0179	0.0302	0.0458	0.0715	0.112
914 (36)	656 119	0.0140	0.0237	0.0359	0.0561	0.0877
1219 (48)	1 167 071	0.00956	0.0162	0.0245	0.0382	0.0597
1524 (60)	1 824 147	0.00710	0.0120	0.0182	0.0284	0.0444

Note: K_c (English units) = K_c (SI units)/3.28, d = in.

Table C.3 Head Loss Coefficients for Square Conduits Flowing Full

$$K_c = \frac{19.60n^2}{R^{4/3}}, \qquad \text{Where } R = \text{Hydraulic Radius (m)}$$

Conduit Size [m × m (ft × ft)]	Flow Area (m²)	Manning Coefficient of Roughness, n			
		0.012	0.014	0.016	0.020
0.61 × 0.61 (2 × 2)	0.372	0.0347	0.0472	0.0616	0.0963
0.91 × 0.91 (3 × 3)	0.828	0.0203	0.0277	0.0361	0.0564
1.22 × 1.22 (4 × 4)	1.488	0.0138	0.0187	0.0245	0.0382
1.52 × 1.52 (5 × 5)	2.310	0.0103	0.0140	0.0182	0.0285
1.83 × 1.83 (6 × 6)	3.349	0.00800	0.0109	0.0142	0.0222
2.13 × 2.13 (7 × 7)	4.537	0.00653	0.00889	0.0116	0.0181
2.44 × 2.44 (8 × 8)	5.954	0.00545	0.00742	0.00970	0.0152
2.74 × 2.74 (9 × 9)	7.508	0.00467	0.00636	0.00831	0.0130
3.05 × 3.05 (10 × 10)	9.303	0.00405	0.00551	0.00720	0.0113

APPENDIX D

Drain Tile and Pipe Specifications

D.1 Clay Drain Tile

The test requirements given in Tables D.1 and D.2 are condensed from ASTM C4-62, *Tentative Specifications for Clay Drain Tile*, but the most current standard should be checked for possible changes.

Drain tile subject to these specifications may be made from clay, shale, fire clay, or mixtures thereof, and burned. The quality of tile selected should be such that the strength exceeds the soil load by a suitable margin. Where subjected to extreme freezing and thawing, extra-quality or heavy-duty tile are recommended.

Size and Minimum Lengths. The nominal sizes of clay drain tile shall be designated by their inside diameter. Tile less than 12 in. in diameter shall be not less than 1 ft in length; 12- to 30-in. tile, not less than their diameter; and tile larger than 30 in., not less than 30 in. in length.

Other Physical Properties. Some of the general physical requirements for the three classes of clay drain tile are given in Table D.2. Drain tile, while dry, shall give a clear ring when stood on end and tapped with a light hammer. They shall also be reasonably straight and smooth on the inside. Drain tile shall be free from cracks and checks extending into the tile in such a manner as to decrease its strength appreciably. They shall be neither chipped nor broken so as to decrease their strength materially or to admit soil into the drain.

D.2 Concrete Drain Tile

The specifications given in Table D.3 are condensed from ASTM C412-83. Standard and extra-quality concrete tile are intended for ordinary soils; special-quality tile are for soils or drainage waters that are markedly acid (pH of 6.0 or lower) or

475

Table D.1   Physical Test Requirements for Clay Drain Tile

Internal Diameter (in.)	Standard		Extra-Quality		Heavy-Duty	
	Average Minimum Strength[a] (lb/ft)	Average Maximum Absorption[b] (%)	Average Minimum Strength[a] (lb/ft)	Average Maximum Absorption[b] (%)	Average Minimum Strength[a] (lb/ft)	Average Maximum Absorption[b] (%)
4,5,6	800	13	1100	11	1400	11
8	800	13	1100	11	1500	11
10	800	13	1100	11	1550	11
12	800	13	1100	11	1700	11
15	870	13	1150	11	1980	11
18			1300	11	2340	11
21			1450	11	2680	11
24			1600	11	3000	11

[a]Average of five tile using the three-edge bearing method.
[b]Average of five tile using the 5-h boiling test.
Note: Tile diameters 14, 16, 27, and 30 are omitted for brevity. Diameter in mm = diameter in in. $\times$ 25.4. N/m = lb/ft $\times$ 14.6.

that contain unusual quantities of soil sulfates, chiefly sodium and magnesium, singly or in combination (assumed to be 3000 ppm or more).

Special-quality tile should have the same wall thickness and strength as extra-quality tile. Average maximum absorption is 8 percent, and closer tolerances are required for wall thicknesses than for extra-quality tile. The 10-min, room-temperature maximum soaking absorption shall be 3 percent for individual tile. The hydrostatic pressure test may be made in lieu of the above 10-min test. For sulfate exposures, sulfate-resistant cement shall be specified.

Table D.2   Distinctive General Physical Properties of Clay Drain Tile

Physical Properties Specified	Standard	Extra-Quality and Heavy-Duty
Number of freezings and thawings (reversals)	36	48
Permissible variation of average diameter below specified diameter (%)	3	3
Permissible variation between maximum and minimum diameters of same tile (% of thickness of wall)	75	65
Permissible variation of average length below specified length (%)	3	3
Permissible variation from straightness (% of length)	3	3
Permissible thickness of exterior blisters, lumps, and flakes that do not weaken tile and are few in number (% of thickness of wall)	20	15
Permissible diameters of above blisters, lumps, and flakes (% of inside diameter)	15	10
General inspection	Rigid	Very rigid

Table D.3   Physical Test Requirements for Concrete Drain Tile

Internal Diameter (in.)	Standard			Extra-Quality		
	Average Minimum Strength[a] (lb/ft)	Average Maximum Absorption[b] (%)	Wall Thickness[c] (in.)	Average Minimum Strength[a] (lb/ft)	Average Maximum Absorption[b] (%)	
4	800	10	1/2	1100	9	
5	800	10	9/16	1100	9	
6	800	10	5/8	1100	9	
8	800	10	3/4	1100	9	
10	800	10	7/8	1100	9	
12	800	10	1	1100	9	
15			1 1/4	1100	9	
18			1 1/2	1200	9	
21			1 3/4	1400	9	
24			2	1600	9	

[a]Average of five tile using the three-edge bearing method.
[b]Average of five tile using the 5-h boiling test.
[c]Minimum diameters shall not be less than the nominal diameters by more than 1/4 in. for 4- and 5-in. tile, 3/8 in. for 6- and 8-in. tile, 1/2 in. for 10- to 14-in. tile, 5/8 in. for 15- to 18-in. tile, and 3/4 in. for 20- to 24-in. tile. No wall thickness is specified for standard quality.
Note: Tile diameters 14, 16, and 20 are omitted for brevity. Diameter in mm = diameter in in. × 25.4. N/m = lb/ft × 14.6.

Size and Minimum Lengths.   Concrete tile less than 12 in. in diameter shall not be less than 1 ft in length, and 12- to 24-in.-diameter tile shall have nominal lengths not less than their diameter.

D.3   Corrugated Plastic Tubing

Specifications for tubing and fittings are given in ASTM F405, F667, and D2412. Pipe stiffness is the slope of the load-deflection curve (kPa or psi) at a specified percentage deflection based on the original inside diameter. The load is applied between two parallel plates at a constant deflection rate of 12.7 mm/min, at a test temperature of 23°C. The test specimen shall be 305 mm long. Minimum tubing stiffness and maximum elongation are given in Table D.4. Heavy-duty tubing is required for leach beds.

Table D.4   Physical Test Requirements for Corrugated Plastic Tubing (76 to 203 mm in Diameter) ASTM F405-76b

Physical Property	Standard [MPa (psi)][b]	Heavy Duty[a] [MPa (psi)][b]
Pipe stiffness at 5% deflection, minimum	0.17 (24)	0.21 (30)
Pipe stiffness at 10% deflection, minimum	0.13 (19)	0.175 (25)
Elongation, maximum %	10	5

[a]Pipe stiffness for 254-, 305-, and 381-mm diameters as per ASTM F667-80.
[b]lb/in. per in.

Tubing Size and Perforations.   Nominal diameters range from 76 to 203 mm in 25.4-mm (1-in.) increments. Perforations shall be cleanly cut and uniformly spaced along the length and circumference of the tubing in a size, shape, and pattern to suit the needs of the user.

Elongation.   A 1.27-m-long specimen shall be tested with the axis vertical using a test load of $5D$ lb, where D is the nominal inside tubing diameter in inches.

Table D.5   Specifications for Tile, Pipe, Hose, and Tubing and Installation

No.	Type and Specification	Specification No.
1	Aluminum sprinkler irrigation tubing, minimum standards	ASAE S263.3[a]
2	Clay drain tile	ASTM[b] C4-Yr[c]
3	Clay drain tile, perforated	ASTM C498
4	Clay pipe, vitrified, perforated	ASTM C700
5	Concrete drain tile	ASTM C412
6	Concrete pipe, perforated	ASTM C444M
7	Concrete, nonreinforced, irrigation pipe systems, design and installation	ASAE S261.7
8	Concrete sewer pipe	ASTM C14M
9	Concrete pipe for irrigation or drainage	ASTM C118M
10	Concrete pipe, reinforced, culvert, storm drain, and sewer	ASTM C76M
11	Concrete tile or pipe, method of testing	ASTM C497M
12	Drain and sewer pipe, bituminized fiber	ASTM D1861-2
13	Drain and sewer pipe, plastic	Commercial standard CS-228[d]
14	Hose and couplings, underground irrigation, thermoplastic pipelines	ASAE 376.1
15	Hose and couplings, self-propelled hose-drag irrigation systems	ASAE S394
16	Irrigation wells, design and construction	ASAE EP400.1
17	Pipe, polyethylene for drip/trickle irrigation laterals	ASAE S435
18	Pipe, polyvinyl chloride, drain, waste, vent, and fittings	ASTM D2665
19	Pipe, plastic, properties by parallel-plate loading	ASTM D2412
20	Tubing, corrugated polyethylene and fittings	ASTM F405, F667
21	Tubing, corrugated polyvinyl chloride and fittings	ASTM F800
22	Tubing, corrugated thermoplastic installation for agricultural drainage or water table control	ASTM F449
23	Tubing, thermoplastic pipe and corrugated tubing installation in septic tank leach fields	ASTM F481

[a]ASAE Standards (1990). "AE Guide to Product and Services." *Agr. Eng.* **71**(3), 77–80 (listing only).
[b]American Society for Testing Materials, 1916 Race Street, Philadelphia, PA 19103.
[c]Specifications include last two digits of calendar year (Yr) when approved. T for tentative and M for metric where appropriate, i.e., C4-90TM.
[d]Available from GPO, Washington, DC 20250.

Loads on Underground Conduits

Underground conduits should be installed such that the load does not cause failure. For concrete and clay tile and corrugated plastic tubing the required strength and stiffness are set forth in current ASTM specifications. Tile should meet the required minimum crushing strength and tubing should have the minimum stiffness at the specified vertical deflection.

Methods of calculating static soil loads are given in Chapter 15. Loads from wheeled vehicles or from superconcentrated loads can be estimated from Fig. E.1 or Table E.2 using the procedure given in Examples E.1 and E.2.

☐ Example E.1

Determine the average load per lineal length and the total load transmitted to a 610-mm (24-in.)-nominal-diameter drain tile installed at a depth of 1.65 m (5.4 ft) from a static concentrated load of 4540 N (1000 lb) directly over the center of the tile.

Solution. Outside diameter (o.d.) = $0.61 + 2(0.1 \times 0.61) = 0.73$ m. Depth to the top of conduit = depth to bottom of tile $- B_c = 1.65 - 0.73 = 0.92$ m (3.0 ft). Read $I_c = 1.0$ for static superload from Fig. E.1. Read $C_t = 20$ percent from curve. Substitute in equation from Fig. E.1.

$$W_t = (1/0.61) \times 1.0 \times 0.20 \times 4540 = 1490 \text{ N/m (100 lb/ft)}$$

$$\text{Total load} = LW_t = 0.61 \times 1490 = 910 \text{ N (200 lb)} \qquad \square$$

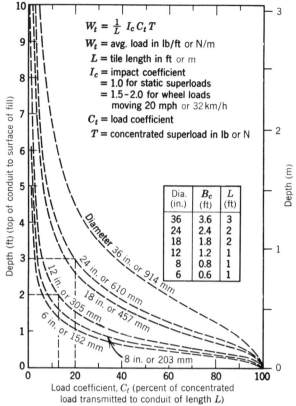

Fig. E.I Concentrated surface load coefficients for rigid pipe. (Revised from M. G. Spangler, C. Mason, and R. Winfrey (1926). *Experimental Determination of Static and Impact Loads Transmitted to Culverts.* Iowa Eng. Expt. Sta. Bull. 79; and A. Marston (1930). *The Theory of External Loads on Closed Conduits in the Light of the Latest Experiments.* Iowa Eng. Expt. Sta. Bull. 96.)

☐ *Example E.2*

Determine the average load per 0.3-m (1-ft) length on a 305-mm (12-in.) tile installed at a depth of 1 m (3.3 ft) if a 4540-N (1000-lb) concentrated load is moving at 32 km/h (20 mph).

Solution. From Fig. E.1 for a depth of 0.6 m (1.0 − 0.4), read $C_t = 13$ percent, and select $I_c = 2.0$ for moving load.

$$W_t = (1/0.305) \times 2.0 \times 0.13 \times 4540 = 3870 \text{ N/m } (260 \text{ lb/lin ft}) \qquad \square$$

Recommended maximum depths for corrugated plastic tubing from static soil loads are given in Table E.1. For ditch conditions the soil load was computed using the equation for flexible conduits, and for wide trenches the projecting load equation for rigid pipe was applied. These loads were substituted in the deflection equation described in Chapter 15.

Table E.1   Recommended Maximum Depths for Tubing Buried in Loose, Fine-Grained Soils in Meters

Nominal Tubing Diameter [mm (in.)]	Tubing Quality	Trench Width at Top of Tubing (m)					
		0.2	0.3	0.4	0.6	0.8	≥1
102 (4)	Standard	[a]	3.9	2.1	1.7	1.6	1.6
	Heavy duty	[a]	[a]	3.0	2.1	1.9	1.9
152 (6)	Standard	[a]	3.1	2.1	1.7	1.6	1.6
	Heavy duty	[a]	[a]	2.9	2.0	1.9	1.9
203 (8)	Standard		3.1	2.2	1.7	1.6	1.6
	Heavy duty		[a]	3.0	2.1	1.9	1.9
254 (10)			[a]	2.8	2.0	1.9	1.9
305 (12)				2.7	2.0	1.9	1.9
381 (15)					2.1	1.9	1.9

[a]Any depth is permissible at this width or less. Minimum side clearance between tubing and trench should be about 0.08 m.

Assumptions: Soil modulus of reaction $E' = 345$ kN/m^2; deflection lag factor $D = 3.4$; vertical deflection 20 percent of nominal diameter; bedding angle factor $K = 0.096$; and soil density of 1.75 g/cm^3. See Chapter 15.

Source: A. D. Fenemor, B. R. Bevier, and G. O. Schwab (1979). "Predictions of Deflection for Corrugated Plastic Tubing." *Trans. ASAE* **22**, 1338–1342.

Table E.2   Wheel Loads Transmitted to Underground Rigid Conduits in Percent[a]

Depth of Backfill Over Top of Tile, H (m)	Trench Width at Top of Drain (m)				
	0.3	0.6	0.9 (%)	1.2	1.8
0.3	17.0	26.0	28.6	29.7	30.2
0.6	8.3	14.2	18.3	20.7	22.7
0.9	4.3	8.3	11.3	13.5	15.8
1.2	2.5	5.2	7.2	9.0	11.5
1.5	1.7	3.3	5.0	6.3	8.3
1.8	1.0	2.3	3.7	4.7	6.2

[a]Includes both live load and impact load transmitted to a 0.3-m length of tile.

Note: Live loads transmitted are practically negligible below a 1.8-m depth.

Source: American Society of Agricultural Engineers (ASAE) (1988). *Design and Construction of Subsurface Drains in Humid Areas.* EP260.4. ASAE, St. Joseph, MI.

APPENDIX F

Conversion Constants

Table F.1 Conversion of Drainage Coefficients

Drainage Coefficient [mm (in.)]	cfs/ac	gpm/ac	L/s per ha	m³/day per ha
1.0 (1/25)	0.0017	0.75	0.116	10
1.6 (1/16)	0.0026	1.18	0.185	16
3.2 (1/8)	0.0052	2.36	0.370	32
6.4 (1/4)	0.0105	4.71	0.741	64
8.5 (1/3)	0.0142	6.29	0.984	85
9.5 (3/8)	0.0157	7.07	1.100	95
12.7 (1/2)	0.0210	9.43	1.470	127
15.9 (5/8)	0.0262	11.79	1.840	159
19.1 (3/4)	0.0315	14.14	2.211	191
22.2 (7/8)	0.0367	16.50	2.569	222
25.4 (1)	0.0420	18.86	2.940	254

Note: 1 cfs/ac = 69.96 L s⁻¹ ha⁻¹ = 6044 m³ day⁻¹ ha⁻¹. 1 L/s = 15.85 gpm.

Table F.2 Conversion of Temperatures from Fahrenheit to Celsius[a]

°F	°C	°F	°C	°F	°C
−20	−28.9	40	4.4	100	37.8
−15	−26.1	45	7.2	105	40.6
−10	−23.3	50	10.0	110	43.3
− 5	−20.6	55	12.8	115	46.1
0	−17.8	60	15.6	120	48.9
5	−15.0	65	18.3	125	51.7
10	−12.2	70	21.1	130	54.4
15	− 9.4	75	23.9	135	57.2
20	− 6.7	80	26.7	140	60.0
25	− 3.9	85	29.4	145	62.8
32	0.0	90	32.2	150	65.6
35	1.7	95	35.0	212	100.0

[a] °F = 9/5 (°C) + 32 or °C = 5/9 (°F − 32).

[1] For length, area, and volume conversions see inside of back cover.

Table F.3   Miscellaneous Conversion Constants

Pressure and Force

1 in. Hg = 3386.4 Pa	1 Pa = 1 N/m²
= 2.04 psi	1 atm = 1013 mb
1 mm Hg = 133.3 Pa	= 101.3 kPa
1 mm water = 9.8 Pa	= 760 mm Hg
1 psi = 51.7 mm Hg	= 33.93 ft water
1 psi = 6.895 kPa	1 mb = 100 Pa
1 lb f = 4.45 N	1 mb = 10.2 mm water
1 lb m = 0.454 kg	1 ft water = 2.985 kPa
1 lb f/ft² = 47.88 Pa	1 kPa = 0.335 ft water

Power and Energy

1 Btu = 1055 J	1 kW = 1.341 hp
1 cal = 4.19 J	1 kWh = 3.6 × 10⁶ J
1 hp = 550 ft-lb/s	
= 746 W	

Volume, Weight, and Mass

1 U.S. gal = 8.34 lb	1 oz = 28.35 g
1 Imp. gal = 10.02 lb	1 lb = 453.6 g
1 ft³ water = 62.4 lb	1 lb/ft³ = 16.02 kg/m³
1 yd³ = 0.7656 m³	1 short ton (t) = 907.18 kg = 2000 lb
1 ft³ = 7.48 gal	1 metric tonne = 1000 kg (Mg)
	= 2205 lb
	1 t/a = 2.24 Mg/ha (metric tonne)
	1 kg = 2.205 lb mass

Velocity

1 fps = 0.305 m/s	1 mph = 1.61 km/h = 0.447 m/s
= 1097 m/h	1 km/h = 0.621 mph

Useful Formulas and Procedures

VOLUME FORMULAS

The average end area formula for computing the volume of storage in a reservoir, earth fill in a dam, or ditch excavation is

$$V = \frac{d}{2}(A_1 + A_2) \tag{G.1}$$

where
V = volume of storage (L^3),
d = distance between end areas (L),
A_1 and A_2 = end area (L^2).

The prismoidal formula is

$$V = \frac{d}{6}(A_1 + 4A_m + A_2) \tag{G.2}$$

where A_m = middle area halfway between the end areas (L^2).

Where preliminary surveys are made by taking slopes in the reservoir area, the storage may be estimated from the approximate formula for a frustrum of a cone,[1]

$$V = A_0 d + \frac{177 d^2 A_0^{1/2}}{S} \tag{G.3}$$

where A_0 = area at spillway crest (L^2),
d = depth of water above spillway crest (L),
S = average slope of reservoir sides and banks, through range of d, in percent.

---

[1]Culp, M. M. (1948). "The Effect of Spillway Storage on the Design of Upstream Reservoirs." *Agr. Eng.* **29**, 344–346.

LAYOUT OF CIRCULAR CURVES

The procedure to be followed in laying out a circular curve is as follows: The transit is first set up at the point of intersection P.I. as indicated in Fig. G.1 and the angle I is measured. Next the tangent distance is calculated with the formula

$$T = R \tan \frac{I}{2} \qquad (G.4)$$

where T = tangent distance (L),
       I = intersection angle in degrees,
       R = radius of curvature (L).

The point of curvature P.C. and the point of tangency P.T. are located by measuring the computed distance T from the P.I. Set up the transit at P.C. and locate stations on the curve by chaining and measuring off deflection angles as computed by the equation

$$e = \frac{c}{100} \cdot \frac{D}{2} = \frac{cD}{200} \qquad (G.5)$$

where e = deflection angle in degrees,
       c = chord length in m or ft,
       D = degree of curve in SI or English units (Chapter 13).

From this equation 100-unit-long chords require deflection angles of $\frac{1}{2}D$, 50-unit stations $\frac{1}{4}D$, and so on. After the first station beyond the P.C. is located, the deflection angle for each succeeding station is the summation of the deflection angles for all previous chord distances. Since most curves are rather flat, the arc distance is nearly equal to the chord length. The total length of the curve is

$$L = 100 \ I/D \ \text{(m)} \qquad (G.6)$$

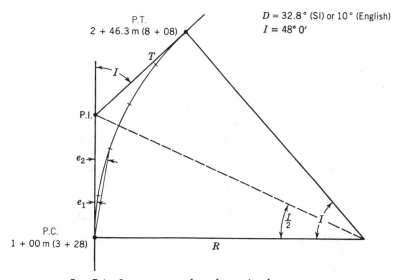

Fig. G.1   Layout procedure for a circular curve.

The design of a circular curve is illustrated by the following problem. Because the degree of curvature was defined in English units originally, computation is simpler than in SI units.

☐ *Example G.1*
___

Design a 32.8-degree (SI) curve for Fig. G.1 if angle I is 48 deg 0 min.

Solution.   From definition of degree of curvature given in Chapter 13,

$$R = 50/\sin 16.4 \text{ deg} = 177.1 \text{ m (581.1 ft)}$$

and from Eq. G.4 for $I/2 = 24$ deg,

$$T = 177.1 \times \tan 24 \text{ deg} = 78.85 \text{ m (258.7 ft)}$$

From Eq. G.5, the deflection angle from Station $1 + 00$ m $(3 + 28$ ft) to $1 + 30$ m $(4 + 27$ ft) is

$$e_1 = 30 \times 32.8/200 = 4.92 \text{ deg}$$

and, similarly, e_2 for Station $1 + 60$ m $(5 + 25$ ft) is $4.92 + 4.92 = 9.84$ deg, and on to P.T., where e is $I/2 = 24$ deg. From Eq. G.6, for Station $1 + 00$ m to P.T.,

$$L = (100 \times 48)/32.8 = 146.3 \text{ m (480 ft)}$$

Station at P.T. is $(1 + 00) + (1 + 46.3) = 2 + 46.3$ m $(8 + 08$ ft). For English units, curves are normally laid out in 50- or 100-ft stations.   ☐

SETTING SLOPE STAKES

In making the location survey prior to construction of a dam or a ditch, center line stakes and slope stakes are set at each station or at more frequent intervals to guide the operator. On level or nearly level topography the offset of the slope stakes from the center line can be easily computed by adding one half the top or bottom width plus the sideslope ratio (z) times the depth. On irregular land the slope stakes are set by trial and error. Although the following procedure, illustrated in Fig. G.2, applies to ditch location, the same method is applicable to earth dam construction. First, the offset distance from the center line is estimated and an elevation for the point determined. If the depth from this point to the bottom of the ditch corresponds to the computed distance to the center line, the slope stake has been set correctly. For example, in Fig. G.2 the slope stake is at an elevation of 52.5 m and the bottom of the ditch is 45.0 m. The computed distance is

$$d = \Delta Ez + w/2 = (52.5 - 45.0) \; 1 + 6/2 = 10.5 \text{ m}$$

where d = distance from center line to edge of fill or ditch (L),
   ΔE = difference in elevation from the stake to the bottom of ditch or top of fill (L),

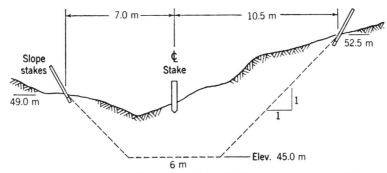

Fig. G.2   Setting slope stakes for a ditch on uneven ground.

z = sideslope ratio (horizontal to vertical),
w = bottom width of ditch or top width of dam (L).

If this distance is not 10.5, a new trial point must be selected, the elevation determined, and the distance from the center line again compared with the computed distance.

APPENDIX H

Filter Design Criteria

Considerable experimentation with the design of filters has been performed by the U.S. Corps of Engineers, U.S. Bureau of Reclamation, and many others. The USBR (1987)[1] recommends the following limits for toe drains in earth dams which will satisfy filter stability criteria and provide an ample increase in permeability from the base to the filter material. These criteria are satisfactory for natural sand and gravel or crushed rock.

$$\frac{D_{15} \text{ (filter)}}{D_{15} \text{ (base)}} \geq 5 \tag{H.1}$$

provided the filter does not contain particles more than 5 percent finer than 0.074 mm,

$$\frac{D_{15} \text{ (filter)}}{D_{85} \text{ (base)}} \leq 5 \tag{H.2}$$

$$\frac{D_{85} \text{ (filter)}}{\text{Maximum size openings in pipe drain}} \geq 2 \tag{H.3}$$

and the distribution size curve of the filter should be roughly parallel to that of the base material. In the above criteria D_{15} and D_{85} are the particle diameters at which 15 and 85 percent, respectively, of the total soil particles that are smaller on a weight basis. In addition to the above criteria the maximum particle size is 76.2 mm (3 in.).

For pipe drains in irrigated land of the West the USBR has developed filter criteria (Chapter 15) simpler than those described above for dams. The SCS

---

[1]U.S. Bureau of Reclamation (USBR) (1987). *Design of Small Dams*, 3rd ed. GPO, Washington, DC.

$(1973)^2$ criteria have modified the limits in the above equations and, in addition, specify that for pipe drains,

$$\frac{D_{50} \text{ (filter)}}{D_{50} \text{ (base)}} = 12 \text{ to } 58 \qquad \text{(H.4)}$$

The right-hand sides of Eqs. H.1 and H.3 are changed to (12 to 40) and ($\geq$½), respectively. The criteria require that no more than 10 percent of the filter material should pass the 25-mm sieve (No. 60).

The following example illustrates a typical design for a thin filter around a perforated pipe drain in a dam.

☐ Example H.1

Determine the filter size limits for the conditions given in Fig. H.1 if the pipe drain openings are 12.7 mm (½ in.) in diameter and the gradation of the foundation soil is that shown in Fig. H.2.

Solution. From available materials the sand and gravel with size distribution curves shown in Fig. H.2 were selected, based on the following calculated values:

Layer	Layer Size (mm) D_{15}	Layer Size (mm) D_{85}	D_{15} Size Requirement (mm) for the Next Coarser Layer (Filter) From Eq. H.1	D_{15} Size Requirement (mm) for the Next Coarser Layer (Filter) From Eq. H.2	Maximum Opening in Pipe Drain from Eq. H.3 (mm)
Soil	0.006	0.1	0.03 to 0.24	$\leq$0.5	0.05
Sand	0.14[a]	2.4[a]	0.70 to 5.6	$\leq$12.0	1.2
Gravel	4.0[a]	50.0[a]	20.0 to —[b]	[b]	25.0

[a]These materials gave distribution curves nearly parallel to that for the soil.
[b]Maximum permissible size is 76.2 mm at D_{100} for any filter material.

[2]U..S. Soil Conservation Service (SCS) (1973). *Drainage of Agricultural Lands.* Water Information Center, Port Washington, NY.

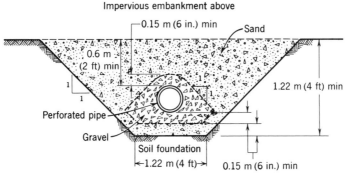

Fig. H.1 Typical filter for a toe drain of a dam. (Redrawn and revised from U.S. Bureau of Reclamation (USBR) (1987). *Design of Small Dams.* GPO, Washington, DC.)

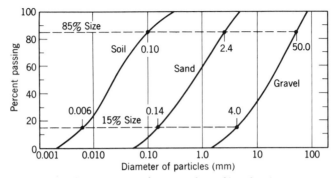

Fig. H.2 Particle-size distribution curves for a toe drain filter for dams. (Redrawn from U.S. Bureau of Reclamation (USBR) (1987). *Design of Small Dams*. GPO, Washington, DC.)

Since the size of openings in the drain pipe is 12.7 mm (½ × 25.4), the sand will not meet the criteria in Eq. H.3, but the gravel with a permissible opening of 25.0 mm is satisfactory. Thus, it was necessary to provide the gravel layer next to the perforated drain. Theoretically, each layer of the filter could be very thin, but practically a reasonable thickness is necessary to make sure that some slight readjustment during or after construction does not disrupt the layer. For earth dams the USBR (1987) recommends a minimum filter thickness of 0.9 m (3 ft), but the minimum thickness of individual layers should be 0.15 m (6 in.). Other minimum criteria are shown in Fig. H.1. □

Computer Programs for Soil and Water Engineering

The following table contains some of the computer programs in use by government agencies, industry, consultants, and research scientists. Many other programs have been developed and readers are encouraged to contact local sources to obtain programs that may be better suited to local conditions or special problems.

Title and Reference	Capabilities	Source	Operating System
AGNPS (Young et al., 1989)	Prediction of runoff and erosion rates for large watersheds	USDA–ARS, North Central Soil Conservation Res. Lab. Morris, MN 56267	MS-DOS
CREAMS (Knisel, 1980)	Calculation of erosion and chemical transport for comparing management systems	USDA–ARS Southeast Watershed Res. Lab. P.O. Box 946 Tifton, GA 31793	MS-DOS Mainframe Source
DESIGNER (Driscoll and Bralts, 1985)	Analysis and design of microirrigation submain units	Department of Agricultural Engineering Michigan State University East Lansing, MI 48823	MS-DOS
DRAINMOD (Skaggs, 1982)	Prediction of the effects of drain depth and spacing on water table elevations and crop yields	USDA–SCS South National Technical Center Fort Worth, TX or state SCS offices	MS-DOS

(Continued on next page.)

Title and Reference	Capabilities	Source	Operating System
Land Leveling (Spectra-Physics, 1981)	Best fit of single plane surface, cut/fill map, total cut	The Spectra-Physics Corporation 5475 Kellenburger Road Dayton, OH 45424–1099	Apple
SCS DESIGN (Hart et al., 1983)	Computation of values for SCS surface irrigation design equations for basins, borders and furrows	Department of Agricultural Engineering University of Arizona Tucson, AZ 85721 or state SCS offices	MS-DOS
SCS Engineering Programs (Rovang et al., 1990)	Design of waste storage, waterways, sediment basins, curve number, peak discharge, stadia reduction and plotting, areas, and volumes	Conservation engineer State SCS offices	MS-DOS
RUSLE (Renard et al., 1991)	Prediction of average annual erosion	USDA–ARS National Soil Erosion Res. Lab. Purdue University West Lafayette, IN 47907	MS-DOS UNIX
SCHEDULER (Shayya et al., 1990)	Scheduling of irrigations using climatic, soil, and crop data	Department of Agricultural Engineering Michigan State University East Lansing, MI 48823	MS-DOS
SRFR (Strelkoff, 1991)	Prediction of surface irrigation flows, infiltration profiles, and efficiencies	U.S. Water Conservation Lab. 4331 East Broadway Phoenix, AZ 85040	MS-DOS
TR55 (USDA–SCS, 1986)	Prediction of storm runoff volume, peak rate of discharge, hydrographs, and storage volumes for small watersheds	USDA–SCS Engineering Division P.O. Box 2890 Washington, DC 20013	MS-DOS UNIX Mainframe
WEPP (Nearing et al., 1989)	Prediction of runoff and erosion rates for hillslopes and small watersheds	USDA–ARS National Soil Erosion Res. Lab. Purdue University West Lafayette, IN 47907	MS-DOS UNIX Source

REFERENCES

Bralts, V. F., D. M. Edwards, and I. Wu (1987). "Drip Irrigation Design and Evaluation Based on the Statistical Uniformity Concept." *Adv. Irrig.* **4**, 67–115.

Driscoll, M. A., and V. F. Bralts (1985). *Microcomputer Drip Irrigation Designer: A User's and Programmer's Guide.* Dept. of Agr. Eng., Michigan State University, East Lansing, MI.

Flanagan, D. C. (ed.) (1990). *WEPP Second Edition, Water Erosion Prediction Project–Hillslope Profile Model Documentation Corrections and Additions.* National Soil Erosion Res. Lab., West Lafayette, IN.

Hart, W. E., H. G. Collins, G. Woodward, and A. S. Humphreys (1983). "Design and Operation of Gravity or Surface Irrigation Systems," In *Design and Operation of Farm Irrigation Systems*, M. E. Jensen (ed.), Monograph No. 3, pp. 501–580. ASAE, St. Joseph, MI.

Knisel, W. G. (ed.) (1980). *CREAMS: A Field-Scale Model for Chemicals, Runoff, and Erosion from Agricultural Management Systems.* Conservation Research Report No. 26. USDA–Science and Education Administration, Washington, DC.

Lane, L. J., and M. A. Nearing (eds.) (1989). *USDA-Water Erosion Prediction Project: Hillslope Profile Model Documentation.* NSERL Report No. 2. USDA–ARS National Soil Erosion Res. Lab., West Lafayette, IN.

Nearing, M. A., G. R. Foster, L. J. Lane, and S. C. Finkner (1989). "A Process-Based Soil Erosion Model for USDA-Water Erosion Prediction Project Technology." *ASAE Trans.* **32**, 1587–1593.

Renard, K. G., G. R. Foster, F. A. Weesies, and J. P. Porter (1991). "RUSLE Revised Universal Soil Loss Equation." *J. Soil Water Cons.* **46**(1), 30–33.

Rovang, R. M., L. P. Herndon, T. E. Radermacher, and K. E. Harward (1990). *Experience Gained in Development of SCS Prototype Engineering Software.* Paper No. 90260, ASAE, St. Joseph, MI.

Shayya, W. H., V. F. Bralts, and T. R. Olmsted (1990). "A General Irrigation Scheduling Package for Microcomputers." *Computers Electronics Agr.* **5**, 197–212.

Skaggs, R. W. (1982). "Field Evaluation of a Water Management Simulation Model." *ASAE Trans.* **25**, 666–674.

Spectra-Physics Construction and Agricultural Division (1981). *Land Leveling Design System for Apple Computers.* Spectra-Physics, Dayton, OH.

Strelkoff, T. (1991). *SRFR—A Model of Surface Irrigation*, Version 20. Proceedings of 1991 National Conference of Irrig. and Drain. Div., ASCE, Honolulu, HI, July 22–26, 676–682.

U.S. Department of Agriculture–Agricultural Research Service (USDA–ARS) (1986). *Agricultural Nonpoint Source Pollution Program User Manual (AGNPS-PC).* Minnesota Pollution Control Agency, Dept. of Soil Science, University of Minnesota, St. Paul, MN.

U.S. Department of Agriculture–Soil Conservation Service (USDA–SCS) (1986). *Urban Hydrology for Small Watersheds.* Technical Release 55. National Technical Information Serv., Springfield, VA.

Young, R. A., C. A. Onstad, D. D. Bosch, and W. P. Anderson (1989). "AGNPS: A Nonpoint Source Pollution Model for Evaluating Agricultural Watersheds." *J. Soil Water Cons.* **44**(2), 168–173.

Index

English – SI Length Conversion Constants

Length	in.	ft	yd	mi	cm	m	km
1 in.	1	0.083	0.027	—	2.54	—	—
1 ft	12	1	0.333	—	30.48	0.305	—
1 yd	36	3	1	—	91.44	0.914	—
1 mi (statute)	—	5280	1760	1	—	1609	1.61
1 cm	0.394	0.033	0.011	—	1	0.1	—
1 m	39.37	3.281	1.094	—	100	1	0.001
1 km	—	3281	1094	0.621	—	1000	1

English – SI Area Conversion Constants

Area	in.^{2}	ft^2	yd^2	ac	cm^2	m^2	ha
1 in.^{2}	1	0.007	—	—	6.45	0.00064	—
1 ft^2	144	1	0.1111	—	—	0.0929	—
1 yd^2	1296	9	1	—	—	0.8361	—
1 ac	—	43 560	4840	1	—	4047	0.405
1 cm^2	0.155	—	—	—	1	0.0001	—
1 m^2	1550	10.76	1.20	—	10 000	1	0.0001
1 ha	—	107 650	11 961	2.47	—	10 000	1

English – SI Volume Conversion Constants

Volume	in.^{3}	ft^3	U.S. gal	L	m^3	ac-ft	ha-m
1 in.^{3}	1	—	0.0043	0.0164	—	—	—
1 ft^3	1728	1	7.481	28.32	0.0283	—	—
1 U.S. gal	231	0.134	1	3.785	0.0038	—	—
1 L	61.02	0.0353	0.2642	1	0.001	—	—
1 m^3	61 022	35.31	264.2	1000	1	0.00081	0.0001
1 ac-ft	—	43 560	325 872	—	1233.4	1	0.1233
1 ha-m	—	353 198	—	10×10^6	10 000	8.108	1

Note: 1 yd^3 = 0.765 m.^{3} 1 m^3 = 1.308 yd.^{3} For conversion constants see Appendix F.